Circular Bioeconomy: Novel Processes and Materials for Food Preservation

Circular Bioeconomy: Novel Processes and Materials for Food Preservation

Editors

Sergio Torres-Giner
Amparo Chiralt
Chelo Gonzalez-Martinez

MDPI

Basel • Beijing • Wuhan • Barcelona • Belgrade • Novi Sad • Cluj • Manchester

Editors

Sergio Torres-Giner
University Institute of Food Engineering—FoodUPV
Polytechnic University of Valencia
Valencia
Spain

Amparo Chiralt
University Institute of Food Engineering—FoodUPV
Polytechnic University of Valencia
Valencia
Spain

Chelo Gonzalez-Martinez
University Institute of Food Engineering—FoodUPV
Polytechnic University of Valencia
Valencia
Spain

Editorial Office
MDPI
St. Alban-Anlage 66
4052 Basel, Switzerland

This is a reprint of articles from the Special Issue published online in the open access journal *Foods* (ISSN 2304-8158) (available at: www.mdpi.com/journal/foods/special_issues/processes_materials_food_preservation).

For citation purposes, cite each article independently as indicated on the article page online and as indicated below:

Lastname, A.A.; Lastname, B.B. Article Title. *Journal Name* **Year**, *Volume Number*, Page Range.

ISBN 978-3-7258-0112-1 (Hbk)
ISBN 978-3-7258-0111-4 (PDF)
doi.org/10.3390/books978-3-7258-0111-4

Cover image courtesy of Sergio Torres-Giner

Contents

About the Editors

Sergio Torres-Giner

Sergio Torres-Giner received a Bachelor's degree in Chemical Engineering in 2003 from the Polytechnic University of Valencia (UPV), Spain. In 2004, he attained an MSc in Process Systems Technology at Cranfield University, England, followed by an MBA in Industrial Management in 2005 at the Catholic University of Valencia "San Vincente Martir", Spain. He completed his Ph.D. in 2010 in Food Science at the University of Valencia, Spain. He currently works as a scientist and a professor at the University Institute of Food Engineering—FoodUPV in the field of macromolecular science, with an interest in the applications of food packaging technologies. He has 20 years of experience working at both public research agencies and industrial R&D organizations. He has published over 100 peer-reviewed scientific papers indexed in JCR, 10 books and encyclopedia chapters, and 4 patents. His research activity has strongly contributed to advancing knowledge on biopolymers and applying such knowledge to food-related products.

Amparo Chiralt

Amparo Chiralt received a Ph.D. in Chemistry, and she is currently a Full Professor of Food Technology at the Polytechnic University of Valencia (UPV). She is the author of 103 book chapters for prestigious publishers, 339 publications indexed in WOS, more than 500 contributions to international congresses, and 10 patents. She has participated in 109 competitive R&D projects (40 as project leader) and 11 knowledge transfer contracts with food companies. She has been included on the list of "Highly Cited Researchers" in Agricultural Science from 2018 to 2023. She has directed 60 doctoral theses and supervised most doctors holding relevant positions at Spanish and foreign academic institutions and companies. She was Vice Chancellor of Postgraduate Studies (5 years) and Research (4 years) at UPV and has been Director of the Doctoral School (2013–2021). Over the last 15 years, her studies have focused on active (antioxidant/antimicrobial) biodegradable materials for food packaging applications using renewable resources. This represents a contribution to the valorization of agro-food waste and reduction in the use of highly polluting oil-derived plastics within the concept of Circular Economy.

Chelo Gonzalez-Martinez

Chelo Gonzalez-Martinez is a Full Professor of Food Technology at the Polytechnic University of Valencia (UPV), Spain. She has been the coordinator of the Master in Food Management and Safety, and director of the PhD program in Food Science, Technology, and Management. She is currently Head of the University Institute of Food Engineering—FoodUPV. She has been involved in more than 28 research projects and she has co-authored 95 articles in *SCI* journals, 15 book chapters, over 150 Congress participations, and 16 doctoral theses. Since 2019, she has been a member of the Education on Food Studies Working Group of the international association IUFoST. She has also participated in several European projects related to the quality and innovation of education in the area of Food Studies. Her research activity is focused on the physicochemical aspects of food and food processing, along with the development and characterization of active coatings and biodegradable packaging materials incorporating natural antimicrobials for their application in food preservation.

Preface

Circular Bioeconomy is a concept becoming an integral part of industrial green technological processes, whereby residues are considered as novel resources and a new sustainable closed model based on high-value applications of biomass is developed. In the field of food technology, the scientific community is undertaking efforts to utilize food processing by-products as highly sustainable resources for the production of food packaging materials and bioactive compounds. Thus, multiple research groups are currently engaged in the complete valorization of food wastes into alternative feedstocks for monomers and intermediates to produce biopolymers and green composites, as well as in the extraction and purification of antioxidants, natural preservatives or antimicrobials to enrich food quality and ensure food safety.

This reprint is divided into ten chapters that compile original research studies published recently in the field of Circular Bioeconomy. In particular, the book gathers different advancements concerning with both the processes and materials obtained from the valorization of food wastes and also the conversion of food residues into innovative solutions for food preservation. The reprint begins with two chapters focused on the preparation and incorporation of nano- and microcapsules of natural antioxidant and antimicrobial additives into biodegradable polymers to develop active packaging systems able to protect and extend the shelf life of perishable foods. These studies also discuss the potential ways to promote the application of nanotechnology for active food packaging development, highlighting the importance of eliminating legislative restrictions and creating a unique global organization to control and monitor nanomaterials. The reprint continues with two more chapters that demonstrate the environmental benefit that can be achieved via the valorization of food processing by-products as highly sustainable resources to produce food packaging materials and active and bioactive compounds. The reprint contains two other chapters devoted to the particular use of lignocellulosic materials obtained from processing by-products and wastes derived from the food and agroforestry industries as cost-effective fillers in biopolymer-based composites. The reprint concludes with four chapters covering the development of innovative packaging systems produced using bio-based and/or biodegradable materials as a novel sustainable solution to extend food shelf life and reduce food waste.

This reprint offers a novel knowledge source for a diverse audience of researchers and professionals working in the areas of food packaging technology, materials science and sustainability. The reprint compiles original and cutting-edge research works accessible to a broad readership that can serve to promote the application of Circular Bioeconomy within both industry and governments.

Sergio Torres-Giner, Amparo Chiralt, and Chelo Gonzalez-Martinez
Editors

foods

Editorial

Circular Bioeconomy: Novel Processes and Materials for Food Preservation

Sergio Torrres-Giner *, Amparo Chiralt and Chelo González-Martínez

University Institute of Food Engineering—FoodUPV, Universitat Politècnica de València (UPV), Camino de Vera s/n, 46022 Valencia, Spain; dchiralt@tal.upv.es (A.C.); cgonza@tal.upv.es (C.G.-M.)
* Correspondence: storresginer@upv.es

Citation: Torrres-Giner, S.; Chiralt, A.; González Martínez, C. Circular Bioeconomy: Novel Processes and Materials for Food Preservation. *Foods* **2023**, *12*, 4341. https://doi.org/10.3390/foods12234341

Received: 8 November 2023
Accepted: 17 November 2023
Published: 1 December 2023

Food preservation is a set of procedures and resources aimed at blocking the action of external and internal agents that may alter the original characteristics of food. It has implications regarding aspects of food quality and food safety since it delays the deterioration of organoleptic and nutritional properties and prevents toxicity and contamination. Alteration agents can be grouped into physical (e.g., mechanical damage, temperature, humidity, air, and light), chemical (e.g., browning or rancidity reactions), and biological (e.g., enzymes, macro-, and microorganisms) agents. The food industry has traditionally made use of a combination of different processes and materials to ensure food preservation. These techniques have been classically grouped into two methods, those based on the use of temperature (e.g., pasteurization, sterilization, refrigeration, and freezing) and non-thermal methods (e.g., dehydration and drying processes, the application of electromagnetic waves, radio frequency, salting, curing, acidification, pickling, glazing, smoking, etc.). However, in the 21st century, food preservation technology is facing two major challenges, namely the requirement to reduce food waste and to use more sustainable packaging materials. On the one hand, according to the United Nations Food and Agricultural Organization (FAO), approximately 1/3 of the total food produced worldwide was lost or wasted in the supply chain. This represents 1.3 billion tons of food with direct costs of nearly USD 750 billion yearly. Aside from its economic impact, food waste is a potent greenhouse gas emitter (mainly methane) contributing to environmental pollution. Moreover, the fast-growing world population is accelerating the increased demand for food, and, subsequently, the larger quantity of food production with its increased generation of food wastes. On the other hand, over 85% of food is distributed processed and/or packaged in the retail market, where plastic materials are present in more than 50% of single-use packaging and 40% of the total. This accounts for an annual plastic consumption of approximately 380 million tons, which is based on a linear economy model where 95% of the plastic packaging material value, that is, $80–120 billion annually, is lost after the first short-use cycle. Therefore, both food waste and plastic food packaging pose a challenge not only from an environmental but also from an economic point of view.

From the above, crafting methods of valorizing food waste and reducing the environmental impact of food plastic packaging are both crucial for developing a sustainable productive model and for achieving the United Nations (UN) Sustainable Development Goals (SDGs) of 2030. For the implementation and achievement of these sustainable development policies, the valorization of food waste under the biorefinery framework together within the circularity of processes and products have gained momentum in recent years. Thus, the shifting of these policies and regulations on how to mitigate and/or utilize food wastes has forced the development of the so-called Circular Bioeconomy, which is now becoming an integral part of industrial green technological processes. It is based on a combination of the principles of both Bioeconomy and Circular Economy, in which biomass and residues are originally regarded as a novel source of raw materials, whereas, at the same time, end products follow a closed model where no waste is generated and the value is recovered. Thus, in the frame of food preservation, the concept of the Circular Bioeconomy can be described as the production of materials and additives from food waste and biomass in a sustainable and integrated/cascaded manner while generating zero waste. Food wastes include both the edible and non-edible parts of food that are generated throughout the whole food supply chain, which are thus potential sources of biopolymers and natural additives. Biopolymers, such as proteins, carbohydrates, and biopolyesters, have special advantages in the domain of biodegradable packaging materials. Moreover, these biopolymers can be formulated with antioxidant and antimicrobial compounds extracted from food waste to develop active and bioactive packaging systems that can contribute to the extension of food shelf life. This Special Issue aims to compile some of the most recent advances in the processes and materials dealing with the valorization of agricultural and food wastes as well as the development of innovative sustainable solutions to improve food preservation and reduce food waste.

In the Circular Bioeconomy context, new food packaging technologies based on biodegradable polymers formulated with natural additives should be developed. As described in the review of Baghi et al. [1], nano- and microcapsules of natural antioxidant and antimicrobial additives can be successfully incorporated into biodegradable polymers to protect and extend the shelf life of perishable foods. Nano- and microencapsulation processes also increase the bioavailability and solubility of these active compounds, providing controlled release and targeted delivery. Furthermore, these additives can additionally improve the thermal and chemical stability of biopolymers. Following this trend, a thymol (TO)-rich activated carbon (AC) nanohybrid structure was prepared by Giannakas and coauthors [2]. This nanohybrid material was, thereafter, extruded with low-density polyethylene (LDPE) to develop novel active packaging films. The authors observed that loadings of 15 wt% of the nanohybrid material improved the barrier properties of LDPE, exhibiting 230% and 1928% higher barriers to water vapor and oxygen, respectively. Moreover, the resultant active film presented high antioxidant activity, showing 44.4% in the 2,2-diphenyl-1-picrylhydrazyl radical (DPPH) assay, a low thymol release rate ($k_2 \approx 1.5\ s^{-1}$), and high antibacterial activity, resulting in a 2-day extension of fresh pork fillet shelf life. Nevertheless, it is also worth noting that these research works highlighted the current restriction of the application of encapsulated antimicrobial and antioxidant agents in biopolymers for active packaging applications by European regulations for food-contact materials and safety aspects. It was envisaged that the application of nanotechnology in active food packaging will be promoted by the elimination of legislative restrictions and the creation of a unique global organization. At the same time, greater control and monitoring of nanomaterials as well as risk assessment and public consumer familiarization will be required for the acceptance of such an innovation.

In addition, the scientific community is undertaking strong efforts for the utilization of food-processing by-products as a highly sustainable resource to produce food packaging materials and active and bioactive compounds. Multiple research groups are currently engaged in the complete valorization of food wastes into alternative feedstocks for monomers and intermediates to produce biopolymers and green composites. Some intense research is also being conducted in the extraction and purification of antioxidants, natural preservatives, and antimicrobials to enrich food quality and provide food safety. In this regard, Freitas et al. [3] demonstrated the potential of ultrasound-assisted extraction (UAE) as an eco-friendly and efficient process to extract bioactive compounds from rice straw. The performance of UAE treatments in combination with conventional heating and stirring extraction methods was evaluated in terms of the extraction kinetics of phenolic compounds as well as the antioxidant and antimicrobial activities of the obtained extracts. Higher extract yields and antioxidant activities of the water-soluble phenolic compounds were observed through the application of ultrasound pretreatment followed by thermal treatment. In particular, extracts with improved antioxidant activities were attained after ultrasonication for 30 min followed by thermal treatment in water reflux at 100 °C for 60 min. The latter conditions favored, once the substrate surface exposed to the extraction was effectively increased, the promotion of covalent bond cleavage between the phenolic compounds and the lignocellulosic fraction. In another example of valorization, Pavon et al. [4] obtained pine resin and gum rosin derivatives and then used them as natural low-cost additives, such as stabilizers, compatibilizers, and/or plasticizers by. Pine resin and gum rosin, the non-volatile fraction of pine resin, are exudated from conifers and tapping pine trees. The collection of these secretions has recently resurged due to cleaning activities being undertaken to reduce fire risk. The authors showed the benefit of incorporating gum rosin and four gum rosin derivatives into thermoplastic starch (TPS) via melt extrusion. It was determined that loadings of 10 wt% of pine resin derivatives can stiffen the structure of TPS from −100 °C to 40 °C, suggesting a good filler-to-matrix cohesion in this temperature range. Moreover, all the biopolymer/gum rosin formulations disintegrated in less than 90 days under composting conditions. These sustainable materials were proposed to be applied for obtaining rigid packaging materials that do not suffer plastic deformation at room temperature and are fully bio-based and biodegradable and thus aligned with the principles of Circular Bioeconomy.

Also following the Circular Bioeconomy approach, lignocellulosic materials obtained from processing by-products and wastes of the food and agroforestry industries are also gaining more attention as cost-effective fillers in biopolymer-based composites due to their natural origin and favored biodegradability. For instance, Terroba-Delicado et al. [5] valorized spent coffee grounds (SCGs), a type of waste generated during the 'fruit-to-cup' transformation of coffee beans. In this research, SCGs obtained from spent coffee grains used in the liquor industry remarkably increased the ductility of compostable polylactide (PLA) pieces. Moreover, the simultaneous addition of chemically modified oligomers of lactic acid (OLAs) contributed to improving the impact strength, and, more notably, the tensile strength of PLA. Thus, the resultant green composite pieces with enhanced mechanical performance are great candidates for food packaging and disposable articles. As shown by Mellinas et al. [6], the cocoa bean shell extract, which is obtained from the residues of the chocolate production process, also represents a great opportunity for waste valorization. The authors developed active films of pectin reinforced with these food waste derived fillers and zinc oxide/zinc nanoparticles (ZnO/Zn-NPs), reporting that both the extract and nanoparticles showed high compatibility with the carbohydrate matrix. As a result, loadings of 3 wt% of nanoparticles enhanced pectin's barrier to oxygen by 50% and the screen to ultraviolet (UV) radiation reached 98%. Moreover, the photocatalytic activity of the nanoparticle-containing pectin films was demonstrated, showing photodegradation efficiency values of nearly 90% after 60 min, which are all essential properties for the packaging of food that is highly sensitive to oxidative degradation.

Finally, the novel packaging materials that are currently being developed according to the Circular Bioeconomy's principles should also provide shelf-life extension to reduce food waste. This was demonstrated, for instance, by Bugatti et al. [7], who developed active cellulose-based trays to preserve ready-to-eat pasta. These trays were made of cellulose coupled with cellulose acetate and coated with layered double hydroxide (LDH) nanofiller containing 4-hydroxybenzoate. The authors evaluated the organoleptic characteristics as well as mold and bacterial growth for up to 30 days of storage at 4 °C, showing that the shelf life of the cooked pasta with tomato sauce was improved. In another study, Hernández-García et al. [8] studied the food packaging performance of films of poly(3-hydroxybutyrate-*co*-3-hydroxyvalerate) (PHBV) reinforced with atomized microfibrillated cellulose (MFC). To this end, the green composite films were applied as lids in trays to package minced pork meat and sunflower oil. It was observed that the PHBV/MFC films successfully preserved the physicochemical and microbiological quality of pork meat for one week of storage at 5 °C, whereas these also prevented sunflower oil oxidation in accelerated oxidative storage conditions for 21 days. The quality and shelf life of pork meat were also evaluated in another research study using multilayer PLA films [9]. The packaged meat fillets were effectively preserved for 7 days in PLA, showing values of food quality and safety in the range of commercial high-barrier multilayers, which are currently based on structures made of non-biodegradable polymers that are extremely difficult to recycle. Finally, Sun et al. [10] demonstrated that films of poly(L-lactide-*co*-butylene fumarate) (PLBF) and poly(L-lactide-*co*-glycolic acid) (PLGA) can successfully extend the shelf life and maintain the quality during cold storage of white mushrooms (*Agaricus bisporus*). These copolymers have the advantage to modify the oxygen (O_2) and carbon dioxide (CO_2) ratio of PLA, resulting in lower permeability values. In particular, the copolymer PLBF film was able to offer a shelf life of up 15 days, which was achieved by creating an optimal gas composition for mushroom preservation, that is, 0.27–0.11% O_2 and 6.33–7.92% CO_2. This new biopolymer packaging material significantly reduced respiration metabolism, membrane lipid peroxidation, and postharvest senescence.

Funding: This research was funded by the Spanish Ministry of Science and Innovation (MICI), grant number PID2021-128749OB-C33.

Acknowledgments: S.T.-G. acknowledges MICI for his Ramón y Cajal contract (RYC2019-027784-I). The guest editors thank all the authors for submitting their work to this Special Issue and for its successful completion and also the reviewers for participating in the peer-review process of the submitted manuscripts. We are also grateful to Kathrine Zhao and the editorial assistants of *Foods* who made the entire Special Issue creation a smooth and efficient process.

Conflicts of Interest: The authors declare no conflict of interest.

References

1. Baghi, F.; Gharsallaoui, A.; Dumas, E.; Ghnimi, S. Advancements in Biodegradable Active Films for Food Packaging: Effects of Nano/Microcapsule Incorporation. *Foods* **2022**, *11*, 760. [CrossRef] [PubMed]
2. Giannakas, A.E.; Karabagias, V.K.; Moschovas, D.; Leontiou, A.; Karabagias, I.K.; Georgopoulos, S.; Karydis-Messinis, A.; Zaharioudakis, K.; Andritsos, N.; Kehayias, G.; et al. Thymol@activated Carbon Nanohybrid for Low-Density Polyethylene-Based Active Packaging Films for Pork Fillets' Shelf-Life Extension. *Foods* **2023**, *12*, 2590. [CrossRef] [PubMed]
3. Freitas, P.A.V.; González-Martínez, C.; Chiralt, A. Application of Ultrasound Pre-Treatment for Enhancing Extraction of Bioactive Compounds from Rice Straw. *Foods* **2020**, *9*, 1657. [CrossRef] [PubMed]
4. Pavon, C.; Aldas, M.; López-Martínez, J.; Hernández-Fernández, J.; Arrieta, M.P. Films Based on Thermoplastic Starch Blended with Pine Resin Derivatives for Food Packaging. *Foods* **2021**, *10*, 1171. [CrossRef]
5. Terroba-Delicado, E.; Fiori, S.; Gomez-Caturla, J.; Montanes, N.; Sanchez-Nacher, L.; Torres-Giner, S. Valorization of Liquor Waste Derived Spent Coffee Grains for the Development of Injection-Molded Polylactide Pieces of Interest as Disposable Food Packaging and Serving Materials. *Foods* **2022**, *11*, 1162. [CrossRef] [PubMed]
6. Mellinas, A.C.; Jiménez, A.; Garrigós, M.C. Pectin-Based Films with Cocoa Bean Shell Waste Extract and ZnO/Zn-NPs with Enhanced Oxygen Barrier, Ultraviolet Screen and Photocatalytic Properties. *Foods* **2020**, *9*, 1572. [CrossRef]
7. Bugatti, V.; Viscusi, G.; Gorrasi, G. Formulation of a Bio-Packaging Based on Pure Cellulose Coupled with Cellulose Acetate Treated with Active Coating: Evaluation of Shelf Life of Pasta Ready to Eat. *Foods* **2020**, *9*, 1414. [CrossRef] [PubMed]
8. Hernández-García, E.; Chiralt, A.; Vargas, M.; Torres-Giner, S. Lid Films of Poly(3-hydroxybutyrate-*co*-3-hydroxyvalerate)/Microfi brillated Cellulose Composites for Fatty Food Preservation. *Foods* **2023**, *12*, 375. [CrossRef] [PubMed]
9. Hernández-García, E.; Vargas, M.; Torres-Giner, S. Quality and Shelf-Life Stability of Pork Meat Fillets Packaged in Multilayer Polylactide Films. *Foods* **2022**, *11*, 426. [CrossRef] [PubMed]
10. Sun, T.; Bian, J.; Wang, Y.; Hu, J.; Yun, X.; Chen, E.; Dong, T. One-Step Synthesis of Poly(L-Lactic Acid)-Based Soft Films with Gas Permselectivity for White Mushrooms (*Agaricus bisporus*) Preservation. *Foods* **2023**, *12*, 586. [CrossRef] [PubMed]

foods

MDPI

Review

Advancements in Biodegradable Active Films for Food Packaging: Effects of Nano/Microcapsule Incorporation

Fatemeh Baghi [1,2], Adem Gharsallaoui [1], Emilie Dumas [1] and Sami Ghnimi [1,2,*]

1 Laboratoire d'Automatique, de Génie des Procédés et de Génie Pharmaceutique, CNRS, University Claude Bernard Lyon 1, 43 Bd 11 Novembre 1918, 69622 Villeurbanne, France; fbaghi@isara.fr (F.B.); adem.gharsallaoui@univ-lyon1.fr (A.G.); emilie.dumas@univ-lyon1.fr (E.D.)

2 Institut Supérieur d'Agriculture et Agroalimentaire Rhône-Alpes (ISARA), 23 Rue Jean Baldassini, CEDEX 07, 69364 Lyon, France

* Correspondence: sami.ghnimi@univ-lyon1.fr or sghnimi@isara.fr; Tel.: +33-(0)4-27-85-86-70

Abstract: Food packaging plays a fundamental role in the modern food industry as a main process to preserve the quality of food products from manufacture to consumption. New food packaging technologies are being developed that are formulated with natural compounds by substituting synthetic/chemical antimicrobial and antioxidant agents to fulfill consumers' expectations for healthy food. The strategy of incorporating natural antimicrobial compounds into food packaging structures is a recent and promising technology to reach this goal. Concepts such as "biodegradable packaging", "active packaging", and "bioactive packaging" currently guide the research and development of food packaging. However, the use of natural compounds faces some challenges, including weak stability and sensitivity to processing and storage conditions. The nano/microencapsulation of these bioactive compounds enhances their stability and controls their release. In addition, biodegradable packaging materials are gaining great attention in the face of ever-growing environmental concerns about plastic pollution. They are a sustainable, environmentally friendly, and cost-effective alternative to conventional plastic packaging materials. Ultimately, a combined formulation of nano/microencapsulated antimicrobial and antioxidant natural molecules, incorporated into a biodegradable food packaging system, offers many benefits by preventing food spoilage, extending the shelf life of food, reducing plastic and food waste, and preserving the freshness and quality of food. The main objective of this review is to illustrate the latest advances in the principal biodegradable materials used in the development of active antimicrobial and antioxidant packaging systems, as well as the most common nano/microencapsulated active natural agents incorporated into these food-packaging materials.

Keywords: biodegradable packaging; active packaging; antimicrobial agent; antioxidant agent; nano/microencapsulation; biopolymers

Citation: Baghi, F.; Gharsallaoui, A.; Dumas, E.; Ghnimi, S. Advancements in Biodegradable Active Films for Food Packaging: Effects of Nano/Microcapsule Incorporation. *Foods* **2022**, *11*, 760. https://doi.org/10.3390/foods11050760

Academic Editors: Sergio Torres-Giner, Amparo Chiralt and Chelo Gonzalez-Martinez

Received: 12 February 2022
Accepted: 3 March 2022
Published: 6 March 2022

1. Introduction

The quality and safety of food products has always been a concern in the food industry. The large number of people affected by foodborne illnesses, about 600 million people in 2015 [1], and the increasing risk of the transmission of these diseases by the growth of international trade elucidates the necessity of improving food safety strategies. Microbial spoilage causes the loss of more than 25% of food products before consumption and wastes a considerable amount of food each year [2]. Using active packaging and antimicrobial additives for food preservation are two important target areas to protect and extend the shelf-life of perishable foods by preserving them from external environmental impacts and contamination [3]. Packaging can reduce or prevent foods from physical damage and spoilage and preserve beneficial constituents and organoleptic properties from the time of packaging to the time of consumption [4]. Different parameters are considered to control the food quality, including water activity, pH, temperature, light, and the partial pressures of oxygen and carbon dioxide. Thus, food packaging can be designed based on these parameters to avoid contamination and deterioration [5]. Despite many advanced technologies in packaging, there is always more demand for novel packaging systems with a higher efficacy, which help maintain the safety and quality of food products in a safe and environmentally sustainable manner.

There are three issues in this regard. First, today's lifestyle calls for packaged foods with an extended shelf life, especially in industrialized countries. Second, consumers are more concerned about good eating habits and the harmful effects of synthetic/chemical additives in food products. They want to consume healthier and more natural foods. Third, the rise in plastic pollution is a global concern. Therefore, the packaging should be designed following these three principles.

Among the different types of food packaging, the conventional way of packaging perishable foodstuffs, which has been extensively used for many years, is the use of plastic films based on petrochemical polymers. Considerable amounts of the polymers produced each year are used in food packaging. It accounts for 36.0, 40.7, and 46.5% of the world, Japan, and South Korea, respectively [6]. The main plastic packaging materials used in food packaging are polyethylene (PE), polyethylene terephthalate (PET), polypropylene (PP), polystyrene (PS), and polyvinyl chloride (PVC), which represent more than 90% of the total volume of plastics used in industry and about 50–70% of the total plastic waste derived from them [7–9].

Flexible packaging plastics are massively used in packaging due to their capacity to extend self-life and facilitate product handling. They take various forms, including plastic films, bags, flexible food packaging plastics (including mono-layered and multi-layered), and other single-use flexible plastics [6].

These materials have various advantages, such as good processing properties, excellent physicochemical characteristics, stability, resilience, and low price [10]. However, they are non-renewable, non-biodegradable, and take a long time to decompose [11]. Only 9% of these materials are recycled, 12% are incinerated, and the rest are discarded into nature [12]. Millions of tons of plastics are wasted, with more than 32 million tons yearly in only the United States [13], leading to a significant amount of plastic in the environment. Plastics accumulate in landfills and ecosystems such as oceans and coasts, causing harmful environmental effects. For example, plastic packaging can break down into microplastics with low degradation rates and find its way into fish stomachs and, therefore, into the marine food chain [14,15]. This environmental pollution poses a serious threat to animals and humans with serious consequences such as contamination of food and lack of essential nutrients [15], brain damage, and behavioral disorders [16], altering human chromosomes and obesity, infertility, and cancer after long exposure [17,18].

Furthermore, these plastic materials depend on the limited petroleum resources [19,20]. Therefore, there is an urgent need to use alternative packaging materials to overcome these drawbacks. Thanks to many recent studies, biopolymers were introduced as biodegradable packaging materials to replace the petrochemical materials. They have several advantages for manufacturing food packaging, including biodegradability and nontoxicity [21,22]. Nevertheless, there are some challenges to coping with this route, such as non-sufficient mechanical and barrier properties compared with conventional fossil-based plastics.

The degradation of biopolymers is performed by microorganisms (e.g., bacteria, fungi) throughout enzymatic catalysis processes [23]. Bio-based polymers can be classified into different categories, as shown in Figure 1. The main biopolymers used in the manufacture of food packaging are bio-based polymers derived from polysaccharides, proteins, and lipids [24]. In recent years, many studies have been conducted using these natural biopolymers extracted from biomass to produce film packaging, which results in understanding their technical characteristics and their applications as sustainable and eco-friendly materials. Their beneficial properties, such as being biodegradable, renewable, and abundant in nature, make them a good choice for use in food packaging. However, certain improvements need to be made in areas such as the mechanical properties, heat transfer, gas and water vapor permeability, and their plasticity to be competitive with plastic films [22]. Microorganisms can also produce bio-based polymers, and new technologies are able to produce synthetic biodegradable polymers, as shown in Figure 1.

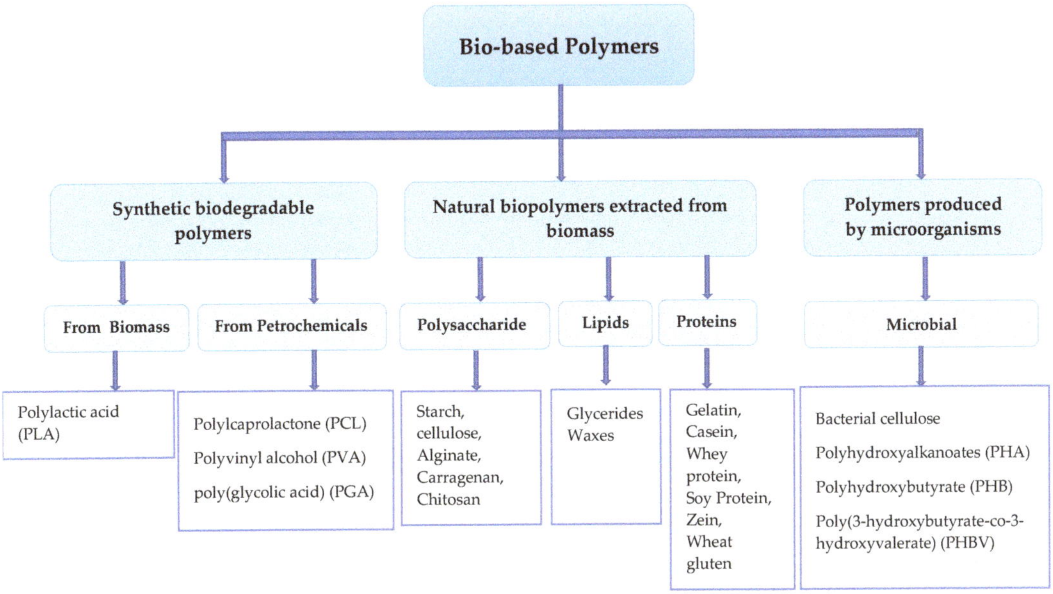

Figure 1. Classification of biodegradable polymers based on their source.

The use of bio-based polymeric materials is part of the progress of food packaging that takes into account environmental concerns. The next step in improving food packaging is to prevent or retard the deterioration of packaged food. Based on the sources used, we can differentiate between first-, second-, and third-generation feedstocks based on the classification of biofuels. The first generation is based on edible materials such as maize, sugarcane, whey, or corn. The second generation includes lignocellulosic and non-edible residues such as municipal waste or side-stream products from food and agriculture industries. The third generation includes biomass from algae [25]. It has been proven that films derived from biopolymers have an excellent potential to contain various additives, such as antimicrobial and antioxidant agents, in their matrix and release them during storage [2,26–28]. This packaging system is called "active packaging" and helps to extend the shelf life of food by absorbing (scavenging compounds) or diffusing (emitting compounds) various compounds. Scavengers have the ability to remove undesirable substances such as oxygen, moisture, ethylene, carbon dioxide, and odor from the internal packaging space. Emitters can release substances that have a positive impact, such as antimicrobial or antioxidant agents, enzymes, nutraceuticals, and aromatic compounds [2,29].

Active packaging can reduce the risk of foodborne pathogens and improve the quality and safety of food products [1,4]. They offer many advantages by reducing, restricting, or inhibiting the growth of spoilage and pathogenic microorganisms in food products and effectively extending the shelf life of packaged content. This strategy avoids adding the antimicrobials directly to the food product and offers the potential to control their release during storage time [27,28,30].

Among the incorporated additives, much attention has been paid to natural antimicrobial agents. Among these, essential oils (EOs), which are aromatic substances that are considered as secondary metabolites and are extracted from different parts of plants and used for many years in traditional medicines, have attracted a great deal of attention in the food industry [31,32].

Despite the promising results of EOs as antimicrobial agents, their volatility, low solubility in water, and sensitivity to oxidation limit their applications in producing food packaging films [3,33]. A promising technology to eliminate these limitations is encapsulation. Encapsulation could be performed through different techniques in which EOs are considered a core material covered by wall materials. Various wall materials, such as carbohydrates, proteins, and lipids, were used to encapsulate active EOs molecules, which improves their solubility and stability and protects them during film making [31,34]. In this context, it is important to choose appropriate wall materials compatible with core properties and the conditions of encapsulation and film formation [35]. Microencapsulation is a process in which active molecules are loaded into micro-size-capsules (>1 μm) protect and isolate them from environmental conditions, masking undesirable flavors, and increasing their solubility and stability [36]. These microcapsules can be modified by reducing their size to obtain nano-size-capsules (less than 100 nm) in order to improve their properties, including bioavailability, solubility, and adsorption [34,37]. Generally, the main outcomes of nanoencapsulation are the reinforcement of biological activity and bioavailability due to the larger surface of interaction with food and, thus, the higher possibility of penetration into the cell membrane of microorganisms, which results in being more effective and supplying controlled release to the food. The strategy of incorporating nanoencapsulated bioactive molecules into film packaging extends the time of food storage by inhibiting the growth of pathogenic and spoilage microorganisms, along with using fewer amounts of antimicrobial agents and avoiding their direct addition to food products.

Considering the above discussion, this review aims to summarize the potential of bio-based polymers and the natural antimicrobial and antioxidant agents in the area of biodegradable active packaging. It also highlights the encapsulation strategies used to overcome the bioactive molecules limitations, and the most common nano/microencapsulated active natural agents incorporated into these food-packaging materials.

2. Biodegradable Food Packaging

More than 40% of the petroleum-based plastic materials produced are used for packaging, and half of those are used for food packaging. About 95% of plastic packaging is discarded after the first use [38]. The degradation of plastics can occur through chemical, physicochemical, or physical processes. It may take place following one or a combination of four mechanisms in nature, namely, photodegradation, hydrolysis, thermo-oxidative degradation, and biodegradation [39]. Photodegradation caused by UV-B radiation often initiates the degradation of common plastics. Degradation then moves forward through hydrolysis and thermo-oxidative degradation. Finally, degradation processes break down plastics into fragments with a lower molecular weight, which can be metabolized by microbes. The duration of degradation depends on the conditions and type of material, but the process is slow, and it may take more than 50 years for full degradation [39,40].

The definition of biodegradable bioplastic according to the European standard EN 13432 is "Under the action of microorganisms in the presence of oxygen, decomposition of an organic chemical compound into carbon dioxide, water and mineral salts, other elements present (mineralization) and appearance of a new biomass. In the absence of oxygen, decomposition into carbon dioxide, methane, mineral salts, and creation of a new biomass" [41]. Plastic biodegradation has been studied by focusing on the isolation of individual microorganisms that are able to decompose biodegradable plastics. It has been proved that bacterial and fungal consortia have a great potential in plastic waste biodegradation and bioremediation [17]. The polymeric chains are broken down into compounds such as oligomers, dimmers, and monomers with a lower molecular weight by enzymatic fractionation [42].

Plastic production has almost doubled during the last decade, and it is predicted that it will increase more, reaching 33 billion tons by 2050. The recent world health crisis caused by the spread of COVID-19 caused more plastic pollution due to the massive consumption of plastics for medical use and personal protective equipment, as well as using plastic shopping bags to prevent cross-contamination. For instance, medical waste has sharply increased by 370% and packaging plastic by 40% [43].

Consequently, plastic pollution has become one of the most critical environmental problems. Plastic has become ubiquitous, and it is a serious global problem for nature, human health, society, and the global economy [44–46]. Plastic pollution causes hazardous impacts on water systems, soil conditions, and air quality. The digestion and entanglement of plastic debris is an acute danger to animal and human life, and plastic waste destroys wildlife habitats [46]. Plastic consumption causes many troubles, including behavior disorders, dietary dilution, obesity, altering human chromosome sequences, infertility, and even cancer [17]. In addition, recent research investigated the effect of plastic consumption and proved the negative impacts on adolescent mental health and behavioral problems [47].

Thus, the growing accumulation of plastic wastes in nature increases the pressure to tackle the problem. Recycling aims to consume fewer raw materials, but it remains technologically and economically a big challenge [48]. Various degrees of recyclability of different products, consumer information about different recycling labels for plastic, appropriate technology, plastic waste management strategies, and risk of contamination make the recycling process complex. Indeed, the collecting, preprocessing, and recycling of plastic waste is complicated and expensive [38]. In fact, less than 10% of all plastic waste was recycled by the end of 2015 [12,49]. Incineration is also used to tackle plastic pollution, but it causes the release of undesirable gases such as dioxins, dioxin-like compounds, carbon monoxide, nitrogen oxides, alkenes, alkanes, aromatic, and chlorinated hydrocarbons into the atmosphere [50].

Therefore, biodegradability is a critical point that should be considered. Indeed, some petroleum-derived compounds, such as aliphatic-aromatic copolymers, have the potential to produce biodegradable polymers with good technical properties and modify their microstructure and composition to meet the particular requirements for different applications [51]. While biodegradability is a useful feature of these petroleum-based polymers, the processes of their production are polluting. Furthermore, fossil resources are limited, and they recharge only after millions of years. Thus, an alternative to fossil-based polymers is essential.

Thanks to extensive research, bioplastics have been introduced as a promising alternative that can be biodegradable and renewable. Figure 2 illustrates a general classification of biodegradable and non-biodegradable fossil and biobased plastics. Hence, there are the following four different groups of plastics:

- Fossil-based and non-biodegradable: refers to classical plastics such as conventional polyethylene (PE) and polystyrene (PS);
- Fossil-based and biodegradable: includes polycaprolactone (PCL), polybutylene succinate (PBS), and poly (butylene adipate-co-terephthalate) (PBAT);
- Bio-based and non-biodegradable: bio-polyethylene (PE) is an example of this group produced from bioethanol fuel, which is produced from sugar cane;
- Bio-based and biodegradable: this group is an interesting choice with high potential to apply in food packaging without environmental impacts, which can be natural or synthetic such as cellulose, starch blends, and polyesters such as PLA and PHA [52].

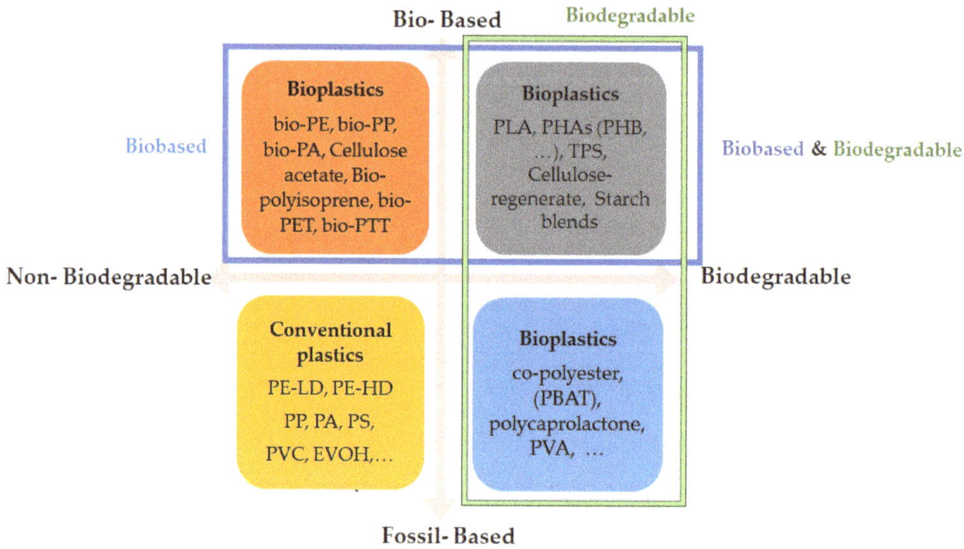

Figure 2. Material coordinate system of bioplastics (adapted from [53]). EVOH: ethylene-vinyl alcohol; PA: polyamide; PBAT: polybuthylene adipate terephthalate; PE: polyethylene; PE-HD: high-density polyethylene; PE-LD: low-density polyethylene; PET: poly(ethylene terephthalate); PHA: polyhydroxyalkanoate; PHB: polyhydroxybutyrate; PLA: polylactic acid; PP: polypropylene; PS: polystyrene; PTT: polytrimethylene terephthalate; PVA: polyvinyl alcohol; PVC: polyvinyl chloride; TPS: thermoplastic starch.

European Bioplastics introduced the term "bioplastics" in 2016 [52]. In order to better understand this concept and clear up any confusion, it is worth noting some definitions related to this notion. The term "biomaterials" is reserved for materials for medical use and is defined as "non-living material used and designed to interact with biological systems". Biopolymers are polymers made by living organisms (proteins, polysaccharides, etc.) or biodegradable synthetic polymers. Agro-polymers are also applied as polymers made by living organisms and derived from agricultural resources (starches, cellulose, etc.). Materials that are derived in whole or in part from biomass resources are called "bio-based." Biomass resources are organic materials that are available on a renewable or recurring basis, such as crop residues, wood residues, grasses, and aquatic plants. Corn ethanol is a well-known example of a bio-based material derived from biomass resources. Bio-based refers technically to any product that contains some amount of bio-based material. The term is typically applied only to materials containing carbon. Bioplastics is a broader term, dedicated to biodegradable and/or bio-based polymers produced by synthesis [52].

The production of bioplastics has increased in recent years. About 2.1 million tons were produced in 2019, according to European Bioplastics, and this amount is predicted to rise to 2.4 million tons around 2024. Biodegradable and non-biodegradable parts account for 58.1% and 41.9% of total bioplastics, respectively (Figure 3). Meanwhile, the total capacity of biodegradable polymers is still quite small compared to petroleum-based plastics.

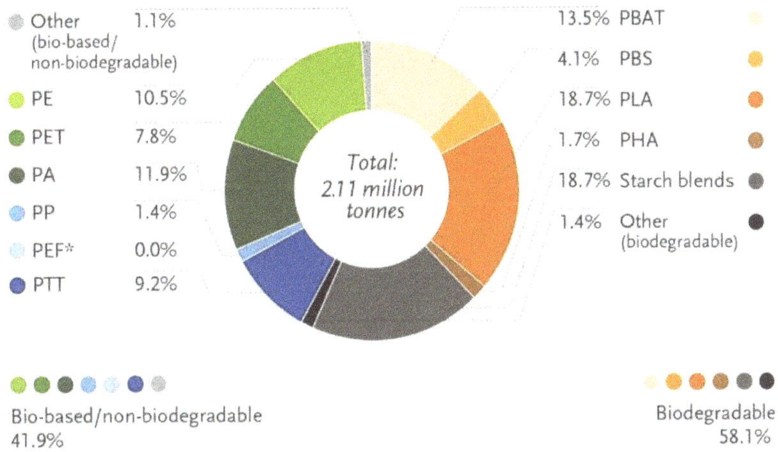

Figure 3. The projected global production of bioplastics in 2024 [53]. * PEF is currently in development and predicted to be available on a commercial scale in 2023.

Polybutylene adipate terephthalate (PBAT), polylactic acid (PLA), polyhydroxyl alkenoate (PHA), and polybutylene succinate (PBS) are principal synthetic bioplastics commercially used to date [52]. Research on bioplastics is in progress to improve their properties, such as film-forming and barrier properties, and broaden their applications. For example, polylactic acid (PLA) is an important example of this group. It is a very versatile polyester and the most prevalent polymer synthesized from biomass. PLA has, besides starch blends, the highest production capacity of the biodegradable materials available [54]. PLA is biodegradable and industrially compostable and is used in food packaging due to its non-toxicity. It is obtained from lactic acid during the fermentation of renewable crops such as sugar beets and corn [23]. PLA is sealable at lower temperatures and acts as both a flavor and an odor barrier for foodstuffs. Its industrial composting starts after only 8 days at temperatures between 50–60 °C and 60% relative humidity (RH). This time can be decreased if PLA is blended with PCL, a biodegradable petroleum-based polymer, and becomes home-compostable [55]. All polymer materials used for packaging can be theoretically substituted with those types synthesized from renewable monomers. However, there is an economic challenge regarding processing costs and also the lack of a collection channel for these bioplastics [56].

Some of the commercially produced biodegradable polymers are shown in Table 1.

Table 1. Overview of major commercial biodegradable polymers.

Biodegradable Polymers	Commercial Name	Company
Polyhydroxyalkanoate (PHA)/ Polyhydroxybutyrate (PHB)	Minerv	Bio-On, Italy
	Biocycle	PHB Industrial, Brazil
	Biomer	Biomer, Germany
	Nodax	Danimer Scientific, USA,
	AmBio	Shenzhen Ecomann, China
	Kaneka	Kaneka Corporation, Japan
	Solon	RWDC Industries, Singapore
	ENMAT	TianAn Biologic Mat., China
	Hydal	Bochemie, Czech Republic
	Green Bio	Tianjin Green-Bio, China
	PHB	Imperial Chemical Industries, UK
	TephaFLEX	TEPHA, USA
	ENMAT	Tinam, China
	PHA	SIRIM, Malaysia

Table 1. *Cont.*

Biodegradable Polymers	Commercial Name	Company
Starch	Solanyl BiomeHT Starch MATER-BI Starch Cardia Starch Starch	Rodenburg, Netherlands Biome Bioplastics, UK Green Home, South Africa Novamont, Italy Biobag, Norway Cardia Bioplastics, Australia Starch Tech Inc., USA Evercorn, Japan
Casein/Whey proteins	Casein Wheylayer	Lactips, France Wheylayer ltd, Germany
polybutylene succinate (PBS)	PBSA Bionolle EnPol, PBSA PBSA PBSA	Highpolymer, Japan Ire chemicals, South Korea Kingfa, China IPC-CAS, China
Polybutylene adipate terephthalate (PBAT)	Ecoflex Biomax MATER-BI Easter Bio	BASF, Germany Dupont, USA Novamont, Italy Eastman Chemicals, USA
Cellulose	CNF Eco, Cartocan MelOx Cellulose NatureFlex TIPA Zelfo Microcel	Toppan, Japan Klabin, Brazil International paper, USA Futamura, Japan TIPA Corp, Israel The Green Factory, France Roquette, France
Poly(lactic acid)	PLA PLA INGEO CPLA PLA L-PLA Bio-Flex NATIVIA PLA PLA PLA Ecovio	Bio4pack, Germany NatureWorks, USA Great River, China Galactic, Belgium Corbion, Netherlands FKuR, Germany Taghleef Industries, UAE Minima Technology, Taiwan Naturabiomat, Austria Natur-Tec, USA BASF, Germany

2.1. *Biopolymers*

Biopolymers are interesting candidates for biodegradable packaging. They are extensively used on the market, especially in food packaging, because they are bio-based, nontoxic, and biodegradable. They are biodegradable materials with a faster generation rate. Biopolymers can be obtained from various resources, such as plants and domestic and marine animals. Based on their origin, they are classified into the following three main groups: polysaccharides, proteins, and lipid-based packaging [57].

2.1.1. Polysaccharide-Based Packaging

Polysaccharides are one of the most important bio-based polymers. Thanks to many favorable properties and being nontoxic, polysaccharides have been used in food packaging. Polysaccharides are the most abundant carbohydrates found in nature. They are long-chain polymeric carbohydrates composed of monosaccharide units bound together by glycosidic linkages. They are made from nearly 40 different monosaccharides with various structures. There are different types of polysaccharides with distinct properties [58]. The following two classes of polysaccharides are available: (i) hetero-polysaccharides, which may contain two or more different monosaccharides such as gellan and xanthan, and (ii) homo-polysaccharides, which contain one type of mono-saccharide such as pullulan, curdlan, levan, and bacterial cellulose with linear chains [59]. Some polysaccharides, including starches, cellulose, alginates, carrageenan, and chitosan, are used for packaging because their properties are appropriate for manufacturing film packaging [4].

Starches

Starches are formed by large glucose units joined by glycosidic bonds with the chemical formula of $(C_6H_{10}O_5)_n$. It is the most common carbohydrate in human diets. Native starch consists of two types of glucose polymers, namely, amylose and amylopectin, and it is produced by most green plants, including corn, wheat, potatoes, rice, and cassava, etc., contributing to 80% of total starch production [60–62]. Starch is widely studied as a natural biodegradable material and has many applications in the fields of drug delivery, tissue engineering, biomedical scaffolds and stents, and food packaging [63]. Starch is non-toxic, abundant, biocompatible, and has good film-forming potential. Indeed, its polyhydroxy structure facilitates modulating its structure and functional properties through chemical or enzymatic reactions [64]. This great potential and the other attractive advantages of starches, such as being edible, available, low cost, and biodegradable, make them a good choice for food packaging. Despite all these properties, there are some limitations to applying native starch for food packaging, such as water susceptibility, brittle mechanical behavior, poor barrier properties, and trivial resistance to extreme processing conditions such as high temperature and shear in the native form. Several methods have been proposed to overcome these limitations. Using plasticizers such as glycerol or polyglycerol [65–67] and other additives, such as cellulose, gelatin, chitosan, and citric acid, can enhance the functionality of starch-based biodegradable materials [64]. Starch film properties could also be modified by using deep eutectic solvent [68] and under reactive extrusion (REx) conditions (high pressure and temperature, and low moisture content) [69]. Blending starch with many synthetic biodegradable polymers, including polylactic acid (PLA), polyvinyl alcohol (PVA), polycaprolactone (PCL), polybutyl succinic acid-butyl adipate (PBSA), and polyadipate butylene terephthalate (PBAT), is also a promising technique for improving mechanical and processing properties as well as the water-resistance of starch-based biodegradable materials [64]. These blends with starches, made by a proportion of about 30–80% of copolymers based on their application, are completely biodegradable [55,58,62,63].

Starch-based films can be used as a monolayer or laminated with other films to improve the barrier properties. Moreover, when combined with flexible polyesters such as PBAT, starch-based films become flexible, and blending with polylactic acid (PLA) improves their thermoforming properties and rigidity [57]. Furthermore, recently, the use of nanoparticles such as nanoclay or zinc in starch films has improved their mechanical properties [70–72].

Many authors have investigated the potential of using starch films in active packaging and reported the antimicrobial activities of different kinds of starch-based films by incorporating antimicrobial agents into them [73]. For instance, adding cinnamon essential oil to cassava starch film preserved bakery products against fungi and extended their shelf-life [70,74,75]. To date, starch-based films are emerging as an excellent alternative to conventional plastic for a variety of food packaging applications. For example, cassava starch-based foams were incorporated with grape stalks for the packaging of English cakes [76]. Starch-based films incorporated with citric pectin and feijoa peel flour (FPF) were used for apple packaging [77]. Cowpea starch-based films incorporated with maqui berry extract were used for salmon packaging [78]. Yam starch-based films incorporated with eugenol were used for pork [79], and pea starch (PS)/polylactic acid (PLA)-based bilayer films for cherry tomatoes [80]. Rye starch films containing rosehip extract (RHE) for chicken breast packaging [81], acetylated cassava thermoplastic starch and green tea blends with linear low-density polyethylene (LLDPE) films for sliced bacon [82], corn starch and gelatin films containing N-α-lauroyl-l-arginine ethyl ester monohydrochloride (LAE) for chicken breast fillets [83], and cassava starch-based films incorporated with zinc-nanoparticles for tomatoes preservation [84].

Cellulose

Cellulose is a polysaccharide with the formula $(C_6H_{10}O_5)_n$, consisting of a linear chain of several hundred to many thousands of $\beta(1 \rightarrow 4)$ linked D-glucose units. Cellulose and its derivatives are the most abundant biopolymers and are found in many resources [85].

These resources include wood, agricultural residues, factory and food wastes, food leftovers, marine organisms, some types of grass, cereal brans and husks, sugarcane bagasse, corn kernels, along with microbial biosynthesis (algae, fungi, bacteria) and peels of different fruits and vegetables [86,87]. Some of its derivatives are cellulose acetate, sulfate, nitrate and carboxymethyl, ethyl and methyl nano-cellulose [88]. They have interesting characteristics such as being edible, biodegradable, lightweight, nontoxic, bioavailable, and having nutritional value and favorable sensory and organoleptic properties such as color, appearance, aroma, flavor, and taste, and they can be easily found in significant quantities at a low cost. They also have a high potential to be used for encapsulating and incorporating various active molecules of antimicrobial and antioxidant materials [89]. It is worth mentioning that cellulose of bacterial origin exhibits extraordinary and differentiated properties compared to other polysaccharide-based polymers. It is receiving extensive attention for applications in the food industry as a thickening and gelling agent, stabilizer, water-binding additive, and also as a food packaging material [90].

Cellulose has excellent physical and mechanical properties and high thermal resistance, but there are some shortcomings, including a high water absorption capacity and insufficient interfacial adhesion. Cellulose has been studied for many years to improve its properties. Cellulose multilayers were assessed to enhance barrier properties, resulting in further tensile strength, as well as grease resistance and barrier to water [91]. The effect of incorporation of carotenoids in cellulose acetate, as natural antioxidants, was studied and showed good protection for foods susceptible to photooxidation [92]. Thereby, due to its outstanding characteristics, cellulose used for food packaging reduces the cost of food packaging and preserves a wide range of foodstuffs in an eco-friendly manner.

Alginate

Alginate is the salt of alginic acid, a bio-based polymer consisting of D-mannuronic and L-guluronic monomers extracted from brown marine algae (brown seaweeds) and some bacteria, such as *Pseudomonas aeruginosa* [93,94]. Alginate is nontoxic, non-antigenic, biocompatible, and biodegradable, and it has the capacity to form a hydrogel when applied in encapsulation. Due to these properties, alginate is used in textiles, cosmetics, and pharmaceutics, and also in food industries as a thickener and stabilizer [95–98]. In the food industry, alginate is extensively used in the preparation of edible coatings because of its film-forming properties, hydrophilicity, and biocompatibility [99,100]. The application of alginate-based coatings in food preservation is limited due to their poor antimicrobial, UV-shielding, and water barrier properties. Some studies were dedicated to overcoming these shortcomings [101,102]. One recent study investigated the effect of the incorporation of phenolic compounds such as thymol into the sodium alginate film for fresh-cut apple slices. This film exhibited remarkable inhibition of the growth of *Staphylococcus aureus* and *Escherichia coli*, as well as a reduction in weight loss and retention of nutrients, and the surface color of apple slices. These thymol/sodium alginate films also showed a high tensile strength and elongation at break, as well as UV–vis light-blocking capability [103].

Carrageenan

Carrageenan is a polysaccharide extracted from marine red algae by hot alkali separation. Akin to other polysaccharides, carrageenan is nontoxic, biocompatible, and biodegradable with a low immunogenicity. Furthermore, carrageenan offers excellent properties compared with other encapsulation materials, such as a better stability of capsules, higher electronegativity, and a better protection of encapsulated contents, allowing them to be considered a good candidate in delivery systems for bioactive ingredients [104,105]. Therefore, it has an extensive range of applications in the food and pharmaceutical fields. Various structures of carrageenan make different biological activities, including antioxidant, antitumor, immunomodulatory, anti-inflammatory, anticoagulant, antiviral, antibacterial, antifungal, and anti-hyperglycemic properties.

Carrageenan has many applications in the food industry, including gelling, thickening, stabilizing, protective coating, and fat substitution [106]. Many authors developed pH-sensitive and antioxidant-packaging films based on carrageenan [107]. Another investigation showed a protective effect of fish oil, encapsulated in carrageenan, against lipid and protein oxidation and an improvement in the shelf life and sensory characteristics of enriched nuggets [108].

Chitosan

Chitosan, a linear polysaccharide with (1-4)-linked 2-amino-deoxy-β-d-glucan derived from chitin, is the second most abundant biopolymer and the most abundant biopolymer of animal origin. It is found in the exoskeletons of crustaceans and insects and in the cell walls of fungi and yeast [109]. It is a multipurpose biopolymer with interesting physicochemical and biological properties that are applied in agriculture, pharmacy, and biomedicine industries [110]. It is used in food packaging due to its low cost, biodegradability, biocompatibility, non-toxicity, film formation, viscosity, and ion binding [111]. The combination of chitosan with other biomaterials, nanometals, and active compounds allows further exploitation of chitosan and modification of some of the shortages of its characteristics, including hydrophilicity, weak mechanical properties, low gas permeability, and low encapsulation efficiency, as well as enhancing the bioavailability and biological properties [112].

The emerging chitosan-gelatin composite films present excellent physical properties, including mechanical, surface hydrophobicity, color, barrier, and thermal characteristics [113]. Further, chitosan films incorporated with ε-polylysine were used for packaging beef fillet. Different features were investigated, and this technique was introduced as an effective way to extend the shelf life of beef fillets and maintain their quality during refrigerated storage [114]. Another study highlighted the antioxidant and antimicrobial activities of chitosan-based packaging incorporated with nanocapsules of *Cinnamodendron dinisii* essential oil with zein as the wall material. This active chitosan-based film was tested on beef and showed preservative effects against spoilage and color changes [115].

2.1.2. Protein-Based Packaging

One of the other important types of bio-based polymers are proteins with two main plant and animal origins. They are made of different polar and nonpolar α-amino acids. There are different sources of plant-origin proteins such as wheat gluten, corn zein, soy, peanut, rice-bran, cotton seed, barley, and sunflower, along with animal origins such as gelatin, collagen, casein, whey, and fish myofibrillar protein [116]. The proteins of plant origin are used more due to their greater availability and lower cost.

Proteins are extensively used as packaging materials due to their good mechanical characteristics, abundance, and high nutritional value. They also have the potential to be incorporating agents for active packaging applications in addition to their nontoxicity and biodegradability [116,117]. However, some shortcomings, such as a high level of sensitivity to moisture and poor water vapor barrier properties, need to be improved. In fact, films that are composed of plant protein isolates have excellent oxygen barrier properties due to their polar nature and crosslinked polymer network; however, the water vapor barrier is low [25]. Using plasticizers, cross-linkers, and other additives through various physical, chemical, and enzymatic methods can be applied for this purpose [4,118]. The most used protein-based polymers are presented below.

Soy Protein

Soy proteins are composed of globulin proteins 7S (β-conglycinin) and 11S (glycinin), which have different structures, functional, and molecular properties. The functional properties of soy products depend on these two components [119].

Soy proteins have many applications, such as adhesives, composites, plastics, and so on, in different industries, including the food industry. Soy proteins are biodegradable and have exceptional film-forming properties. Furthermore, the effectiveness of incorporating antimicrobial compounds into soy protein films has been reported; therefore, soy protein can be used for producing film packaging [120–122]. Nevertheless, there are some limitations, such as low mechanical and thermal properties, poor processability, and water sensitivity, which can effectively be modified through laminating, coating with other polymers, plasticizing, nanoparticle reinforcing, or blending methods [123]. As an example, Ref. [124] investigated the mechanical and barrier properties of soy protein isolate-based films coated with polylactic acid and demonstrated improvements in the mechanical and water barrier properties of the film. Furthermore, using cellulose nanocrystals in soy protein films showed better film-forming, tensile strength, barrier properties, and water resistance. This film was evaluated for packaging pork and strawberries and showed fewer total viable counts and total volatile basic nitrogen of stored pork meat and extended the shelf-life of strawberries [125]. A new strategy of incorporating polyethyleneimine as a plasticizer and lignin-silver nanoparticles as a green reinforcer into soy protein films has made it possible to manufacture tough, strong, and UV-shielding soy protein-based composite films [126].

Wheat Gluten

Wheat gluten is a by-product of the wheat starch industry. It comprises mainly two water-insoluble proteins called glutenin and gliadin. Gliadins are composed of low-molecular-weight proteins ranging from 30,000 to 80,000 Da and glutenin containing high-molecular-weight proteins ranging from 80,000 to several million Da. The functional and structural characteristics of these two proteins determine the functional properties of wheat gluten. Generally, the films produced from glutenin showed better barrier properties than the films from gliadins or whole gluten [91,127,128]. Owing to its viscoelastic properties, lower solubility, and biodegradability, wheat gluten has many food and nonfood applications. Wheat gluten films are used in food packaging because of their oxygen barrier properties. However, they present a poor water vapor barrier that can be modified by adding plasticizers, coatings, and blends with hydrophobic polymers [129]. In this regard, Ref. [130] reported that wheat gluten/silica hybrid coating films could improve the inherent moisture sensitivity of this protein with a four-fold reduction in water vapor transmission rate. In addition, the films prepared from the blending of three thermoplastic gluten and polycaprolactone (PCL), with and without chrome octanoate as a food-grade catalyst, were recommended as potential shape-memory food packaging materials [131].

Wheat gluten films also have the potential for combination with antimicrobial materials to protect contents from spoilage [132]. The use of carvacrol as an antimicrobial agent and montmorillonite as a filler in wheat gluten-coated papers improved the antimicrobial efficiency against *Escherichia coli* [133].

Corn Zein

Zein is a class of prolamin protein found in the endosperm of corn (maize) mainly composed of α-zein, β-zein, and γ-zein [134]. It is extracted from corn gluten meal, a byproduct of starch production, through ethanolic extraction. Zein can form tough, glossy, hydrophobic, greaseproof coatings, and it is considered an antimicrobial agent. Zein is an abundant, renewable material with a low cost. Thus, it has a variety of applications in the food, chemical, environmental, medical, metallurgical, pharmaceutical, and biotechnology industries [135,136].

Zein is a hydrophobic and thermoplastic material, soluble in alcohol but insoluble in water that has film-forming properties and has been used commercially as an edible film for nuts and confectionery products for many years [4,137]. Zein films are manufactured via different processes, including the casting of solutions of zein, thermoplastic processing, and blown extrusion. The mechanical and thermal characteristics of zein are affected by the processes by which they are produced [137,138]. They offer high tensile strength and low water vapor permeability, and the water resistance of zein films can be improved by lamination with fatty acids because of their ability to form hydrophobic interactions with fatty acids [139]. Some mechanical properties of zein films need to be improved, such as fragility and flexibility, by adding plasticizers or blending with other biopolymers to remove these drawbacks [140–142]. A recent study assessed the mechanical properties of gelatin/zein composite films and introduced them as good candidates for high moisture food packaging applications due to the great improvement in water insolubility, water vapor permeability, and mechanical properties compared with conventional gelatin and zein films [143].

Casein and Whey Proteins

Milk proteins are composed of the following two dominant proteins: casein in the form of micelles, which accounts for approximately 80% of the total proteins, and the remaining 20% are whey proteins. Casein, the main protein of milk, has four types, namely, αs1-, αs2-, β-, and κ-casein [144]. Casein micelles are composed of large numbers of casein molecules, which are stabilized by calcium-phosphate bridging [145]. It can be used in food packaging because it can produce a flexible and transparent film with good gas and lipid barrier properties at low relative humidity. It also shows good emulsifying behaviors along with the mechanical and thermal stability of the caseinate.

Whey protein is a by-product of the cheese-making process or a by-product of acid casein production. Roughly 70% consists of a mixture of β-lactoglobulin (β-Lg) and α-lactalbumin (α-La). The other components of whey protein are protease peptones, bovine serum albumin, immunoglobulins, and other minor proteins [146].

Globally, milk proteins have high nutritional value, which provides much of the protein and amino acid dietary needs for the human diet. They are natural animal products with excellent functional and sensory properties and are safe for food applications [147]. Their functional and structural characteristics make them versatile materials as a sustainable source of biopolymers. Their most important properties are their nutritional value, biodegradability, biocompatibility, safety, thermal stability, emulsification, gelation, foaming, and water-binding capacities [148]. Thanks to these features, they are used in packaging and also in active packaging with incorporated bioactive and nutraceutical compounds [13,149]. The film-forming capacity and biodegradability of milk proteins make them a promising alternative to conventional non-biodegradable packaging [150]. Composite films containing milk protein-based biopolymers degrade in a very short time, with 100% degradation on day one without harmful residues. However, there are economic feasibility challenges to be overcome in the coming years [151].

Despite the potential of using milk proteins as a biopolymer for packaging, they exhibit some disadvantages compared to other conventional polymer materials, including their poor mechanical and barrier properties against gases, aromas, and water vapor. These drawbacks can be overcome in several ways, including blending with other edible biopolymers to form composite materials [152], multilayering films by coatings with nano- or micro-dimensions of more than one type of biopolymer, and also adding synthetic plasticizers [144,153]. An improvement in the bactericidal properties of the casein and chitosan blend in film packaging due to the presence of ionic interaction between both macromolecules was proven by [154].

It is worth noting that milk proteins constitute a range of biological activity components. For example, casein is an important source of antimicrobial peptides [155,156]. Furthermore, milk proteins can be used as macro and nano carriers for the encapsulation of nutraceuticals, vitamins, minerals, and antimicrobial compounds due to their biocompatibility, ease of controlling the release, and dispersibility of the encapsulated compounds [157]. Using antimicrobial agents in casein and whey protein film packaging has been investigated in many studies. The inhibitory effect on *Staphylococcus aureus* of casein-based films containing *Zataria multiflora* as the essential oil has been proved [158]. Further, the antimicrobial potential of cheese whey bioactive proteins and peptides such as lactoferrin in the development of antimicrobial edible film composites was reported by [159].

Gelatin

Gelatin is produced from the partial hydrolysis of collagen. In the past, gelatin has been manufactured from bovine and porcine skin and bone. In recent years, fish and poultry have also been used to produce gelatin. Production of fish gelatin is increasing due to vegetarianism and the potential of halal and kosher markets [160]. Gelatin has gel-forming properties that offer many applications in food, photography, cosmetics, and pharmaceutics. More notably, it is widely used in food to provide elasticity, viscosity, and stability. Further, it has a linear structure and limited monomer composition, leading to excellent film-forming properties [161]. Gelatin is used as an emulsifier, foaming agent, colloid stabilizer, biodegradable film-forming material, micro-encapsulating agent, and source of bioactive peptides when enzymatically hydrolyzed [162].

Gelatin films possess good mechanical, functional, and barrier properties, depending on the film's formulations and processing conditions. However, they are sensitive to moisture and show poor barrier properties against water vapor, which can be improved by using plasticizers or cross-linking agents or combining with other biopolymers such as soy protein isolate, oils, fatty acids, and certain polysaccharides [163–166].

Adding antimicrobial and antioxidant compounds into gelatin films also conferred more preservation of foodstuffs. Recently, the effect of curcuma in gelatin film was investigated by [167] and revealed remarkable antimicrobial activity against foodborne pathogenic bacteria, *E. coli*, and *L. monocytogenes*, and showed strong antioxidant activity along with improvement of the UV protection, water vapor barrier, and mechanical properties. In addition, the antimicrobial and antioxidant properties of the gelatin films were increased by embedding different kinds of functional nanoparticles such as quercetin, lactoferrin, and chitosan nanofiber into a gelatin-based film [168].

In conclusion, all these protein-based biopolymers are being studied massively due to their promising functions for the production of biodegradable active packaging with many beneficial economic and environmental attributes, enhancing the safety and quality of food products.

2.1.3. Lipids-Based Packaging

Lipids are classified into different types, such as phospholipids, phosphatides, mono-, di-, and tri-glycerides, terpenes, cerebrosides, fatty alcohol, and fatty acids. They are extracted from various natural resources of animal, insect, and plant origins. Among them, glycerides or waxes are mainly used in the production of films. These compounds are mainly nonpolar with a high hydrophobicity, insoluble in aqueous media, and soluble in organic solvents. They are used for coatings and edible films based on different biodegradable materials, including polysaccharides and proteins, due to their excellent moisture barrier properties [13,169]. They have a glossy appearance, minimize moisture loss, and reduce the cost of packaging films [170]. Lipids are also used as carriers for various bioactive compound-delivery in the food and pharmaceutical industries [171,172].

Packaging films using lipids have been studied for many years, and it is moving forward because of the capacity of lipids-based polymer to produce effective active edible films. Palm wax was incorporated into fish gelatin film used in food packaging, resulting in better UV barrier, mechanical, and water resistance properties [173]. Lipids can also be combined with other biopolymers in order to modify their poor hydrophobic characteristics. Recently, carnauba wax was added into a starch-based film to enhance its hydrophobicity and reduce its water solubility and improve its thermal stability and moisture, light and water vapor barrier properties [174].

2.1.4. Microorganism-Based Packaging

Microorganisms are another source to produce biopolymers. Many microorganisms are used to produce polymer materials, including *Alcaligenes, Bacillus, Azotobacter, Rhizobium*, and *Halobacterium* [57,175]. For instance, some polymers are produced from microorganisms such as polyhydroxyalkanoates (PHA), poly(β-hydroxybutyrate) (PHB), poly(3-hydroxybutyrate-co-3-hydroxyvalerate) (PHBV), levan, curdlan, gellan, dextran, bacterial cellulose, and microbial polysaccharides including exopolysaccharides, capsular polysaccharides, and xanthan. They are applied in the medical, agricultural, and packaging fields. Among them, PHAs and bacterial cellulose are the most produced polymers [22,55]. PHA polymers are resistant to heat and ultraviolet light. PHAs are durable for high-speed processing and can withstand high temperatures during storage [25]. Genetically modified microorganisms were also introduced to produce biodegradable polymers with renewable sources. However, genetic engineering is controversial, and it is believed that it is harmful to human health.

3. Active Biodegradable Packaging Films

Packaging is a barrier from gases, moisture, dust, or light, but conventional packaging is not totally effective for preventing the spoilage of food products [28] by chemical or biological reactions. Chemical spoilage mostly happens by oxidation processes of food ingredients, and biological spoilage is caused by enzymes, viral and parasitic activity, and microbial contamination [176]. Among them, microbial contamination is one of the main reasons for food deterioration. Many preservation techniques have been applied for a long time, such as fermentation, drying, thermal processing, freezing, refrigeration, modified atmosphere, irradiation, and, more notably, adding antimicrobial agents to the food [177].

In this regard, antimicrobial agents can be mixed with food to inhibit microbial growth. In such cases, a considerable quantity of antimicrobial compounds would be required, which may have an undesirable impact on the quality, taste, organoleptic properties, and appearance of foods. Moreover, as many of these materials have low stability, there is the possibility of neutralization of antimicrobial agents quickly at the same time by the active substances in food or during processing. Antimicrobial agents can also be added directly to the surface of food, and their effectiveness can be limited by diffusing them into the mass of food, which can cause the same unfavorable results as their combination with food [176].

Therefore, in order to overcome these problems and extend the preservation period of foods along with maintaining their quality, their sensory property, and freshness, some modern strategies are needed. Intelligent packaging and active packaging are two promising technologies that promote the shelf life of food products with minimal adverse effects. In intelligent packaging, various sensors and indicators are employed, including time-temperature indicators, gas indicators, humidity sensors, optical, calorimetric, and electrochemical biosensors for detecting defects, quality monitoring, and following the packaged contents to control the storage conditions during all steps of food processing to the consumption step [29].

Active packaging applies absorption and diffusion mechanisms to protect the foodstuffs and extend their shelf life. Active packaging can act as a scavenger and absorb various materials such as carbon dioxide, oxygen, and ethanol, which cause the spoilage of foods. Active packaging can contain various active molecules such as antimicrobials, antioxidants, and even coloring and nutritional agents, which can be emitted from the packaging into the foods during the storage time [178].

Active and intelligent packaging are already used for food packaging in the USA, Japan, Australia, and other countries. California-based Nature Fresh Farms is a company that uses active packaging for vegetables such as brussels sprouts and cucumbers, with plans to widen active packaging use for their products. Some other companies, including Uvesa, ElPozo Alimentación, BTSA. Biotecnologías Aplicadas, Monteloeder, Nurel, Bandesur Alcalá, and SP group, collaborated in an active packaging project to extend the shelf-life of meat with active packaging (Avanza-S project). Moreover, an edible coating developed by Eden Agritech, a startup company in Thailand, extended the shelf-life of fresh-cut fruit by up to three times [179].

However, strict European regulations for food contact materials and the safety aspects of the application of these innovative technologies have restricted their applications. Knowledge about consumer acceptance, economic aspects, and environmental impacts is required for more developments and applications [180]. Indeed, the development and application of active and intelligent packaging systems needs both legislative regulation and consumer acceptance. In this regard, a European study was conducted within the EU FAIR R&D program framework, called the Actipak project (January 1999–December 2001). It contributed to establishing and implementing active and intelligent packaging in the current relevant regulations for packaged food in Europe and initiated amendments to European legislation for food-contact materials. The aim of this project was the classification of active packaging, including antimicrobial release films, and the evaluation of antimicrobial films on cheese, meat, and fruit [180]. At that time, there was no European regulation that specifically covered the use of active and intelligent packaging. Later, in 2009, the issuing of European Regulation (EC) No. 450/2009 helped faster market penetration in Europe [181].

3.1. Antimicrobial Active Packaging

Antimicrobial packaging can be made from different polymers, including petrochemical-based polymers and bio-based materials. Meanwhile, a wide range of antimicrobial agents, chemical or natural, can also be applied to manufacture antimicrobial packaging films. The necessity of producing healthy, high-quality foods and minimizing the side effects of chemical agents, along with less environmental impacts, causes researchers to use natural antimicrobial agents and biodegradable materials to make antimicrobial biodegradable packaging films [176].

Antimicrobial agents incorporated into biopolymeric compounds are released slowly into the food surface over time. In this way, an active concentration of antimicrobial agents is present on food for a longer time without the risk of neutralization and negative effects on food. This method extends the lag phases for the growth of microorganisms and prevents food spoilage. Keeping the balance between the release rate of antimicrobial agents and the food spoilage kinetics is a key point in optimizing the effectiveness of active packaging [179].

It is not possible to find a single antimicrobial agent that can suppress all types of microorganisms and prevent all kinds of food spoilage. However, many different biopolymers have potential applications in active packaging.

Table 2 summarizes the recent studies related to food preservation by using antimicrobial bio-based films. There are several antimicrobial compounds authorized based on the standards of each country.

3.1.1. Natural Antimicrobial Agents of Plant Origin

Antimicrobial agents can be found in nature, especially in plants. A wide range of antimicrobial compounds are synthetized naturally in plants to protect against microorganisms and other predators.

There is a preference to use these agents in food because of the easy regulation process to extract them and consumer demand to reduce the use of synthetic food additives due to their undesirable effects. The most important natural antimicrobial agents are extracted from spices and herbs such as rosemary, cloves, horseradish, mustard, cinnamon, sage, oregano, basil, marjoram, savory, thyme, and many others. They are mainly in the form of essential oils (EOs) [182].

Table 2. Antimicrobial bio-based packaging for food applications.

Food	Antimicrobial Agents	Bio-Based Polymer	Target Microorganisms	Main Findings	References
Cheese	Essential oils from the following two spices: *Rosmarinus officinalis* and *Laurus nobilis*	Zein nanofibers	*Staphylococcus aureus* and *Listeria monocytogenes*	Both showed antimicrobial activity, with higher effects from *Laurus nobilis* than *Rosmarinus officinalis*.	[183]
Cheese	Moringa oil	Chitosan	*Listeria monocytogenes* and *Staphylococcus aureus*	High antibacterial activity against *Listeria monocytogenes* and *Staphylococcus aureus* at 4 °C and 25 °C for 10 days, without any effect on the sensory quality of cheese.	[184]
Soft (minas frescal) cheese	Nisin	Starch/halloysite/nanocomposite films	*Listeria monocytogenes*	After 4 days, antimicrobial nanocomposite films with 2 g/100 g nisin significantly reduced the initial counts of the bacterium and those with 6 g/100 g nisin completely inhibited *L. monocytogenes*.	[185]
Cheddar cheese	Nisin-silica liposomes	Chitosan	*Listeria monocytogenes*	Anti-Listeria activity without effect on the sensory properties of cheese.	[186]
Fresh cheese and apple juice	Nisin	pullulan nanofibers	*Leuconostoc mesenteroides* *L. monocytogenes* *Salmonella Typhimurium*	Bactericidal effect against *L. monocytogenes*, *L. mesenteroides*, and *S. typhimurium* in apple juice after 20, 48, and 48 h, respectively.	[187]
Chicken meat	Tea tree oil (TTO)liposome	Chitosan	*Salmonella enteritidis and Salmonella typhimurium*	Almost no impact on the sensory properties. In total, 5 $\log_{10}$ reductions of *Salmonella* were observed in chicken meat by TTO liposomes/chitosan nanofibers treatment for 4 days at 12 °C and 25 °C.	[188]
Fish	Bacteriocin 7293 (Bac7293), a novel bacteriocin from *Weissella hellenica* BCC 7293	Poly (lactic acid)/sawdust particle biocomposite film	Gram-positive: *Listeria monocytogenes*, *Staphylococcus aureus* Gram-negative: *Pseudomonas aeruginosa*, *Aeromonas hydrophila*, *Escherichia coli*, *Salmonella Typhimurium*	Growth inhibition on both Gram-positive and Gram-negative bacteria.	[189]

Table 2. *Cont.*

Food	Antimicrobial Agents	Bio-Based Polymer	Target Microorganisms	Main Findings	References
Fish	Essential oil from *Plectranthus amboinicus*	Chitosan	*Bacillus subtilis*, *Escherichia coli*, *Staphylococcus aureus*, *Salmonella typhimurium*, *Klebsiella pneumoniae*, *Pseudomonas aeruginosa*	Improvement in tensile strength, opacity, and water vapor barrier with antimicrobial efficiency against foodborne pathogens.	[190]
Fish and chicken	*Amaranthus* leaf extract (ALE)	Polyvinyl alcohol (PVA) and gelatin	Gram-positive: *Bacillus cereus* and *Staphylococcus aureus* Gram-negative: *Escherichia coli* and *Pseudomonas fluorescence*	Better protection against UV light and reduced water solubility and water vapor permeability, and improvement of mechanical properties. Inhibition of microbial growth and minimization of oxidative rancidity in 12 days shelf life compared with 3 days shelf life for neat film.	[191]
Fish fillets	Curcumin and nisin	Electrospun nisin/curcumin (NCL) nanomats	Lactic acid bacteria (LAB) and Total Mesophilic Aerobic (TMAB)	On the 4th day, the count of TMAB in the samples coated with NCL mats was 3.28 log CFU g^{-1} compared to 6.61 log CFU g^{-1} in control samples.	[192]
Chicken breast fillets	Virgin olive oilgrape seed oiland savory essential oil	Gelatin-pectin	*Staphylococcus aureus*, *Salmonella typhimurium* *Fluorescence pseudomonas*	Savory essential oil presented more antimicrobial activity. The mixture of them in film showed antimicrobial activity against mentioned bacteria for 12 days storage.	[193]
Chicken breast fillets	Carvacrol (0.75% *w/w*) and citral (1.0% *w/w*)	Sago starch and guar gum	*Bacillus cereus* *Escherichia coli*	The tensile strength of films reduced while elongation at break increased, and the film showed good antimicrobial activity.	[194]
	Laurus nobilis essential oil and *Rosmarinus officinalis* essential oil	Polyvinyl alcohol (PVOH)	*Listeria monocytogenes*	Inhibition of the lipid oxidation together with antimicrobial activity.	[195]
Lamb meat	2% rosemary oil	Cellulose nanofiber/whey protein matrix containing titanium dioxide particles (1% TiO_2)	*Escherichia coli* *Salmonella enteritidis* *Listeria. monocytogenes* *Staphylococcus aureus*	The active packaging significantly reduced microbial growth, lipid oxidation, and lipolysis of the lamb meat during storage.	[196]

Table 2. *Cont.*

Food	Antimicrobial Agents	Bio-Based Polymer	Target Microorganisms	Main Findings	References
Strawberries	Cinnamon	Polybutylene adipate terephthalate (PBAT) films loaded ith cellulose nanofibers (CNF)	*Salmonella enterica* subsp. *enterica serovar Choleraesuis* and *Listeria monocytogenes*	The active film showed a high thermal stability with decreasing water vapor permeability. Strawberries had lower weight loss after 15 days of storage, better freshness preservation without fungal attack, and antimicrobial activity against bacteria.	[197]
Cherry tomatoes	Cinnamon	Chitosan as the outer layer and the mixture of sodium alginate and the amphiphilic starch as the intermediate layer	*Escherichia coli* *Staphylococcus aureus*,	This active film showed more freshness and lower weight loss rate within two weeks compared to polyethylene films. The inhibition growth rates for *E. coli* and *S. aureus* were 36% and 30%, respectively, and soil biodegradability rate was 70% in 28 days.	[198]
Cucumber	Clove oil	Chitosan	*Escherichia coli*	Maintained the color and flavor of cucumber for more than 4 days and until 4.97 $\log_{10}$, reductions of *E. coli* biofilm in population.	[199]
Strawberries	*Thyme*	Porous polylactic acid (PLA) nanofibers and coated with poly(vinyl alcohol)/poly(ethylene glycol) (PVA/PEG) blends	*Escherichia Coli* *Staphylococcus aureus*	Strawberries packed with this film exhibited better freshness and more than 99% antimicrobial activity against mentioned bacteria.	[200]
Strawberries	*Citral Litsea* (L.) *cubeba* essential oil	Polyvinyl acetate (PVA)	*Escherichia coli* *Staphylococcus aureus* *Aspergillus niger*	The broad-spectrum, direct, and indirect (gas phase) antimicrobial activity was observed against bacteria and fungi.	[201]
Vegetable products	Cinnamon and oregano	Cellulose	*Listeria grayi* *Listeria monocytogenes*	Cinnamon and oregano essential oils inhibited the growth of both bacteria in the vapor phase. The packaging with cellulose stickers impregnated with cinnamon reduced the *Listeria* count on frozen vegetable samples.	[202]
Fruit	Cinnamon	Zein	*Escherichia coli*	Improvement of barriers and mechanical properties of zein film with antimicrobial effect on *E. coli* and fruit samples.	[203]

Table 2. *Cont.*

Food	Antimicrobial Agents	Bio-Based Polymer	Target Microorganisms	Main Findings	References
-	*Zataria multiflora* and *Cinnamon zeylanicum* essential oils	Soy Protein Isolate (SPI)/Gelatin	*Staphylococcus aureusBacillus cereus Listeria monocytogenes. Salmonella typhimurium Escherichia coli*	This active film incorporated with 20% *Z. multiflora* reduced 100% of *S. aureus*, *B. cereus*, and *L. monocytogenes*. The reduction for *E. coli* and *S. typhimurium* were 70% and 63%, respectively.	[204]
-	Lavender essential oil	Starch, furcellaran, and gelatin	*Escherichia coli Staphylococcus aureus*	Increase film thickness and decrease water absorption and degree of swelling of the film with increasing concentration of oils. Additionally, the film showed both antioxidant and antimicrobial activity.	[205]
-	Rosemary mint essential oil, nisin and lactic acid	Chitosan, pectin, and starch	*Bacillus subtilis*, *Escherichia coli*, *Listeria monocytogenes*	Rosemary and nisin improved water barrier properties, tensile strength, and thermal stability, as well as microstructural heterogeneity and opacity. The film also showed inhibitory activity against all mentioned bacteria and antioxidant activity.	[206]
-	Rosemary essential oil	Chitosan	*Listeria monocytogenes*, *Pseudomonas putida Streptococcus agalactiae*, *Escherichia coli*, and *Lactococcus lactis*	Antimicrobial activity with a better effect on Gram-positive bacteria (i.e., *L. monocytogenes*, *S. agalactiae*)	[207]
-	Rosemary essential oil	Glycerol, gelatin, chitosan, and pectin	*Bacillus subtilis*, *Staphylococcus aureus*, *Enterococcus aerogenes*, *Enterococcus faecalis* and *Escherichia coli*	Optimization of the mixture with 10.0% of chitosan, 24.3% of gelatin, 0.5% of pectin, and 65.2% of glycerol. Inhibition of the growth of the mentioned microorganisms.	[208]
-	*Glycyrrhiza glabra* L. root essential oil (GGEO)	Carboxymethyl cellulose–polyvinyl alcohol (CMC-PVA)	*Gram-positive: Listeria monocytogenes*, *Staphylococcus aureus* *Gram-negative: Escherichia coli Salmonella Typhimurium*	Better inhibitory effects against the Gram-positive bacteria compared with Gram-negative bacteria.	[209]
-	Carvacrol (0.75% *w*/*w*) and citral (1.0% *w*/*w*)	Sago starch (SS) and guar gum	*Bacillus cereus Escherichia coli*	The tensile strength of films reduced while elongation at break increased, and the film showed good antimicrobial activity.	[194]

Essential oils (EOs) are the second metabolite and naturally volatile organic compounds produced from the extraction of different parts of plants, including flowers (jasmine, rose, violet, and lavender), herbs, buds (clove), leaves (thyme, eucalyptus, salvia), fruits (anis, star anise), twigs, bark (cinnamon), zest (citrus), seeds (cardamom), wood (sandal), rhizomes, and roots (ginger) [210]. They are present in cavities, secretory cells, epidermic cells, canals, or glandular trichomes. Essential oils present antibacterial, antifungal, anti-inflammatory, anesthetic, insecticidal, and antiviral properties. Thereby, they have many applications in the cosmetic, perfumery, pharmaceutic, beverage, feed, and food industries [211,212]. They are produced by various methods, such as solvent extraction, supercritical fluid extraction, hydro distillation, solvent-free (microwave) extraction, physical expression (cold pressing), ultrasound-assisted extraction, and combinations of extraction methods [182,210,213]. Essential oils are composed of a mixture of about 20–80 individual components at different concentrations, most of which are terpenes, terpenoids, aldehydes, isoflavonoid acids, and ketones. The antimicrobial activity of EOs depends on the chemical structure of these individual components. There are about 3000 essential oils; approximately 300 are used for food, cosmetics, fragrance, and medical applications. EOs are naturally hydrophobic and lipophilic compounds, with a density that is often lower than water. They are soluble in organic solvents such as ether, alcohol, and fixed oils but immiscible with water. They can be isolated from the aqueous phase via decantation [33,210].

Essential oils have gained a growing interest in the food industry and packaging due to their antimicrobial and antioxidant properties applied to the preservation of foodstuffs, including fruits, vegetables, fish, meat, and dairy products [3]. As they are natural compounds, they can be an excellent alternative to synthetic food additives. Their hydrophobicity allows them to permeate through the lipids of the cell membrane of bacteria, disrupting the structure and finally inhibiting them [33].

The antimicrobial effect of EOs has been studied in detail, and it depends on their chemical structure. It is confirmed that the presence of hydroxyl groups in phenolic compounds such as carvacrol and thymol has a key role in their activities against bacteria [214]. However, EOs are unstable, volatile compounds, which prevents their direct use on foods, as that would be ineffective with uncontrolled release. In fact, they start to degrade easily through oxidation, volatilization, heating, and light if they are not preserved from external factors [33,215]. Furthermore, EOs are aromatics that can have undesirable sensory effects. Encapsulation of essential oils into carriers such as liposomes, polymeric particles, and solid lipid nanoparticles is indispensable [3,216].

Many studies have been conducted on the antimicrobial activities of EOs on different microorganisms with promising results in inhibiting the growth of different types of pathogens, which has attracted huge attention in food packaging [217]. In this regard, a recent study conducted by [202] evaluated the antimicrobial activities of cinnamon, oregano, and carvacrol against *L. grayi* and *L. monocytogenes* in vitro. The results showed their effectiveness in inhibiting the growth of *L. grayi* as well as *L. monocytogenes* in the vapor phase. Chitosan films prepared by incorporation of apricot kernel essential oil, which is a strong antioxidant and antimicrobial agent, in different concentrations, showed excellent antimicrobial and antioxidant properties as compared to neat chitosan films. They successfully inhibited fungal growth on packaged bread slices, improved water resistance, and enhanced the water vapor barrier property [218].

3.1.2. Natural Antimicrobial Agents of Animal Origin

Some animal proteins and enzymes can act on both Gram-positive and Gram-negative bacteria by destroying their cell membranes [219]. Antimicrobial peptides are a large group of antimicrobial agents of animal origin, including pleurocidin, lactoferrin, defensins, and protamine [220]. Some enzymes such as lactoperoxidase and lysozyme, plus some polysaccharides and lipids from animal sources such as chitosan, as well as some fatty acids of animal origin such as eicosatetraenoic acid and docosahexaenoic acid, could also effectively act against Gram-positive and Gram-negative bacteria and fight foodborne microorganisms [219]. Chitosan, as the main part of these groups, was introduced in the previous section. Herein, we introduce some of the other most applicable antimicrobial substances of animal origin.

Pleurocidin

Pleurocidin is a novel antimicrobial peptide found in the skin mucous secretions of the winter flounder (*Pseudopleuronectes americanus*). It inhibits the growth of both Gram-positive and Gram-negative foodborne microorganisms. Its effect on some microorganisms, including *Saccharomyces cerevisiae, E. coli O157:H7, L. monocytogenes, Vibrio parahemolyticus*, and *Penicillium expansum*, was confirmed [221] for food preservation purposes. Incorporation of pleurocidin in poly-vinyl alcohol in order to preserve its bioactivity leads to the inhibition of foodborne pathogens, and it shows better efficiency compared to free pleurocidin samples tested with real foods such as apple cider [222].

Lactoferrin

Lactoferrin is an 80 kDa glycoprotein containing around 700 amino acids. It is a significant part of the non-heme iron-binding glycoprotein of our organism and belongs to the transferrin class. It is present in most mammalian body fluids such as milk [223]. Anti-inflammatory, antibacterial, antiviral, antioxidant, antitumor, anti-obesity, antibiotic, and immunomodulatory activities are some properties that have been reported to date. Lactoferrin regulates the free iron level in body fluids and causes bacteriostatic properties, which are beneficial for health [224,225].

It exhibits broad-spectrum antimicrobial properties against bacteria, viruses, and fungi [226]. Its bio preservative potential against Salmonella *enterica* and *Escherichia coli O157:H7* as two common dairy pathogens was examined, and it showed a considerable reduction in their growth with only lactoferrin concentrations at or above 112.5 mg/mL in the milk [227]. Lactoferrin's antimicrobial activity makes it a potential candidate for use in food preservation. Recently, immobilization of lactoferrin in gelatin films improved their optical, mechanical, barrier, and preservative properties [168].

Lactoperoxidase

Lactoperoxidase is a heme-containing glycoprotein composed of 608 amino acids with a molecular mass of approximately 78 kDa [228]. It is an antimicrobial enzyme that is secreted in the epithelial cells of the mammary gland and is found in large amounts in cow's milk [229]. It can destroy bacteria such as *Salmonella, Shigella, Pseudomonas*, and coliforms. Thereby, it has applications in the food industry for the preservation of unprocessed milk, poultry, fish, and meat [230–233]. Due to its antimicrobial activity, lactoperoxidase has potential use in food packaging. Its incorporation into defatted soybean meal, a by-product left after crushing soybeans for oil, showed its effectiveness in sliced ham preservation without changing its sensory properties considerably [232].

Lysozyme

Lysozyme is another important enzyme widespread throughout the animal and plant kingdoms that shows protective biological functions [234]. Lysozyme is a basic protein in many living organisms, found in body fluids and tissues; therefore, there are many sources of lysozyme [235]. One of the most convenient sources is chicken eggs, the richest in lysozyme. It has interesting characteristics, including antimicrobial activity along with protection against infection, with the effect of strengthening the immune system [236,237]. Despite its interesting properties, the use of free lysozyme in packaging is restricted because it is unstable. Food protease can hydrolyze it and inactivate its catalytic activity [238,239]. The modification of lysozyme via heat, chemicals, and hydrolysis enhances its biological activities and broadens its applications [240]. According to [241], lysozyme coating in packaging films can reduce the water vapor permeability, oxygen permeability, and oil permeability without unwanted effects on the appearance of the films. The effect of lysozyme-N-succinyl chitosan on extending the shelf life of strawberries was proved [242]. Other authors assessed exploiting lysozyme nanofibers as antimicrobial and antioxidant-reinforcing additives by incorporating lysozyme nanofibers into the pullulan matrix as an eco-friendly edible film [243]. Recently, [244] incorporated lysozyme into cold plasma-activated polylactic acid pouches and assessed its effectiveness with pear juice and rice milk-based smoothies. The authors highlighted that activated packages could inhibit *Listeria monocytogenes*. The color of pouches manufactured with this method was also better and more stable [244]. Other studies demonstrated that the synergy of low-dose lysozyme enhanced the inhibition of eugenol–casein nanoparticles against *Staphylococcus aureus* and *Bacillus* sp. by 5.83-fold and 5.53-fold, respectively [245].

3.1.3. Antimicrobial Agent Produced by Microorganisms

Some microorganisms and their derivatives are able to inhibit the growth of other microorganisms [246]. Bacteriocins, produced by some bacteria such as lactic acid bacteria (LAB), are peptides with antimicrobial activity against various non-pathogenic and sometimes pathogenic bacteria [247]. They have recently received much attention due to their antimicrobial activity, especially bacteriocins generated from lactic acid in food and drug systems, most of which are classified as safe agents by the American Food and Drug Administration (FDA) [248].

Current bacteriocins that have the potential for food biopreservation are nisin, leucocin, lactocin 705, enterocin 4, and enterocin produced by various microorganisms [247]. Bacteriocins can be incorporated into packaging films to inhibit spoilage microorganisms and extend the shelf life of foods [247]. The incorporation of bacteriocin from lactic acid bacteria into active food packaging was successful in meat and meat products [249]. The active film obtained from polyvinyl alcohol blended with casein hydrolysates and nisin exhibited antimicrobial activity by reducing the growth of spoilage bacteria and then extending the shelf-life of the packaged food product [250]. The concentrated bacteriocin from lactic acid bacteria was also effective in reducing the growth of *Listeria monocytogenes* in ready-to-eat sliced emulsion-type sausage [251].

Pediocins are bacteriocins produced by *Pediococcus* spp., a group of homofermentative Gram-positive bacteria belonging to the *Lactobacillaceae* family [252]. It is an active peptide that specifically inhibits the growth of *Listeria monocytogenes* [253]. The activity of pediocin can be limited by the presence of salt, heat, or proteolytic enzymes, along with pH and storage time [254]. Pediocin has many applications in the food industry, including the preservation of fermented sausages, vegetables, and dairy products [255]. It can be used directly on food products as a preservative or incorporated into food packaging in order to prevent the spoilage of food products [256]. According to [257], adding nisin and pediocin to starch film inhibited the growth of *L. monocytogenes* and *C. perfringens*.

Reuterin is also an antimicrobial compound produced by *Lactobacillus* spp. strains in the presence of glycerol and is considered an antimicrobial agent against several pathogenic and spoilage microorganisms. Its effectiveness was evaluated for many foodstuffs, including meat, dairy products, beef sausages, and cooked pork, resulting in inhibition of *L. monocytogenes*, *S. aureus*, *E. coli O157:H7*, and so forth. Despite pediocin, reuterins act in a wide range of pH, which makes them a good candidate for biopreservation of many foods. It exhibits broad inhibitory activities against Gram-positive and Gram-negative bacteria, bacterial spores, molds, yeasts, and protozoa [258]. According to [258], reuterin at a concentration of 6.9 mM showed a fungicidal effect in yogurt.

3.1.4. Natural Antimicrobial of Algal and Mushrooms Origin

Mushrooms, with about 140,000 species, have antimicrobial and antioxidant capacities. In addition to their nutritional value, they have preservative and medical properties [259]. Methanol, ethanol, acetone, dichloromethane, and chloroform extracts of many different mushrooms showed inhibitory activity towards pathogenic microorganisms such as *E. coli*, *S. aureus*, *B. subtilis*, *C. albicans*, and *P. aeruginosa* and so forth [246,260]. Antimicrobial and anti-fungal effects of two wild edible mushrooms from the Kashmir valley, *Morchella esculenta* and *Verpa bohemica*, against *E. coli* and *Aspergillus fumigates* were assessed by [261]. Besides, the antimicrobial activity of *Auricularia auricular-judae* mushroom has been confirmed against *Staphylococcus aureus, Bacillus subtilis, Escherichia coli, Pseudomonas aeruginosa, Klebsiella pneumoniae*, yeast (*Candida albicans*), and dermatophytic pathogens [262].

Algae also have antimicrobial activity due to their polyphenolic content [263]. Antimicrobial activity of various algae-derived compounds was reported, such as the effectiveness of *Himanthalia elongata, Saccharina latissima, Laminaria digitate, Padina*, and *Dictyota* against *L. monocytogenes, Salmonella, Enterococcus faecalis, P. aeruginosa, B. cereus*, and *E. coli* [246]. Attempts to exploit the antimicrobial and antioxidant properties of algae-derived compounds led to their being determined as good candidates for pharmaceutical applications as well as food coatings and films [246]. Polyelectrolyte structured films prepared from cationic starch and sodium alginate showed good thermal, antimicrobial, and surface properties [264].

3.2. Antioxidant Active Packaging

Higher than normal concentrations of free radicals during oxidative stress are very harmful to the body as they damage or modify the major components of a cell and cause many diseases. Oxidation is also one of the major causes of food spoilage. Antioxidants are radical scavengers that trap free radicals and help delay or prevent oxidation [265]. Indeed, antioxidants minimize oxygen in the food system by reducing the capacity of active antioxidants. Therefore, they retard both lipid oxidation and protein denaturation. They can also interrupt the oxidation chain reaction and prevent the oxidation reaction along with inactivation of enzyme activity, resulting in blocking catalyzed oxidation reactions [266].

Polyphenols, flavonoids, vitamins C, E, and terpenoids have been reported as natural and synthetic antioxidant agents [267]. Natural antioxidants, compared to synthetic antioxidants, exhibit benefits such as stronger antioxidant effects, better consumer acceptability, potential health benefits, and safety [268].

Two synthetic antioxidants, butylated hydroxyanisole (BHA) and butylated hydroxytoluene (BHT), have been extensively studied in active packaging. However, health awareness and the ever-increasing demand for using natural molecules are encouraging industries to use natural ingredients [269].

In this regard, recently, great consideration has been given to natural antioxidants. Natural antioxidants, also called green antioxidants, are composed of simple phenols, phenolic acids, ascorbic acid, tocopherols, carnosic acid, rutin, carotenoids, flavonoids, vitamins, and anthocyanins [270]. Antioxidant activity is widely observed in fruits, spices, and herbs, such as essential oils [271]. Natural sources of antioxidants comprise blueberries, strawberries, blackberries, rosemary, turmeric, saffron, ginger, chili, green tea, coffee bean, catnip, sage, kenaf seed, roselle seed, thyme, potato peel and sugar beet pulp, spices and herbs, and so forth [270,272–274]. The most important group of natural antioxidants are phenolic compounds due to their strong free-radical scavenging effect [266].

Recently, some strategies for incorporating active components such as oxygen scavengers and antioxidants into food packaging have been introduced for food preservation [179]. However, antioxidants may reduce or lose their activities due to interaction with the other ingredients or during processing [26]. The incorporation of antioxidants in the packaging matrix can help to overcome these defects, retard food spoilage, and extend the shelf life of packaged foods [179,275]. Antioxidants applied to food packaging should meet some criteria, such as being low cost, being non-toxic, having high activity at low concentrations, having strong permeability, and showing good stability without affecting the quality of the food [266]. In addition, the proper antioxidants for packaging applications should be chosen based on molecular size, polarity, and release properties of antioxidant compounds [276]. Some forms of antioxidant packaging are already used in the food industry, such as sachets, pads, or labels for foods in which iron and ferrous oxide are common active ingredients [277].

In the past few years, there have been extensive studies on active packaging with antioxidant compounds. A recent study assessed the antioxidant activity of green tea extract for meat preservation. It indicated the potential use of green tea extract in active films [278]. Additionally, polyfunctional starch/tea polyphenol nanofibrous films also demonstrated higher antioxidant activity and optimum mechanical and hydrophobic properties [279]. Moreover, Ref. [280] evaluated the antioxidant activity of chitosan films enriched with *Artemisia campestris* extracts. *Artemisia campestris* is a widespread species of plant in the sunflower family that exhibits antioxidant and UV–Vis barrier properties with good thermal stability. Ref. [281] investigated the influence of *Syzygium cumini* seed extract (SCSE) incorporation in sodium alginate/gum Arabic (SG) films. They highlighted that the addition of SCSE into SG films declined the thermal stability, elongation at break, tensile strength, and moisture content and improved the scavenging activity, opacity, solubility, and water vapor permeability. Ref. [282] produced sodium alginate (SA) maltodextrins (MD) based functional films incorporated with phenolic extract of *Azolla pinnata* leaves fern (AF). They demonstrated that the addition of AF extract to SA.MD films increased the thickness and enhanced the scavenging properties. Furthermore, the film's solubility, swelling degree, and water vapor permeability were decreased. Ref. [283] fabricated carboxymethyl cellulose (CMC) based novel functional films containing Chinese chives root extract (CRE) and revealed that a higher extract concentration decreased the tensile strength, water solubility, swelling degree, and water vapor permeability. Furthermore, the addition of CRE into CMC exhibited good antioxidant and antimicrobial activity of films.

Antioxidant packaging as an emerging active packaging has great potential for use in food packaging and preserving food in a safe and eco-friendly manner. This technology can consist of biopolymers or nanofillers to improve mechanical or physical properties, resulting in various antioxidant-active packaging for different foodstuffs.

4. Active Food Packaging with Nano/Microencapsulated Ingredients

The limitations of using most of these natural antioxidant and antimicrobial agents are their easy degradation, low water solubility, low bioavailability, and undesirable tastes [284]. Encapsulation is a promising technology for overcoming these limitations and enhancing the physical, chemical, and thermal stability along with masking the unwanted taste, increasing the bioavailability and solubility of natural antimicrobial and antioxidant compounds, and providing the possibility of controlled release and targeted delivery [284,285]. Indeed, this innovative technology keeps the sensitive bioactive natural compounds from damage under harsh conditions such as high temperature or pressure, light, oxidation, and neutralization because of unwanted interaction with the other ingredients or foods throughout processing and during storage periods. Thanks to encapsulation techniques, natural antimicrobial and antioxidant compounds are covered by wall material in order to protect their activities against external impacts [286,287].

Encapsulation of antimicrobial agents is highly useful for food applications, notably for active packaging with high breadth potential due to the vast variety of natural antimicrobial compounds and many possible wall materials for encapsulating them. It is a wide search field that has received huge attention recently, specifically the characterization of natural antimicrobial compounds and the choice of compatible shell materials.

Encapsulation is a process by which one substance in solid, liquid, or gaseous states is entrapped/coated in another material, called wall material. This process can produce different scales of particles, namely, millimeters, micrometers (microencapsulation), and nanometers (nanoencapsulation) [288]. The aim of encapsulation is to produce capsules in nano or micro size. Nano/microcapsules are composed of a core, which is surrounded by a monolithic shell (Figure 4).

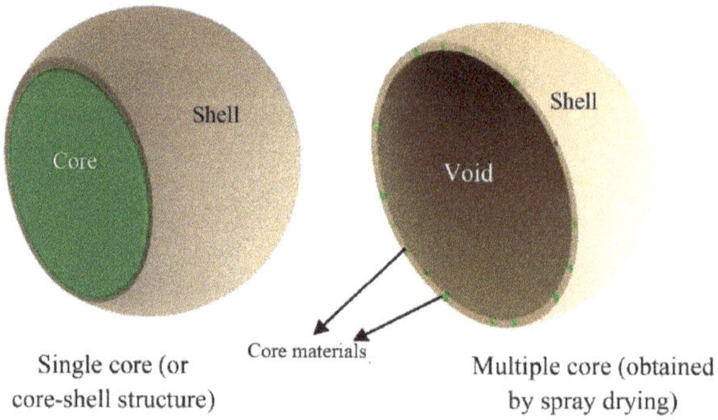

Figure 4. Types of encapsulated particles, in which core materials are natural antimicrobials coated with wall materials as a shell (Reproduced with permission from [285]).

The core, which is mainly a bioactive agent, is centered in capsules, and it is also called the internal phase, encapsulant, payload phase, or fill. The shell that coats the core is considered a wall, coating, envelope, membrane, carrier, encapsulating agent, external phase, or matrix [289,290]. The micro/nanocapsules can be classified into four categories depending on whether they have a single- or multi-core and a shell. They can be found in the form of single-core with single-wall, or single-core coated by multi-wall, as well as multi-core with single-wall, or multi-core enveloped by multi-wall [291]. The main outcome of encapsulation is to overcome the physical instability and protect active molecules from degradation in addition to controlling the release of active molecules [292]. Indeed, encapsulation contributes to preserving the active molecule against harsh conditions such as high temperature, pressure, or pH through processing. Moreover, the possibility of controlled release is another important benefit of encapsulation. In this regard, encapsulation could minimize the concentration of active agents needed in different applications by controlling their release and preserving them against harsh conditions.

More specially, the use of encapsulation techniques in the food industry confers several advantages, as mentioned by [291]. Their main advantages are as follows:

- The preservation of sensitive molecules during processing conditions such as phenolic compounds with antimicrobial and antioxidant activity;
- Encapsulation at nano and micro sizes enhances the bioavailability of active molecules;
- The prevention of the alteration of the sensory properties of food by some bioactive agents with unpleasant aroma and taste. Essential oils and oil fish are two examples of active molecules with extreme aromas that can alter the taste of food. Encapsulation prevents the change in taste by covering the molecules and reducing the necessary concentration;
- Controlled release of active compounds to improve food quality and safety;
- The final product of the encapsulation process is mostly a fine powder. It offers several benefits, such as improvement of stability and flowability. It is easier to handle and store the active molecules. In addition, agglomeration and change in density can be reduced by encapsulation.

Different wall materials could be used for the encapsulation, depending on the nature of active molecules. The wall materials should meet some characteristics comprising resistance to high shears and mechanical stresses as well as the ability to incorporate the active molecules with the minimum thickness [293]. In addition, the wall material structures play a key role in encapsulation efficiency and the durability of micro/nanocapsules [294], and their compatibility with the core material is an important element for choosing a wall material. Stronger wall material or a double layer should be chosen for the encapsulation of highly sensitive molecules or under harsh conditions [34]. The apt shell in encapsulation exhibits some criteria, including the stabilization of the core, film-forming ability, flexibility, stability, low density, non-hygroscopicity, the capacity for controlled release, and no unwanted reactions with the payload [285]. Considering all these characteristics, various biopolymers have been applied in the encapsulation of food ingredients [291], including polysaccharides (gum arabic, modified starches, maltodextrins, alginates, pectin, carrageenan, cellulose derivatives, chitosan, and cyclodextrins), fats and waxes (hydrogenated vegetable oils, beeswax, lecithin, medium-chain triglycerides, and glyceryl behenate), proteins (gelatin, whey protein, sodium caseinate, soy protein, gluten, caseins, zein, and silk fibroin), and synthetic compounds (paclitaxel, mPEG500-b-p, polyacrylonitrile, polycaprolactone, polylactic acid, poly methyl methacrylate (PMMA), acrolein, and glycidyl methacrylate epoxy polymers). The natural materials are mostly cheap, abundant, non-toxic, and compatible with food formulations. There are some parameters that should be taken into account for designing the encapsulation process, including capsule size, final physical state, stability, and physical condition of the release [291,295].

The encapsulated particles are divided into three main classes that differ in size, namely, macrocapsules (with a size of more than 5000 μm), microcapsules (0.2–5000 μm), and nanocapsules (with a size of less than 0.2 μm). According to EU regulation, nanoparticles are attributed to the particles when at least 50% of the particles have a size of ≤100 nm [296], whereas pharmaceutical science considers nano size for particles with a size of less than 1000 nm.

Herein, the various techniques of microencapsulation along with nanoencapsulation are mentioned briefly. Microencapsulation techniques include spray drying as a more common technique, spray chilling, spray cooling, extrusion, fluidized bed coating, centrifugal extrusion, freeze-drying, coacervation, interfacial polymerization, molecular inclusion, and so forth [291,295].

Nanoencapsulation is more recent than microencapsulation. The different nanoencapsulation systems for loading natural antimicrobials are summarized in Figure 5, including nanoemulsions, nanoliposomes, solid lipid nanoparticles (SLNs), nanostructured lipid carriers (NLCs), biopolymeric nanoparticles, and equipment-based nanoencapsulation methods, which have commonly been applied to design the best delivery systems for antimicrobial substances [297]. The selection of the proper technology for each process depends on the physicochemical features, particle size, release type, delivery method, and economic and environmental aspects [291].

In nanoencapsulation, the size of particles is decreased to a nanometer by two basic methods, including top-down (using high-energy devices such as high-pressure-homogenizer, ultrasound, and milling) and bottom-up methods (using low-energy techniques such as spontaneous emulsification or anti-solvent precipitation) [298]. Nanocapsules with a nano-size have a higher ratio of surface to volume. Therefore, they offer higher solubility, bioavailability, and adsorption [34]. Nanoencapsulation exhibits several benefits such as homogeneity, improvement in physical and chemical properties, and shelf stability, in addition to better encapsulation efficiency and boosting the functional properties of active molecules [299]. Besides, the nanocapsules are more controllable. Furthermore, the release of micro-sized particles occurs more slowly, and over longer periods of time compared to nanocapsules. Consequently, a smaller number of nanoparticles is required. Furthermore, according to numerous studies, the emergence of nanotechnology in food packaging, in particular the use of nanocapsules in active packaging, offers interesting benefits such as more efficacy by inhibiting the growth of microorganisms, better thermal protection of bioactive compounds, and better release control [300].

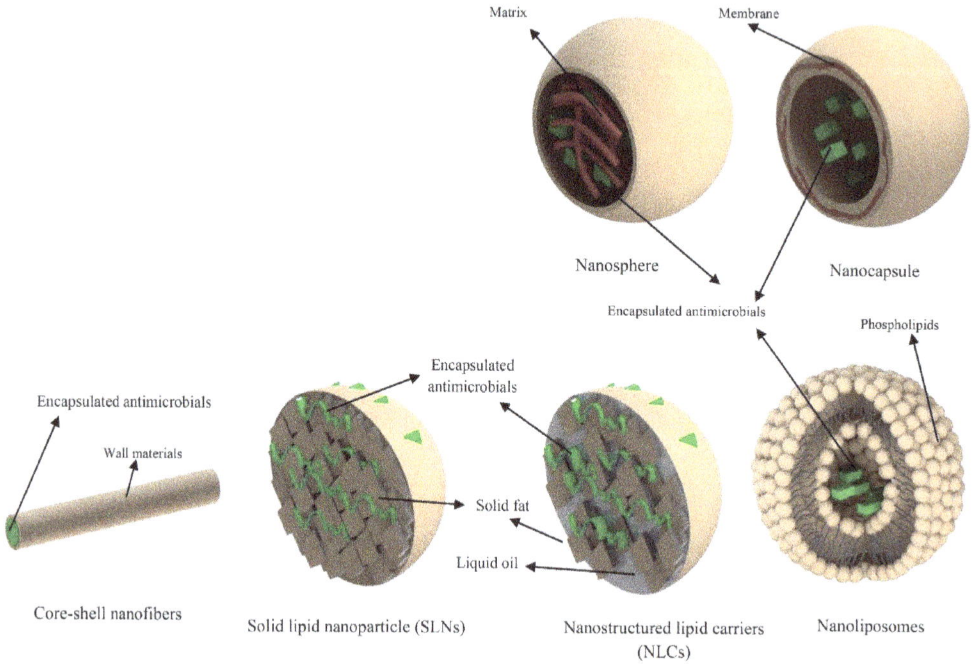

Figure 5. Different nanoencapsulation systems for loading of green natural antimicrobials (Reproduced with permission from [285].

Nanoencapsulation is progressing at a rapid pace in the food sector. The nanoencapsulation of food ingredients aims to produce value-added food products with the possibility of applying this technology in active packaging to boost the preservation of foodstuffs. The incorporation of encapsulated bioactive compounds in the matrix of film packaging is a promising strategy for food preservation, particularly extending the shelf life of fresh foods. Capsules produced through a stable encapsulation process gain remarkable characteristics such as more physiochemical stability and better bioavailability, which show more interest as a delivery system [301,302].

Controlled release packaging (CRP) is a relatively new term that first appeared in the literature in 2005 [303]. A controlled release of bioactive molecules from a film into food is performed during storage periods. The migration of capsules containing antimicrobial or antioxidant compounds can be achieved by direct contact between food and active film packaging or by gas phase diffusion from the packaging layer onto the food surface [304]. The uniqueness of CRP is that it focuses on the kinetics and mechanism of controlled release, what to release, when and how to trigger the release, how much to release, and how fast to release [305]. CRP using antimicrobials may be used for short- or intermediate-term microbial inhibition of highly perishable foods such as fresh meats, seafood, fruit, and vegetables. CRP using antioxidants may be used for long-term retardation of lipid oxidation for shelf-stable foods such as ready-to-eat meals containing fatty components susceptible to oxidation [306].

The control of the release rates of antimicrobial and antimicrobial agents is an emerging area of active food packaging. In fact, rapid release causes consumption of the antimicrobial within a short time, after which the minimum concentration required for the microbial growth inhibition is not maintained. On the other hand, spoilage reactions on the surface may start if the release of the antimicrobial agent from the packaging film is too slow [307].

Different materials and methods may be used in order to apply this strategy by means of various biopolymers and bioactive agents. To date, this technology introduces active antimicrobial packaging with a high potential to minimize the risk of food spoilage and contamination by suppressing the activities of targeted microorganisms. For example, encapsulated essential oils with antimicrobial properties are one of the interesting candidates to be applied for active food packaging. According to [308], lemon myrtle essential oil, encapsulated into cellulose acetate nanofibers, inhibited the growth of 100% *Escherichia coli* and *Staphylococcus aureus* even at the lowest loading concentration of lemon myrtle essential oil of 2 wt%. The antimicrobial activity of cinnamon oil was investigated against foodborne bacteria in free and encapsulated forms, and both exhibit antimicrobial activity. However, the encapsulated form showed antimicrobial action with a lower concentration of 57–125 ppm, compared with non-encapsulated cinnamon, with a concentration of 125–500 ppm. Controlled release of cinnamon capsules leads to better efficiency because the cinnamon concentration in the culture medium reaches its maximum after 4–6 h of incubation when the cells are in the exponential growth phase.

Luteolin, a flavone with potent antioxidant and antimicrobial activities, was encapsulated in oil-in-water nanoemulsions and incorporated into chitosan-matrix [309]. The authors compared films with encapsulated luteolin (CS-LLNEs), films without luteolin (CS-LL), and free luteolin films (CS) as controls. CS-LLNEs film showed slower controlled release to exert antioxidant activity for up to 10 days. In addition, CS-LLNEs film showed better water vapor and oxygen barriers and mechanical properties in comparison with CS-LL film.

Active packaging based on hydroxypropyl methylcellulose containing carvacrol nanoemulsions was developed to extend the shelf-life of wheat bread. The designed system has satisfactory antioxidant activity and good antibacterial activity against *S. aureus* and *E. coli* [310]. Grape seed extract-carvacrol microcapsules were incorporated into chitosan to extend the shelf-life of refrigerated salmon packages. The microcapsules improved the antimicrobial activity of the chitosan film and increased the shelf-life of refrigerated salmon to 4–7 days [311]. The multilayered pectin edible coating was designed with encapsulated *trans*-cinnamaldehyde to extend the shelf life of fresh-cut cantaloupe. Coated fruits lasted longer (7–9 days) compared to uncoated controls (4 days) at 4 °C [312].

Eventually, biodegradable film packaging containing capsules has great potential to meet the needs of three main goals of developed packaging science, comprising sustainability, safety, and convenience [31,313]. Recent progress in the incorporation of nano/microencapsulated natural agents into active food packaging systems is summarized in Table 3.

Table 3. Incorporation of nano/microencapsulated natural agents in active food packaging systems.

Packaging Material and Encapsulated Antimicrobial System	Purpose	References
Active packaging based on hydroxypropyl methylcellulose containing carvacrol nanoemulsions	Development of active packaging system to extend the shelf life of wheat bread. The designed system has a satisfactory antioxidant activity, good antibacterial activity against *S. aureus* and *E. coli.*	[310]
Edible coating fabricated with chitosan, pectin, and encapsulated *trans*-cinnamaldehyde	Designing a multilayered edible coating with antimicrobial agents to extend the shelf life of fresh-cut cantaloupe stored at 4 °C	[312]
Alginate coating containing nano-emulsified basil oil	Development of a coating system against the following spoilage fungi: *Penicillium chrysogenum* and *Aspergillus flavus*	[314]
Active packaging based on hydroxypropyl methylcellulose containing oregano essential oil nanoemulsions	Higher antimicrobial activity against all tested bacterial strains, particularly *S. typhimurium*	[315]
Starch-carboxy methyl cellulose films containing rosemary essential oil (REO)-loaded benzoic acid-chitosan (BA-CS) nanogel	Using of encapsulated REO into BA-CS nanogel in film structure to obtain immediately (REO) and gradual (nanogel) antimicrobial effect against *S. aureus*	[316]
Polylactide films containing essential oils/nanoparticles	Inhibiting the growth of *L. monocytogenes* and *S. typhimurium* on contaminated cheese	[10]
Active packaging containing cinnamon-loaded nanophytosomes into electrospun nanofiber	Higher antimicrobial activity and improving the shelf life of shrimp	[317]
Active packaging based on cellulose nanocrystals (CNCs) reinforced chitosan, containing thyme-oregano, thyme-tea tree, and thyme-peppermint nanoemulsions	Development of active antifungal packaging for rice preservation. Chitosan-based nanocomposite films loaded essential oils mixtures showed significant antifungal activity against *Aspergillus niger*, *Aspergillus flavus*, *Aspergillus parasiticus*, and *Penicillium chrysogenum*, reducing their growth by 51–77%.	[318]
Encapsulation of gallic acid into lentil flour-based nanofibers by electrospinning technology and use of these nanofibers as active packaging materials	Enhancement of the oxidative stability of walnuts present in active packages with encapsulated gallic. The reduction in oxidation of walnuts with lower peroxide, *p*-anisidine, and TOTOX values was observed.	[319]
Active packaging film based on chitosan with grape seed extract-carvacrol microcapsules	Development of active film to extend the shelf-life of refrigerated salmon. The microcapsules improved the antimicrobial activity of the chitosan film and increased the shelf-life of refrigerated salmon to 4–7 days.	[311]

5. Conclusions and Future Developments

In recent years, sustainable packaging has been adopted by the food industry, either as a repackaging of environmental policies, as a marketing strategy in response to social pressures, or as a genuine attempt to grapple with the commercial, social, and environmental issues associated with plastic packaging. This review paper summarizes the current state of biodegradable and active packaging biopolymers as well as major antioxidant and antimicrobial natural agents incorporated into the polymers as nano-microcapsules. More than 300 sources were reviewed, including scientific literature and commercial information.

In conclusion, this review shows that biodegradable active biopolymers are generally characterized by low water vapor barrier properties; hence, the modification of their structure or other techniques can be convenient for modulating this property. These efforts can lead to the production of biodegradable active biopolymers with good barrier and mechanical properties. At the same time, more efforts should be implemented to decrease their price and their competition with food by increasing the use of second- and third-generation feedstock. Another challenge is the lack of a collection channel for these bioplastics. Further studies are necessary to establish biodegradable monolayer or multilayer materials as a viable alternative to commonly used fossil fuel-based materials for food products. A techno-economic analysis (TEA) combined with a life cycle assessment (LCA) should be implemented to estimate and quantify costs, emissions, and energy intensity associated with material acquisition, processing, transport, and end-of-life treatment of biodegradable packaging production.

At the same time, nano-microencapsulation of natural antimicrobial and antioxidant agents improves their physical, chemical, and thermal stability along with increasing their bioavailability and solubility, which provides the possibility of their controlled release and targeted delivery. However, this review highlighted that strict European regulations for food contact materials and safety aspects restricted the application of many encapsulated antimicrobial and antioxidant agents in the releasing systems. Eliminating legislative restrictions and creating a unique global organization would allow for more specific and exact legislation for food-contact materials. Meanwhile, the use of nanotechnology in the antimicrobial and antioxidant food packaging films requires the control and monitoring of nanomaterials as well as risk assessment, which requires information about their toxicity and exposure [296]. In line with this, the European Food Safety Authority (EFSA) has recently produced guidance for the risk assessment of nanoscience and nanotechnology applications in the food and feed chain [320]. Moreover, an effort should be made to familiarize the public consumer with active food packaging and increase their acceptance of such an innovation. This review contributes to increasing the knowledge of the available sustainable biodegradable active packaging and highlights the increasing number of investigations and the increasing industrial interest in the area, as well as further developments needed.

Author Contributions: Conceptualization, F.B. and S.G.; methodology, F.B.; validation, S.G., A.G., and E.D.; formal analysis, S.G.; investigation, F.B.; data curation, F.B.; writing—original draft preparation, F.B; writing—review and editing, S.G. and E.D.; supervision, A.G., S.G. and E.D.; project administration, S.G. All authors have read and agreed to the published version of the manuscript.

Funding: The Terra Isara foundation funded this research, grant number E-F0601-00.

Conflicts of Interest: The authors declare no conflict of interest. The funders had no role in the design of the study; in the collection, analyses, or interpretation of data; in the writing of the manuscript, or in the decision to publish the results.

References

1. Bahrami, A.; Delshadi, R.; Assadpour, E.; Jafari, S.M.; Williams, L. Antimicrobial-loaded nanocarriers for food packaging applications. *Adv. Colloid Interface Sci.* **2020**, *278*, 102140. [CrossRef]
2. Chawla, R.; Sivakumar, S.; Kaur, H. Antimicrobial edible films in food packaging: Current scenario and recent nanotechnological advancements—A review. *Carbohydr. Polym. Technol. Appl.* **2021**, *2*, 100024. [CrossRef]

3. Sharma, S.; Barkauskaite, S.; Jaiswal, A.K.; Jaiswal, S. Essential oils as additives in active food packaging. *Food Chem.* **2021**, *343*, 128403. [CrossRef] [PubMed]
4. Ibarra, G.V.; Sendón, R.; Rodríguez-Bernaldo de Quirós, A. Chapter 29—Antimicrobial Food Packaging Based on Biodegradable Materials. In *Antimicrobial Food Packaging*; Barros-Velázquez, J., Ed.; Academic Press: San Diego, CA, USA, 2016; pp. 363–384. ISBN 978-0-12-800723-5. [CrossRef]
5. López-Carballo, G.; Gómez-Estaca, J.; Catalá, R.; Hernández-Muñoz, P.; Gavara, R. 3-Active antimicrobial food and beverage packaging. In *Emerging Food Packaging Technologies*; Yam, K.L., Lee, D.S., Eds.; Woodhead Publishing Series in Food Science, Technology and Nutrition; Woodhead Publishing: Sawston, UK, 2012; pp. 27–54. [CrossRef]
6. Ahamed, A.; Veksha, A.; Giannis, A.; Lisak, G. Flexible packaging plastic waste-environmental implications, management solutions, and the way forward. *Curr. Opin. Chem. Eng.* **2021**, *32*, 100684. [CrossRef]
7. Kunwar, B.; Cheng, H.N.; Chandrashekaran, S.R.; Sharma, B.K. Plastics to fuel: A review. *Renew. Sustain. Energy Rev.* **2016**, *54*, 421–428. [CrossRef]
8. Alojaly, H.; Benyounis, K.Y. Packaging With Plastics and Polymeric Materials. In *Reference Module in Materials Science and Materials Engineering*; Elsevier: Amsterdam, The Netherlands, 2020; ISBN 978-0-12-803581-8. [CrossRef]
9. Zhang, W.; Zhang, Y.; Cao, J.; Jiang, W. Improving the performance of edible food packaging films by using nanocellulose as an additive. *Int. J. Biol. Macromol.* **2021**, *166*, 288–296. [CrossRef]
10. Ahmed, J.; Hiremath, N.; Jacob, H. Antimicrobial efficacies of essential oils/nanoparticles incorporated polylactide films against *L. monocytogenes* and *S. typhimurium* on contaminated cheese. *Int. J. Food Prop.* **2017**, *20*, 53–67. [CrossRef]
11. Artham, T.; Doble, M. Biodegradation of Aliphatic and Aromatic Polycarbonates. *Macromol. Biosci.* **2008**, *8*, 14–24. [CrossRef]
12. Geyer, R.; Jambeck, J.R.; Law, K.L. Production, use, and fate of all plastics ever made. *Sci. Adv.* **2017**, *3*, e1700782. [CrossRef]
13. Mohamed, S.A.A.; El-Sakhawy, M.; El-Sakhawy, M.A.-M. Polysaccharides, Protein and Lipid -Based Natural Edible Films in Food Packaging: A Review. *Carbohydr. Polym.* **2020**, *238*, 116178. [CrossRef]
14. Pereira, J.M.; Rodríguez, Y.; Blasco-Monleon, S.; Porter, A.; Lewis, C.; Pham, C.K. Microplastic in the stomachs of open-ocean and deep-sea fishes of the North-East Atlantic. *Environ. Pollut.* **2020**, *265*, 115060. [CrossRef]
15. Roman, L.; Kastury, F.; Petit, S.; Aleman, R.; Wilcox, C.; Hardesty, B.D.; Hindell, M.A. Plastic, nutrition and pollution; relationships between ingested plastic and metal concentrations in the livers of two Pachyptila seabirds. *Sci. Rep.* **2020**, *10*, 18023. [CrossRef]
16. Mattsson, K.; Johnson, E.V.; Malmendal, A.; Linse, S.; Hansson, L.-A.; Cedervall, T. Brain damage and behavioural disorders in fish induced by plastic nanoparticles delivered through the food chain. *Sci. Rep.* **2017**, *7*, 11452. [CrossRef]
17. Ali, S.S.; Elsamahy, T.; Al-Tohamy, R.; Zhu, D.; Mahmoud, Y.A.-G.; Koutra, E.; Metwally, M.A.; Kornaros, M.; Sun, J. Plastic wastes biodegradation: Mechanisms, challenges and future prospects. *Sci. Total Environ.* **2021**, *780*, 146590. [CrossRef]
18. Kershaw, P. *Sources, Fate and Effects of Microplastics in the Marine Environment: A Global Assessment*; Technical Report; International Maritime Organization: London, UK, 2015; Available online: http://localhost:8080/xmlui/handle/123456789/735 (accessed on 18 April 2021).
19. Abioye, O.P.; Abioye, A.A.; Afolalu, S.A.; Ongbali, S.O. A Review of Biodegradable Plastics in Nigeria. *Int. J. Mech. Eng. Technol. IJMET* **2018**, *9*, 10. Available online: http://eprints.covenantuniversity.edu.ng/12417/#.YH6elC0itN0 (accessed on 20 April 2021).
20. Hopewell, J.; Dvorak, R.; Kosior, E. Plastics recycling: Challenges and opportunities. *Philos. Trans. R. Soc. B Biol. Sci.* **2009**, *364*, 2115–2126. [CrossRef]
21. Rai, P.; Mehrotra, S.; Priya, S.; Gnansounou, E.; Sharma, S.K. Recent advances in the sustainable design and applications of biodegradable polymers. *Bioresour. Technol.* **2021**, *325*, 124739. [CrossRef]
22. Udayakumar, G.P.; Muthusamy, S.; Selvaganesh, B.; Sivarajasekar, N.; Rambabu, K.; Banat, F.; Sivamani, S.; Sivakumar, N.; Hosseini-Bandegharaei, A.; Show, P.L. Biopolymers and composites: Properties, characterization and their applications in food, medical and pharmaceutical industries. *J. Environ. Chem. Eng.* **2021**, *9*, 105322. [CrossRef]
23. Zhong, Y.; Godwin, P.; Jin, Y.; Xiao, H. Biodegradable polymers and green-based antimicrobial packaging materials: A mini-review. *Adv. Ind. Eng. Polym. Res.* **2020**, *3*, 27–35. [CrossRef]
24. Jeevahan, J.; Chandrasekaran, M. Nanoedible films for food packaging: A review. *J. Mater. Sci.* **2019**, *54*, 12290–12318. [CrossRef]
25. Reichert, C.L.; Bugnicourt, E.; Coltelli, M.-B.; Cinelli, P.; Lazzeri, A.; Canesi, I.; Braca, F.; Martínez, B.M.; Alonso, R.; Agostinis, L.; et al. Bio-Based Packaging: Materials, Modifications, Industrial Applications and Sustainability. *Polymers* **2020**, *12*, 1558. [CrossRef]
26. Mastromatteo, M.; Mastromatteo, M.; Conte, A.; Del Nobile, M.A. Advances in controlled release devices for food packaging applications. *Trends Food Sci. Technol.* **2010**, *21*, 591–598. [CrossRef]
27. Jideani, V.A.; Vogt, K. Antimicrobial Packaging for Extending the Shelf Life of Bread-A Review. *Crit. Rev. Food Sci. Nutr.* **2016**, *56*, 1313–1324. [CrossRef]
28. Khaneghah, M.A.; Hashemi, S.M.B.; Eş, I.; Fracassetti, D.; Limbo, S. Efficacy of Antimicrobial Agents for Food Contact Applications: Biological Activity, Incorporation into Packaging, and Assessment Methods: A Review. *J. Food Prot.* **2018**, *81*, 1142–1156. [CrossRef]
29. Firouz, S.M.; Mohi-Alden, K.; Omid, M. A critical review on intelligent and active packaging in the food industry: Research and development. *Food Res. Int.* **2021**, *141*, 110113. [CrossRef]
30. Khaneghah, M.A.; Hashemi, S.M.B.; Limbo, S. Antimicrobial agents and packaging systems in antimicrobial active food packaging: An overview of approaches and interactions. *Food Bioprod. Process.* **2018**, *111*, 1–19. [CrossRef]

31. Ribeiro-Santos, R.; Andrade, M.; Sanches-Silva, A. Application of encapsulated essential oils as antimicrobial agents in food packaging. *Curr. Opin. Food Sci.* **2017**, *14*, 78–84. [CrossRef]
32. Unalan, I.; Boccaccini, A.R. Essential oils in biomedical applications: Recent progress and future opportunities. *Curr. Opin. Biomed. Eng.* **2021**, *17*, 100261. [CrossRef]
33. Dhifi, W.; Bellili, S.; Jazi, S.; Bahloul, N.; Mnif, W. Essential Oils' Chemical Characterization and Investigation of Some Biological Activities: A Critical Review. *Medicines* **2016**, *3*, 25. [CrossRef]
34. Delshadi, R.; Bahrami, A.; Tafti, A.G.; Barba, F.J.; Williams, L.L. Micro and nano-encapsulation of vegetable and essential oils to develop functional food products with improved nutritional profiles. *Trends Food Sci. Technol.* **2020**, *104*, 72–83. [CrossRef]
35. Gómez, B.; Barba, F.J.; Domínguez, R.; Putnik, P.; Bursać Kovačević, D.; Pateiro, M.; Toldrá, F.; Lorenzo, J.M. Microencapsulation of antioxidant compounds through innovative technologies and its specific application in meat processing. *Trends Food Sci. Technol.* **2018**, *82*, 135–147. [CrossRef]
36. Jurić, S.; Jurić, M.; Siddique, M.A.B.; Fathi, M. Vegetable Oils Rich in Polyunsaturated Fatty Acids: Nanoencapsulation Methods and Stability Enhancement. *Food Rev. Int.* **2020**, *380*, 1–38. [CrossRef]
37. Alehosseini, E.; Jafari, S.M. Micro/nano-encapsulated phase change materials (PCMs) as emerging materials for the food industry. *Trends Food Sci. Technol.* **2019**, *91*, 116–128. [CrossRef]
38. Rochman, C.M.; Browne, M.A.; Halpern, B.S.; Hentschel, B.T.; Hoh, E.; Karapanagioti, H.K.; Rios-Mendoza, L.M.; Takada, H.; Teh, S.; Thompson, R.C. Classify plastic waste as hazardous. *Nature* **2013**, *494*, 169–171. [CrossRef]
39. Andrady, A.L. Microplastics in the marine environment. *Mar. Pollut. Bull.* **2011**, *62*, 1596–1605. [CrossRef]
40. Webb, H.K.; Arnott, J.; Crawford, R.J.; Ivanova, E.P. Plastic Degradation and Its Environmental Implications with Special Reference to Poly(ethylene terephthalate). *Polymers* **2013**, *5*, 1–18. [CrossRef]
41. *EN 13432 Packaging—Requirements for Packaging Recoverable through Composting and Biodegradation—Test Scheme and Evaluation Criteria for the Final Acceptance of Packaging*; European Committee for Standardisation: Brussels, Belgium, 2000. Available online: https://www.en-standard.eu/bs-en-13432-2000-packaging.-requirements-for-packaging-recoverable-through-composting-and-biodegradation.-test-scheme-and-evaluation-criteria-for-the-final-acceptance-of-packaging/ (accessed on 11 February 2021).
42. Ali, S.S.; Elsamahy, T.; Koutra, E.; Kornaros, M.; El-Sheekh, M.; Abdelkarim, E.A.; Zhu, D.; Sun, J. Degradation of conventional plastic wastes in the environment: A review on current status of knowledge and future perspectives of disposal. *Sci. Total Environ.* **2021**, *771*, 144719. [CrossRef]
43. Gorrasi, G.; Sorrentino, A.; Lichtfouse, E. Back to plastic pollution in COVID times. *Environ. Chem. Lett.* **2021**, *19*, 1–4. [CrossRef]
44. Horodytska, O.; Cabanes, A.; Fullana, A. Plastic Waste Management: Current Status and Weaknesses. In *The Handbook of Environmental Chemistry*; Springer: Berlin/Heidelberg, Germany, 2019; pp. 1–18. [CrossRef]
45. Helinski, O.K.; Poor, C.J.; Wolfand, J.M. Ridding our rivers of plastic: A framework for plastic pollution capture device selection. *Mar. Pollut. Bull.* **2021**, *165*, 112095. [CrossRef]
46. Hamilton: Solving Plastic Pollution through Accountability. Available online: https://scholar.google.com/scholar_lookup?title=Solving%20Plastic%20Pollution%20Through%20Accountability&publication_year=2019&author=A.%20Hamilton&author=R.%20Scheer&author=T.%20Stakes&author=S.%20Allan (accessed on 21 April 2021).
47. Xu, H.; Sheng, J.; Wu, X.; Zhan, K.; Tao, S.; Wen, X.; Liu, W.; Cudjoe, O.; Tao, F. Moderating effects of plastic packaged food on association of urinary phthalate metabolites with emotional symptoms in Chinese adolescents. *Ecotoxicol. Environ. Saf.* **2021**, *216*, 112171. [CrossRef]
48. Larrain, M.; Van Passel, S.; Thomassen, G.; Van Gorp, B.; Nhu, T.T.; Huysveld, S.; Van Geem, K.M.; De Meester, S.; Billen, P. Techno-economic assessment of mechanical recycling of challenging post-consumer plastic packaging waste. *Resour. Conserv. Recycl.* **2021**, *170*, 105607. [CrossRef]
49. Liu, C.; Zhang, X.; Medda, F. Plastic credit: A consortium blockchain-based plastic recyclability system. *Waste Manag.* **2021**, *121*, 42–51. [CrossRef]
50. Mattila, H.; Virtanen, T.; Vartiainen, T.; Ruuskanen, J. Emissions of polychlorinated dibenzo-p-dioxins and dibenzofurans in flue gas from co-combustion of mixed plastics with coal and bark. *Chemosphere* **1992**, *25*, 1599–1609. [CrossRef]
51. Ahmadzadeh, S.; Khaneghah, M.A. Role of Green Polymers in Food Packaging. *Encycl. Renew. Sustain. Mater.* **2020**, *2*, 305–319. [CrossRef]
52. Lindström, T.; Österberg, F. Evolution of biobased and nanotechnology packaging—A review. *Nord. Pulp Pap. Res. J.* **2020**, *35*, 491–515. [CrossRef]
53. EUBIO_Admin. Market. *Eur. Bioplastics E.V.* Available online: https://www.european-bioplastics.org/market/ (accessed on 19 July 2021).
54. Hawthorne, L.M.; Beganović, A.; Schwarz, M.; Noordanus, A.W.; Prem, M.; Zapf, L.; Scheibel, S.; Margreiter, G.; Huck, C.W.; Bach, K. Suitability of Biodegradable Materials in Comparison with Conventional Packaging Materials for the Storage of Fresh Pork Products over Extended Shelf-Life Periods. *Foods* **2020**, *9*, 1802. [CrossRef]
55. Ivanković, A.; Zeljko, K.; Talić, S.; Bevanda, A.M.; Lasić, M. Biodegradable Packaging in The Food Industry. *Arch. Für Leb.* **2017**, *68*, 23–52. [CrossRef]
56. Kobayashi, S. Lipase-catalyzed polyester synthesis A green polymer chemistry. *Proc. Jpn. Acad. Ser. B* **2010**, *86*, 338–365. [CrossRef]

57. Van den Oever, M.; Molenveld, K.; Zee, M.; Bos, H. *Bio-Based and Biodegradable Plastics—Facts and Figures. Focus on Food Packaging in The Netherlands*; (Wageningen Food & Biobased Research; No. 1722); Wageningen Food & Biobased Research: Wageningen, The Netherlands, 2017. [CrossRef]
58. Ghuttora, N. Increase the Usage of Biopolymers and Biodegradable Polymers for Sustainable Environment. Available online: http://www.theseus.fi/handle/10024/121984 (accessed on 25 April 2021).
59. Öner, E.T. Microbial Production of Extracellular Polysaccharides from Biomass. In *Pretreatment Techniques for Biofuels and Biorefineries*; Fang, Z., Ed.; Green Energy and Technology; Springer: Berlin/Heidelberg, Germany, 2013; pp. 35–56. ISBN 978-3-642-32735-3. [CrossRef]
60. Shevkani, K.; Singh, N.; Bajaj, R.; Kaur, A. Wheat starch production, structure, functionality and applications—A review. *Int. J. Food Sci. Technol.* **2017**, *52*, 38–58. [CrossRef]
61. Deng, N.; Deng, Z.; Tang, C.; Liu, C.; Luo, S.; Chen, T.; Hu, X. Formation, structure and properties of the starch-polyphenol inclusion complex: A review. *Trends Food Sci. Technol.* **2021**, *112*, 667–675. [CrossRef]
62. Oyeyinka, S.A.; Akintayo, O.A.; Adebo, O.A.; Kayitesi, E.; Njobeh, P.B. A review on the physicochemical properties of starches modified by microwave alone and in combination with other methods. *Int. J. Biol. Macromol.* **2021**, *176*, 87–95. [CrossRef]
63. Żołek-Tryznowska, Z.; Holica, J. Starch films as an environmentally friendly packaging material: Printing performance. *J. Clean. Prod.* **2020**, *276*, 124265. [CrossRef]
64. Cheng, H.; Chen, L.; McClements, D.J.; Yang, T.; Zhang, Z.; Ren, F.; Miao, M.; Tian, Y.; Jin, Z. Starch-based biodegradable packaging materials: A review of their preparation, characterization and diverse applications in the food industry. *Trends Food Sci. Technol.* **2021**, *114*, 70–82. [CrossRef]
65. Mali, S.; Grossmann, M.V.E.; Garcia, M.A.; Martino, M.N.; Zaritzky, N.E. Microstructural characterization of yam starch films. *Carbohydr. Polym.* **2002**, *50*, 379–386. [CrossRef]
66. Nawab, A.; Alam, F.; Haq, M.A.; Lutfi, Z.; Hasnain, A. Mango kernel starch-gum composite films: Physical, mechanical and barrier properties. *Int. J. Biol. Macromol.* **2017**, *98*, 869–876. [CrossRef]
67. Nordin, N.; Othman, S.H.; Rashid, S.A.; Basha, R.K. Effects of glycerol and thymol on physical, mechanical, and thermal properties of corn starch films. *Food Hydrocoll.* **2020**, *106*, 105884. [CrossRef]
68. Adamus, J.; Spychaj, T.; Zdanowicz, M.; Jędrzejewski, R. Thermoplastic starch with deep eutectic solvents and montmorillonite as a base for composite materials. *Ind. Crops Prod.* **2018**, *123*, 278–284. [CrossRef]
69. Thirmizir, M.Z.A.; Ishak, Z.A.M.; Salim, M.S. Compatibilization and Crosslinking in Biodegradable Thermoplastic Polyester Blends. In *Reactive and Functional Polymers Volume Two: Modification Reactions, Compatibility and Blends*; Gutiérrez, T.J., Ed.; Springer International Publishing: Cham, Switzerland, 2020; pp. 23–89. ISBN 978-3-030-45135-6. [CrossRef]
70. Nafchi, A.M.; Alias, A.K.; Mahmud, S.; Robal, M. Antimicrobial, rheological, and physicochemical properties of sago starch films filled with nanorod-rich zinc oxide. *J. Food Eng.* **2012**, *113*, 511–519. [CrossRef]
71. Vaezi, K.; Asadpour, G.; Sharifi, H. Effect of ZnO nanoparticles on the mechanical, barrier and optical properties of thermoplastic cationic starch/montmorillonite biodegradable films. *Int. J. Biol. Macromol.* **2019**, *124*, 519–529. [CrossRef]
72. Ali, A.; Xie, F.; Yu, L.; Liu, H.; Meng, L.; Khalid, S.; Chen, L. Preparation and characterization of starch-based composite films reinfoced by polysaccharide-based crystals. *Compos. Part B Eng.* **2018**, *133*, 122–128. [CrossRef]
73. de Souza, A.C.; Dias, A.M.A.; Sousa, H.C.; Tadini, C.C. Impregnation of cinnamaldehyde into cassava starch biocomposite films using supercritical fluid technology for the development of food active packaging. *Carbohydr. Polym.* **2014**, *102*, 830–837. [CrossRef] [PubMed]
74. Avila-Sosa, R.; Palou, E.; Jiménez Munguía, M.T.; Nevárez-Moorillón, G.V.; Navarro Cruz, A.R.; López-Malo, A. Antifungal activity by vapor contact of essential oils added to amaranth, chitosan, or starch edible films. *Int. J. Food Microbiol.* **2012**, *153*, 66–72. [CrossRef] [PubMed]
75. Souza, A.C.; Goto, G.E.O.; Mainardi, J.A.; Coelho, A.C.V.; Tadini, C.C. Cassava starch composite films incorporated with cinnamon essential oil: Antimicrobial activity, microstructure, mechanical and barrier properties. *LWT-Food Sci. Technol.* **2013**, *54*, 346–352. [CrossRef]
76. Engel, J.B.; Ambrosi, A.; Tessaro, I.C. Development of biodegradable starch-based foams incorporated with grape stalks for food packaging. *Carbohydr. Polym.* **2019**, *225*, 115234. [CrossRef]
77. Sganzerla, W.G.; Rosa, G.B.; Ferreira, A.L.A.; da Rosa, C.G.; Beling, P.C.; Xavier, L.O.; Hansen, C.M.; Ferrareze, J.P.; Nunes, M.R.; Barreto, P.L.M.; et al. Bioactive food packaging based on starch, citric pectin and functionalized with Acca sellowiana waste by-product: Characterization and application in the postharvest conservation of apple. *Int. J. Biol. Macromol.* **2020**, *147*, 295–303. [CrossRef]
78. Baek, S.-K.; Kim, S.; Song, K.B. Cowpea starch films containing maqui berry extract and their application in salmon packaging. *Food Packag. Shelf Life* **2019**, *22*, 100394. [CrossRef]
79. Cheng, J.; Wang, H.; Kang, S.; Xia, L.; Jiang, S.; Chen, M.; Jiang, S. An active packaging film based on yam starch with eugenol and its application for pork preservation. *Food Hydrocoll.* **2019**, *96*, 546–554. [CrossRef]
80. Zhou, X.; Yang, R.; Wang, B.; Chen, K. Development and characterization of bilayer films based on pea starch/polylactic acid and use in the cherry tomatoes packaging. *Carbohydr. Polym.* **2019**, *222*, 114912. [CrossRef]
81. Hassan, M.; Dave, K.; Chandrawati, R.; Dehghani, F.; Gomes, V.G. 3D printing of biopolymer nanocomposites for tissue engineering: Nanomaterials, processing and structure-function relation. *Eur. Polym. J.* **2019**, *121*, 109340. [CrossRef]

82. Panrong, T.; Karbowiak, T.; Harnkarnsujarit, N. Thermoplastic starch and green tea blends with LLDPE films for active packaging of meat and oil-based products. *Food Packag. Shelf Life* **2019**, *21*, 100331. [CrossRef]
83. Moreno, O.; Atarés, L.; Chiralt, A.; Cruz-Romero, M.C.; Kerry, J. *Starch-Gelatin Antimicrobial Packaging Materials to Extend the Shelf Life of Chicken Breast Fillets*; Elsevier: Amsterdam, The Netherlands, 2018; Volume 97, pp. 483–490. [CrossRef]
84. Fadeyibi, A.; Osunde, Z.D.; Egwim, E.C.; Idah, P.A. Performance evaluation of cassava starch-zinc nanocomposite film for tomatoes packaging. *J. Agric. Eng.* **2017**, *48*, 137–146. [CrossRef]
85. Hu, L.; Du, H.; Liu, C.; Zhang, Y.; Yu, G.; Zhang, X.; Si, C.; Li, B.; Peng, H. Comparative Evaluation of the Efficient Conversion of Corn Husk Filament and Corn Husk Powder to Valuable Materials via a Sustainable and Clean Biorefinery Process. *ACS Sustain. Chem. Eng.* **2019**, *7*, 1327–1336. [CrossRef]
86. Liu, W.; Du, H.; Liu, H.; Xie, H.; Xu, T.; Zhao, X.; Liu, Y.; Zhang, X.; Si, C. Highly Efficient and Sustainable Preparation of Carboxylic and Thermostable Cellulose Nanocrystals via FeCl3-Catalyzed Innocuous Citric Acid Hydrolysis. *ACS Sustain. Chem. Eng.* **2020**, *8*, 16691–16700. [CrossRef]
87. Zhao, G.; Du, J.; Chen, W.; Pan, M.; Chen, D. Preparation and thermostability of cellulose nanocrystals and nanofibrils from two sources of biomass: Rice straw and poplar wood. *Cellulose* **2019**, *26*, 8625–8643. [CrossRef]
88. Liu, Y.; Ahmed, S.; Sameen, D.E.; Wang, Y.; Lu, R.; Dai, J.; Li, S.; Qin, W. A review of cellulose and its derivatives in biopolymer-based for food packaging application. *Trends Food Sci. Technol.* **2021**, *112*, 532–546. [CrossRef]
89. Pooresmaeil, M.; Behzadi Nia, S.; Namazi, H. Green encapsulation of LDH(Zn/Al)-5-Fu with carboxymethyl cellulose biopolymer; new nanovehicle for oral colorectal cancer treatment. *Int. J. Biol. Macromol.* **2019**, *139*, 994–1001. [CrossRef]
90. Cazón, P.; Vázquez, M. Bacterial cellulose as a biodegradable food packaging material: A review. *Food Hydrocoll.* **2021**, *113*, 106530. [CrossRef]
91. Wang, W.; Gu, F.; Deng, Z.; Zhu, Y.; Zhu, J.; Guo, T.; Song, J.; Xiao, H. Multilayer surface construction for enhancing barrier properties of cellulose-based packaging. *Carbohydr. Polym.* **2021**, *255*, 117431. [CrossRef]
92. Assis, R.Q.; Pagno, C.H.; Stoll, L.; Rios, P.D.; Rios, A.d.O.; Olivera, F.C. Active food packaging of cellulose acetate: Storage stability, protective effect on oxidation of riboflavin and release in food simulants. *Food Chem.* **2021**, *349*, 129140. [CrossRef]
93. Blanco-Cabra, N.; Paetzold, B.; Ferrar, T.; Mazzolini, R.; Torrents, E.; Serrano, L.; LLuch-Senar, M. Characterization of different alginate lyases for dissolving Pseudomonas aeruginosa biofilms. *Sci. Rep.* **2020**, *10*, 9390. [CrossRef]
94. Günter, E.A.; Popeyko, O.V.; Belozerov, V.S.; Martinson, E.A.; Litvinets, S.G. Physicochemical and swelling properties of composite gel microparticles based on alginate and callus cultures pectins with low and high degrees of methylesterification. *Int. J. Biol. Macromol.* **2020**, *164*, 863–870. [CrossRef]
95. Lee, K.Y.; Mooney, D.J. Alginate: Properties and biomedical applications. *Prog. Polym. Sci.* **2012**, *37*, 106–126. [CrossRef]
96. Xu, M.; Qin, M.; Cheng, Y.; Niu, X.; Kong, J.; Zhang, X.; Huang, D.; Wang, H. Alginate microgels as delivery vehicles for cell-based therapies in tissue engineering and regenerative medicine. *Carbohydr. Polym.* **2021**, *266*, 118128. [CrossRef]
97. Parreidt, S.T.; Müller, K.; Schmid, M. Alginate-Based Edible Films and Coatings for Food Packaging Applications. *Foods* **2018**, *7*, 170. [CrossRef]
98. Zhang, H.; Zhang, F.; Yuan, R. Chapter 13—Applications of natural polymer-based hydrogels in the food industry. In *Hydrogels Based on Natural Polymers*; Chen, Y., Ed.; Elsevier: Amsterdam, The Netherlands, 2020; pp. 357–410. ISBN 978-0-12-816421-1. [CrossRef]
99. Abdel Aziz, M.S.; Salama, H.E. Effect of vinyl montmorillonite on the physical, responsive and antimicrobial properties of the optimized polyacrylic acid/chitosan superabsorbent via Box-Behnken model. *Int. J. Biol. Macromol.* **2018**, *116*, 840–848. [CrossRef]
100. Aziz, M.S.A.; Salama, H.E.; Saad, G.R. Diglycidyl ether of bisphenol A/chitosan-graft-polyaniline composites with electromagnetic interference shielding properties: Synthesis, characterization, and curing kinetics. *Polym. Eng. Sci.* **2019**, *59*, 372–381. [CrossRef]
101. Salama, H.E.; Aziz, A.M.S. Development of active edible coating of alginate and aloe vera enriched with frankincense oil for retarding the senescence of green capsicums. *LWT* **2021**, *145*, 111341. [CrossRef]
102. Bennacef, C.; Desobry-Banon, S.; Probst, L.; Desobry, S. Advances on alginate use for spherification to encapsulate biomolecules. *Food Hydrocoll.* **2021**, *118*, 106782. [CrossRef]
103. Chen, J.; Wu, A.; Yang, M.; Ge, Y.; Pristijono, P.; Li, J.; Xu, B.; Mi, H. Characterization of sodium alginate-based films incorporated with thymol for fresh-cut apple packaging. *Food Control* **2021**, *126*, 108063. [CrossRef]
104. Dong, Y.; Wei, Z.; Xue, C. Recent advances in carrageenan-based delivery systems for bioactive ingredients: A review. *Trends Food Sci. Technol.* **2021**, *112*, 348–361. [CrossRef]
105. Khotimchenko, M.; Tiasto, V.; Kalitnik, A.; Begun, M.; Khotimchenko, R.; Leonteva, E.; Bryukhovetskiy, I.; Khotimchenko, Y. Antitumor potential of carrageenans from marine red algae. *Carbohydr. Polym.* **2020**, *246*, 116568. [CrossRef] [PubMed]
106. Zia, K.M.; Tabasum, S.; Nasif, M.; Sultan, N.; Aslam, N.; Noreen, A.; Zuber, M. A review on synthesis, properties and applications of natural polymer based carrageenan blends and composites. *Int. J. Biol. Macromol.* **2017**, *96*, 282–301. [CrossRef] [PubMed]
107. Liu, Y.; Qin, Y.; Bai, R.; Zhang, X.; Yuan, L.; Liu, J. Preparation of pH-sensitive and antioxidant packaging films based on κ-carrageenan and mulberry polyphenolic extract. *Int. J. Biol. Macromol.* **2019**, *134*, 993–1001. [CrossRef] [PubMed]
108. Pourashouri, P.; Shabanpour, B.; Heydari, S.; Raeisi, S. Encapsulation of fish oil by carrageenan and gum tragacanth as wall materials and its application to the enrichment of chicken nuggets. *LWT* **2021**, *137*, 110334. [CrossRef]

109. Priyadarshi, R.; Rhim, J.-W. Chitosan-based biodegradable functional films for food packaging applications. *Innov. Food Sci. Emerg. Technol.* **2020**, *62*, 102346. [CrossRef]
110. Yu, J.; Wang, D.; Geetha, N.; Khawar, K.M.; Jogaiah, S.; Mujtaba, M. Current trends and challenges in the synthesis and applications of chitosan-based nanocomposites for plants: A review. *Carbohydr. Polym.* **2021**, *261*, 117904. [CrossRef] [PubMed]
111. Mujtaba, M.; Morsi, R.E.; Kerch, G.; Elsabee, M.Z.; Kaya, M.; Labidi, J.; Khawar, K.M. Current advancements in chitosan-based film production for food technology; A review. *Int. J. Biol. Macromol.* **2019**, *121*, 889–904. [CrossRef]
112. Wang, H.; Ding, F.; Ma, L.; Zhang, Y. Edible films from chitosan-gelatin: Physical properties and food packaging application. *Food Biosci.* **2021**, *40*, 100871. [CrossRef]
113. Wang, H.; Qian, J.; Ding, F. Emerging Chitosan-Based Films for Food Packaging Applications. *J. Agric. Food Chem.* **2018**, *66*, 395–413. [CrossRef]
114. Alirezalu, K.; Pirouzi, S.; Yaghoubi, M.; Karimi-Dehkordi, M.; Jafarzadeh, S.; Mousavi Khaneghah, A. Packaging of beef fillet with active chitosan film incorporated with ϵ-polylysine: An assessment of quality indices and shelf life. *Meat Sci.* **2021**, *176*, 108475. [CrossRef]
115. Xavier, L.O.; Sganzerla, W.G.; Rosa, G.B.; da Rosa, C.G.; Agostinetto, L.; de Lima Veeck, A.P.; Bretanha, L.C.; Micke, G.A.; Dalla Costa, M.; Bertoldi, F.C.; et al. Chitosan packaging functionalized with Cinnamodendron dinisii essential oil loaded zein: A proposal for meat conservation. *Int. J. Biol. Macromol.* **2021**, *169*, 183–193. [CrossRef]
116. Dubey, N.K.; Dubey, R. Chapter 27—Edible films and coatings: An update on recent advances. In *Biopolymer-Based Formulations;* Pal, K., Banerjee, I., Sarkar, P., Kim, D., Deng, W.-P., Dubey, N.K., Majumder, K., Eds.; Elsevier: Amsterdam, The Netherlands, 2020; pp. 675–695. ISBN 978-0-12-816897-4. [CrossRef]
117. Tang, X.Z.; Kumar, P.; Alavi, S.; Sandeep, K.P. Recent Advances in Biopolymers and Biopolymer-Based Nanocomposites for Food Packaging Materials. *Crit. Rev. Food Sci. Nutr.* **2012**, *52*, 426–442. [CrossRef]
118. Zhang, H.; Mittal, G. Biodegradable protein-based films from plant resources: A review. *Environ. Prog. Sustain. Energy* **2010**, *29*, 203–220. [CrossRef]
119. Ciannamea, E.M.; Stefani, P.M.; Ruseckaite, R.A. Properties and antioxidant activity of soy protein concentrate films incorporated with red grape extract processed by casting and compression molding. *LWT* **2016**, *74*, 353–362. [CrossRef]
120. Eswaranandam, S.; Hettiarachchy, N.S.; Johnson, M.G. Antimicrobial Activity of Citric, Lactic, Malic, or Tartaric Acids and Nisin-incorporated Soy Protein Film Against Listeria monocytogenes, *Escherichia coli* O157:H7, and *Salmonella gaminara*. *J. Food Sci.* **2004**, *69*, FMS79–FMS84. [CrossRef]
121. Soares, R.M.D.; Scremin, F.F.; Soldi, V. Thermal Stability of Biodegradable Films Based on Soy Protein and Corn Starch. *Macromol. Symp.* **2005**, *229*, 258–265. [CrossRef]
122. Su, J.-F.; Huang, Z.; Liu, K.; Fu, L.-L.; Liu, H.-R. Mechanical Properties, Biodegradation and Water Vapor Permeability of Blend Films of Soy Protein Isolate and Poly (vinyl alcohol) Compatibilized by Glycerol. *Polym. Bull.* **2007**, *58*, 913–921. [CrossRef]
123. Tian, H.; Guo, G.; Fu, X.; Yao, Y.; Yuan, L.; Xiang, A. Fabrication, properties and applications of soy-protein-based materials: A review. *Int. J. Biol. Macromol.* **2018**, *120*, 475–490. [CrossRef]
124. Rhim, J.-W.; Lee, J.H.; Ng, P.K.W. Mechanical and barrier properties of biodegradable soy protein isolate-based films coated with polylactic acid. *LWT-Food Sci. Technol.* **2007**, *40*, 232–238. [CrossRef]
125. Xiao, Y.; Liu, Y.; Kang, S.; Xu, H. Insight into the formation mechanism of soy protein isolate films improved by cellulose nanocrystals. *Food Chem.* **2021**, 129971. [CrossRef]
126. Li, J.; Li, X.; Zhang, F.; Zhang, W.; Li, J. Facile design of tough, strong, and UV-shielding soy protein-based composite films. *Ind. Crops Prod.* **2021**, *166*, 113474. [CrossRef]
127. Jansens, K.J.A.; Vo Hong, N.; Telen, L.; Brijs, K.; Lagrain, B.; Van Vuure, A.W.; Van Acker, K.; Verpoest, I.; Van Puyvelde, P.; Goderis, B.; et al. Effect of molding conditions and moisture content on the mechanical properties of compression molded glassy, wheat gluten bioplastics. *Ind. Crops Prod.* **2013**, *44*, 480–487. [CrossRef]
128. Wang, J.; Wei, Z.; Li, L.; Bian, K.; Zhao, M. Characteristics of enzymatic hydrolysis of thermal-treated wheat gluten. *J. Cereal Sci.* **2009**, *50*, 205–209. [CrossRef]
129. Tunc, S.; Angellier, H.; Cahyana, Y.; Chalier, P.; Gontard, N.; Gastaldi, E. Functional properties of wheat gluten/montmorillonite nanocomposite films processed by casting. *J. Membr. Sci.* **2007**, *289*, 159–168. [CrossRef]
130. Rovera, C.; Türe, H.; Hedenqvist, M.S.; Farris, S. Water vapor barrier properties of wheat gluten/silica hybrid coatings on paperboard for food packaging applications. *Food Packag. Shelf Life* **2020**, *26*, 100561. [CrossRef]
131. Gutiérrez, T.J.; Mendieta, J.R.; Ortega-Toro, R. In-depth study from gluten/PCL-based food packaging films obtained under reactive extrusion conditions using chrome octanoate as a potential food grade catalyst. *Food Hydrocoll.* **2021**, *111*, 106255. [CrossRef]
132. Türe, H.; Gällstedt, M.; Hedenqvist, M.S. Antimicrobial compression-moulded wheat gluten films containing potassium sorbate. *Food Res. Int.* **2012**, *45*, 109–115. [CrossRef]
133. Mascheroni, E.; Guillard, V.; Gastaldi, E.; Gontard, N.; Chalier, P. Anti-microbial effectiveness of relative humidity-controlled carvacrol release from wheat gluten/montmorillonite coated papers. *Food Control* **2011**, *22*, 1582–1591. [CrossRef]
134. Zhang, X.; Dong, C.; Hu, Y.; Gao, M.; Luan, G. Zein as a structural protein in gluten-free systems: An overview. *Food Sci. Hum. Wellness* **2021**, *10*, 270–277. [CrossRef]
135. Shukla, R.; Cheryan, M. Zein: The industrial protein from corn. *Ind. Crops Prod.* **2001**, *13*, 171–192. [CrossRef]

136. Zhou, L.; Wang, Y. Physical and antimicrobial properties of zein and methyl cellulose composite films with plasticizers of oleic acid and polyethylene glycol. *LWT* **2021**, *140*, 110811. [CrossRef]
137. Lai, H.-M.; Padua, G.W. Properties and Microstructure of Plasticized Zein Films. *Cereal Chem.* **1997**, *74*, 771–775. [CrossRef]
138. Wang, Y.; Rakotonirainy, A.M.; Padua, G.W. Thermal Behavior of Zein-based Biodegradable Films. *Starch -Stärke* **2003**, *55*, 25–29. [CrossRef]
139. Ghanbarzadeh, B.; Oromiehi, A.R. Biodegradable biocomposite films based on whey protein and zein: Barrier, mechanical properties and AFM analysis. *Int. J. Biol. Macromol.* **2008**, *43*, 209–215. [CrossRef]
140. Huo, W.; Wei, D.; Zhu, W.; Li, Z.; Jiang, Y. High-elongation zein films for flexible packaging by synergistic plasticization: Preparation, structure and properties. *J. Cereal Sci.* **2018**, *79*, 354–361. [CrossRef]
141. Xu, H.; Zhang, G. Synergistic Effect of Oleic Acid and Glycerol on Zein Film Plasticization. *J. Agric. Food Chem.* **2012**, *60*, 10075–10081. [CrossRef]
142. Kashiri, M.; López-Carballo, G.; Hernández-Muñoz, P.; Gavara, R. Antimicrobial packaging based on a LAE containing zein coating to control foodborne pathogens in chicken soup. *Int. J. Food Microbiol.* **2019**, *306*, 108272. [CrossRef]
143. Ahammed, S.; Liu, F.; Wu, J.; Khin, M.N.; Yokoyama, W.H.; Zhong, F. Effect of transglutaminase crosslinking on solubility property and mechanical strength of gelatin-zein composite films. *Food Hydrocoll.* **2021**, *116*, 106649. [CrossRef]
144. Daniloski, D.; Petkoska, A.T.; Lee, N.A.; Bekhit, A.E.-D.; Carne, A.; Vaskoska, R.; Vasiljevic, T. Active edible packaging based on milk proteins: A route to carry and deliver nutraceuticals. *Trends Food Sci. Technol.* **2021**, *111*, 688–705. [CrossRef]
145. Horne, D.S. Chapter 6—Casein micelle structure and stability. In *Milk Proteins*, 3rd ed.; Boland, M., Singh, H., Eds.; Academic Press: Cambridge, MA, USA, 2020; pp. 213–250. ISBN 978-0-12-815251-5. [CrossRef]
146. Fitzsimons, S.M.; Mulvihill, D.M.; Morris, E.R. Denaturation and aggregation processes in thermal gelation of whey proteins resolved by differential scanning calorimetry. *Food Hydrocoll.* **2007**, *21*, 638–644. [CrossRef]
147. Boland, M. Chapter 21—Milk proteins: The future. In *Milk Proteins*, 3rd ed.; Boland, M., Singh, H., Eds.; Academic Press: Cambridge, MA, USA, 2020; pp. 715–730. ISBN 978-0-12-815251-5. [CrossRef]
148. Martins, J.T.; Bourbon, A.I.; Pinheiro, A.C.; Fasolin, L.H.; Vicente, A.A. Protein-Based Structures for Food Applications: From Macro to Nanoscale. *Front. Sustain. Food Syst.* **2018**, *2*, 77. [CrossRef]
149. Moughan, P.J. Chapter 17—Milk proteins: A rich source of bioactives for developing functional foods. In *Milk Proteins*, 3rd ed.; Boland, M., Singh, H., Eds.; Academic Press: Cambridge, MA, USA, 2020; pp. 633–649. ISBN 978-0-12-815251-5. [CrossRef]
150. Bonilla, J.; Paiano, R.B.; Lourenço, R.V.; Bittante, A.M.Q.B.; Sobral, P.J.A. Biodegradation of Films Based on Natural and Synthetic Biopolymers Using an Aquatic System from Active Sludge. *J. Polym. Environ.* **2021**, *29*, 1380–1395. [CrossRef]
151. Aguirre-Joya, J.A.; De Leon-Zapata, M.A.; Alvarez-Perez, O.B.; Torres-León, C.; Nieto-Oropeza, D.E.; Ventura-Sobrevilla, J.M.; Aguilar, M.A.; Ruelas-Chacón, X.; Rojas, R.; Ramos-Aguiñaga, M.E.; et al. Chapter 1—Basic and Applied Concepts of Edible Packaging for Foods. In *Food Packaging and Preservation*; Grumezescu, A.M., Holban, A.M., Eds.; Handbook of Food Bioengineering; Academic Press: Cambridge, MA, USA, 2018; pp. 1–61. ISBN 978-0-12-811516-9. [CrossRef]
152. Makhijani, K.; Kumar, R.; Sharma, S.K. Biodegradability of Blended Polymers: A Comparison of Various Properties. *Crit. Rev. Environ. Sci. Technol.* **2015**, *45*, 1801–1825. [CrossRef]
153. Tkaczewska, J. Peptides and protein hydrolysates as food preservatives and bioactive components of edible films and coatings—A review. *Trends Food Sci. Technol.* **2020**, *106*, 298–311. [CrossRef]
154. Ponce, A.; Roura, S.I.; Moreira, M.R. Chapter 37—Casein and Chitosan Polymers: Use in Antimicrobial Packaging. In *Antimicrobial Food Packaging*; Barros-Velázquez, J., Ed.; Academic Press: San Diego, CA, USA, 2016; pp. 455–466. ISBN 978-0-12-800723-5. [CrossRef]
155. Haque, E.; Chand, R.; Kapila, S. Biofunctional Properties of Bioactive Peptides of Milk Origin. *Food Rev. Int.* **2008**, *25*, 28–43. [CrossRef]
156. Zhao, Q.; Shi, Y.; Wang, X.; Huang, A. Characterization of a novel antimicrobial peptide from buffalo casein hydrolysate based on live bacteria adsorption. *J. Dairy Sci.* **2020**, *103*, 11116–11128. [CrossRef]
157. Bhosale, S.; Fulpagare, Y.; Desale, R. International Journal of Advanced Research in Biological Sciences Nanoliposomes: Applications in Food and Dairy Industry. *Int. J. Adv. Res. Biol. Sci.* **2019**, *6*, 79–84.
158. Broumand, A.; Emam-Djomeh, Z.; Hamedi, M.; Razavi, S.H. Antimicrobial, water vapour permeability, mechanical and thermal properties of casein based Zataraia multiflora Boiss. Extract containing film. *LWT-Food Sci. Technol.* **2011**, *44*, 2316–2323. [CrossRef]
159. Dinika, I.; Verma, D.K.; Balia, R.; Utama, G.L.; Patel, A.R. Potential of cheese whey bioactive proteins and peptides in the development of antimicrobial edible film composite: A review of recent trends. *Trends Food Sci. Technol.* **2020**, *103*, 57–67. [CrossRef]
160. Uranga, J.; Leceta, I.; Etxabide, A.; Guerrero, P.; de la Caba, K. Cross-linking of fish gelatins to develop sustainable films with enhanced properties. *Eur. Polym. J.* **2016**, *78*, 82–90. [CrossRef]
161. Weng, W.; Zheng, H. Effect of transglutaminase on properties of tilapia scale gelatin films incorporated with soy protein isolate. *Food Chem.* **2015**, *169*, 255–260. [CrossRef] [PubMed]
162. Gómez-Guillén, M.C.; Giménez, B.; López-Caballero, M.E.; Montero, M.P. Functional and bioactive properties of collagen and gelatin from alternative sources: A review. *Food Hydrocoll.* **2011**, *25*, 1813–1827. [CrossRef]
163. Gómez-Guillén, M.C.; Pérez-Mateos, M.; Gómez-Estaca, J.; López-Caballero, E.; Giménez, B.; Montero, P. Fish gelatin: A renewable material for developing active biodegradable films. *Trends Food Sci. Technol.* **2009**, *20*, 3–16. [CrossRef]

164. Suderman, N.; Isa, M.I.N.; Sarbon, N.M. The effect of plasticizers on the functional properties of biodegradable gelatin-based film: A review. *Food Biosci.* **2018**, *24*, 111–119. [CrossRef]
165. Limpisophon, K.; Tanaka, M.; Osako, K. Characterisation of gelatin–fatty acid emulsion films based on blue shark (Prionace glauca) skin gelatin. *Food Chem.* **2010**, *122*, 1095–1101. [CrossRef]
166. Jiménez, A.; Fabra, M.J.; Talens, P.; Chiralt, A. Effect of re-crystallization on tensile, optical and water vapour barrier properties of corn starch films containing fatty acids. *Food Hydrocoll.* **2012**, *26*, 302–310. [CrossRef]
167. Roy, S.; Rhim, J.-W. Preparation of antimicrobial and antioxidant gelatin/curcumin composite films for active food packaging application. *Colloids Surf. B Biointerfaces* **2020**, *188*, 110761. [CrossRef]
168. Tavassoli, M.; Sani, M.A.; Khezerlou, A.; Ehsani, A.; McClements, D.J. Multifunctional nanocomposite active packaging materials: Immobilization of quercetin, lactoferrin, and chitosan nanofiber particles in gelatin films. *Food Hydrocoll.* **2021**, *118*, 106747. [CrossRef]
169. Khwaldia, K.; Arab-Tehrany, E.; Desobry, S. Biopolymer Coatings on Paper Packaging Materials. *Compr. Rev. Food Sci. Food Saf.* **2010**, *9*, 82–91. [CrossRef]
170. Akoh, C.C. *Food Lipids: Chemistry, Nutrition, and Biotechnology*, 4th ed.; CRC Press: Boca Raton, FL, USA, 2017; ISBN 978-1-4987-4487-4.
171. Azar, N.F.A.; Pezeshki, A.; Ghanbarzadeh, B.; Hamishehkar, H.; Mohammadi, M. Nanostructured lipid carriers: Promising delivery systems for encapsulation of food ingredients. *J. Agric. Food Res.* **2020**, *2*, 100084. [CrossRef]
172. Costa-Fernandez, S.; Mattos, J.K.R.; Scheunemann, G.S.; Salata, G.C.; Chorilli, M.; Watanabe, I.-S.; de Araujo, G.L.B.; Santos, M.F.; Ishida, K.; Lopes, L.B. Nanostructured lipid carriers containing chitosan or sodium alginate for co-encapsulation of antioxidants and an antimicrobial agent for potential application in wound healing. *Int. J. Biol. Macromol.* **2021**, *183*, 668–680. [CrossRef]
173. Syahida, N.S.; Ismail-Fitry, M.R.; Ainun, Z.M.A.; Nur Hanani, Z.A. Effects of palm wax on the physical, mechanical and water barrier properties of fish gelatin films for food packaging application. *Food Packag. Shelf Life* **2020**, *23*, 100437. [CrossRef]
174. de Oliveira Filho, J.G.; Bezerra, C.C.D.O.N.; Albiero, B.R.; Oldoni, F.C.A.; Miranda, M.; Egea, M.B.; de Azeredo, H.M.C.; Ferreira, M.D. New approach in the development of edible films: The use of carnauba wax micro- or nanoemulsions in arrowroot starch-based films. *Food Packag. Shelf Life* **2020**, *26*, 100589. [CrossRef]
175. National Academies of Sciences, Engineering, and Medicine. *Genetically Engineered Crops: Experiences and Prospects*; National Academies Press: Washington, DC, USA, 2016; ISBN 978-0-309-43738-7. [CrossRef]
176. Baldevraj, R.S.M.; Jagadish, R.S. 14-Incorporation of chemical antimicrobial agents into polymeric films for food packaging. In *Multifunctional and Nanoreinforced Polymers for Food Packaging*; Lagarón, J.-M., Ed.; Woodhead Publishing: Sawston, UK, 2011; pp. 368–420. ISBN 978-1-84569-738-9. [CrossRef]
177. Sung, S.-Y.; Sin, L.T.; Tee, T.-T.; Bee, S.-T.; Rahmat, A.R.; Rahman, W.A.W.A.; Tan, A.-C.; Vikhraman, M. Antimicrobial agents for food packaging applications. *Trends Food Sci. Technol.* **2013**, *33*, 110–123. [CrossRef]
178. Ibrahim, S.; Elsayed, H.; Hasanin, M. Biodegradable, Antimicrobial and Antioxidant Biofilm for Active Packaging Based on Extracted Gelatin and Lignocelluloses Biowastes. *J. Polym. Environ.* **2021**, *29*, 472–482. [CrossRef]
179. Kuai, L.; Liu, F.; Chiou, B.-S.; Avena-Bustillos, R.J.; McHugh, T.H.; Zhong, F. Controlled release of antioxidants from active food packaging: A Review. *Food Hydrocoll.* **2021**, 106992. [CrossRef]
180. Kruijf, N.D.; Beest, M.V.; Rijk, R.; Sipiläinen-Malm, T.; Losada, P.P.; Meulenaer, B.D. Active and intelligent packaging: Applications and regulatory aspects. *Food Addit. Contam.* **2002**, *19*, 144–162. [CrossRef]
181. Dainelli, D. 8-Global legislation for active and intelligent packaging materials. In *Global Legislation for Food Contact Materials*; Baughan, J.S., Ed.; Woodhead Publishing Series in Food Science, Technology and Nutrition; Woodhead Publishing: Sawston, UK, 2015; pp. 183–199. ISBN 978-1-78242-014-9. [CrossRef]
182. Mendonca, A.; Jackson-Davis, A.; Moutiq, R.; Thomas-Popo, E. Chapter 14—Use of Natural Antimicrobials of Plant Origin to Improve the Microbiological Safety of Foods. In *Food and Feed Safety Systems and Analysis*; Ricke, S.C., Atungulu, G.G., Rainwater, C.E., Park, S.H., Eds.; Academic Press: Cambridge, MA, USA, 2018; pp. 249–272. ISBN 978-0-12-811835-1. [CrossRef]
183. Göksen, G.; Fabra, M.J.; Ekiz, H.I.; López-Rubio, A. Phytochemical-loaded electrospun nanofibers as novel active edible films: Characterization and antibacterial efficiency in cheese slices. *Food Control* **2020**, *112*, 107133. [CrossRef]
184. Lin, L.; Gu, Y.; Cui, H. Moringa oil/chitosan nanoparticles embedded gelatin nanofibers for food packaging against Listeria monocytogenes and Staphylococcus aureus on cheese. *Food Packag. Shelf Life* **2019**, *19*, 86–93. [CrossRef]
185. Meira, S.M.M.; Zehetmeyer, G.; Scheibel, J.M.; Werner, J.O.; Brandelli, A. Starch-halloysite nanocomposites containing nisin: Characterization and inhibition of Listeria monocytogenes in soft cheese. *LWT-Food Sci. Technol.* **2016**, *68*, 226–234. [CrossRef]
186. Cui, H.Y.; Wu, J.; Li, C.Z.; Lin, L. Anti-listeria effects of chitosan-coated nisin-silica liposome on Cheddar cheese. *J. Dairy Sci.* **2016**, *99*, 8598–8606. [CrossRef]
187. Soto, K.M.; Hernández-Iturriaga, M.; Loarca-Piña, G.; Luna-Bárcenas, G.; Mendoza, S. Antimicrobial effect of nisin electrospun amaranth: Pullulan nanofibers in apple juice and fresh cheese. *Int. J. Food Microbiol.* **2019**, *295*, 25–32. [CrossRef]
188. Cui, H.; Bai, M.; Li, C.; Liu, R.; Lin, L. Fabrication of chitosan nanofibers containing tea tree oil liposomes against Salmonella spp. in chicken. *LWT* **2018**, *96*, 671–678. [CrossRef]
189. Woraprayote, W.; Pumpuang, L.; Tosukhowong, A.; Zendo, T.; Sonomoto, K.; Benjakul, S.; Visessanguan, W. Antimicrobial biodegradable food packaging impregnated with Bacteriocin 7293 for control of pathogenic bacteria in pangasius fish fillets. *LWT* **2018**, *89*, 427–433. [CrossRef]

190. Vishnu Priya, N.; Vinitha, U.G.; Meenakshi Sundaram, M. Preparation of chitosan-based antimicrobial active food packaging film incorporated with Plectranthus amboinicus essential oil. *Biocatal. Agric. Biotechnol.* **2021**, *34*, 102021. [CrossRef]
191. Kanatt, S.R. Development of active/intelligent food packaging film containing Amaranthus leaf extract for shelf life extension of chicken/fish during chilled storage. *Food Packag. Shelf Life* **2020**, *24*, 100506. [CrossRef]
192. Meral, R.; Alav, A.; Karakas, C.; Dertli, E.; Yilmaz, M.T.; Ceylan, Z. Effect of electrospun nisin and curcumin loaded nanomats on the microbial quality, hardness and sensory characteristics of rainbow trout fillet. *LWT* **2019**, *113*, 108292. [CrossRef]
193. Khah, M.D.; Ghanbarzadeh, B.; Roufegarinejad Nezhad, L.; Ostadrahimi, A. Effects of virgin olive oil and grape seed oil on physicochemical and antimicrobial properties of pectin-gelatin blend emulsified films. *Int. J. Biol. Macromol.* **2021**, *171*, 262–274. [CrossRef] [PubMed]
194. Dhumal, C.V.; Ahmed, J.; Bandara, N.; Sarkar, P. Improvement of antimicrobial activity of sago starch/guar gum bi-phasic edible films by incorporating carvacrol and citral. *Food Packag. Shelf Life* **2019**, *21*, 100380. [CrossRef]
195. Göksen, G.; Fabra, M.J.; Pérez-Cataluña, A.; Ekiz, H.I.; Sanchez, G.; López-Rubio, A. Biodegradable active food packaging structures based on hybrid cross-linked electrospun polyvinyl alcohol fibers containing essential oils and their application in the preservation of chicken breast fillets. *Food Packag. Shelf Life* **2021**, *27*, 100613. [CrossRef]
196. Alizadeh-Sani, M.; Mohammadian, E.; McClements, D.J. Eco-friendly active packaging consisting of nanostructured biopolymer matrix reinforced with TiO_2 and essential oil: Application for preservation of refrigerated meat. *Food Chem.* **2020**, *322*, 126782. [CrossRef]
197. Montero, Y.; Souza, A.G.; Oliveira, É.R.; dos Santos Rosa, D. Nanocellulose functionalized with cinnamon essential oil: A potential application in active biodegradable packaging for strawberry. *Sustain. Mater. Technol.* **2021**, *29*, e00289. [CrossRef]
198. He, X.; Li, M.; Gong, X.; Niu, B.; Li, W. Biodegradable and antimicrobial CSC films containing cinnamon essential oil for preservation applications. *Food Packag. Shelf Life* **2021**, *29*, 100697. [CrossRef]
199. Cui, H.; Bai, M.; Rashed, M.M.A.; Lin, L. The antibacterial activity of clove oil/chitosan nanoparticles embedded gelatin nanofibers against Escherichia coli O157:H7 biofilms on cucumber. *Int. J. Food Microbiol.* **2018**, *266*, 69–78. [CrossRef]
200. Min, T.; Sun, X.; Yuan, Z.; Zhou, L.; Jiao, X.; Zha, J.; Zhu, Z.; Wen, Y. Novel antimicrobial packaging film based on porous poly(lactic acid) nanofiber and polymeric coating for humidity-controlled release of thyme essential oil. *LWT* **2021**, *135*, 110034. [CrossRef]
201. Thielmann, J.; Theobald, M.; Wutz, A.; Krolo, T.; Buergy, A.; Niederhofer, J.; Welle, F.; Muranyi, P. Litsea cubeba fruit essential oil and its major constituent citral as volatile agents in an antimicrobial packaging material. *Food Microbiol.* **2021**, *96*, 103725. [CrossRef]
202. Tao, R.; Sedman, J.; Ismail, A. Antimicrobial activity of various essential oils and their application in active packaging of frozen vegetable products. *Food Chem.* **2021**, *360*, 129956. [CrossRef]
203. Kaur, M.; Santhiya, D. UV-shielding antimicrobial zein films blended with essential oils for active food packaging. *J. Appl. Polym. Sci.* **2021**, *138*, 49832. [CrossRef]
204. Raeisi, M.; Mohammadi, M.A.; Coban, O.E.; Ramezani, S.; Ghorbani, M.; Tabibiazar, M.; Khoshbakht, R.; Noori, S.M.A. Physicochemical and antibacterial effect of Soy Protein Isolate/Gelatin electrospun nanofibres incorporated with Zataria multiflora and Cinnamon zeylanicum essential oils. *J. Food Meas. Charact.* **2021**, *15*, 1116–1126. [CrossRef]
205. Jamróz, E.; Juszczak, L.; Kucharek, M. Investigation of the physical properties, antioxidant and antimicrobial activity of ternary potato starch-furcellaran-gelatin films incorporated with lavender essential oil. *Int. J. Biol. Macromol.* **2018**, *114*, 1094–1101. [CrossRef]
206. Akhter, R.; Masoodi, F.A.; Wani, T.A.; Rather, S.A. Functional characterization of biopolymer based composite film: Incorporation of natural essential oils and antimicrobial agents. *Int. J. Biol. Macromol.* **2019**, *137*, 1245–1255. [CrossRef]
207. Abdollahi, M.; Rezaei, M.; Farzi, G. Improvement of active chitosan film properties with rosemary essential oil for food packaging. *Int. J. Food Sci. Technol.* **2012**, *47*, 847–853. [CrossRef]
208. Yeddes, W.; Djebali, K.; Aidi Wannes, W.; Horchani-Naifer, K.; Hammami, M.; Younes, I.; Saidani Tounsi, M. Gelatin-chitosan-pectin films incorporated with rosemary essential oil: Optimized formulation using mixture design and response surface methodology. *Int. J. Biol. Macromol.* **2020**, *154*, 92–103. [CrossRef]
209. Bahrami, A.; Fattahi, R. Biodegradable carboxymethyl cellulose–polyvinyl alcohol composite incorporated with Glycyrrhiza Glabra L. essential oil: Physicochemical and antibacterial features. *Food Sci. Nutr.* **2021**, *9*, 4974–4985. [CrossRef]
210. Asbahani, A.E.; Miladi, K.; Badri, W.; Sala, M.; Addi, E.H.A.; Casabianca, H.; Mousadik, A.E.; Hartmann, D.; Jilale, A.; Renaud, F.N.R.; et al. Essential oils: From extraction to encapsulation. *Int. J. Pharm.* **2015**, *483*, 220–243. [CrossRef]
211. Elshafie, H.S.; Camele, I. An Overview of the Biological Effects of Some Mediterranean Essential Oils on Human Health. *BioMed Res. Int.* **2017**, *2017*, e9268468. [CrossRef]
212. Tariq, S.; Wani, S.; Rasool, W.; Shafi, K.; Bhat, M.A.; Prabhakar, A.; Shalla, A.H.; Rather, M.A. A comprehensive review of the antibacterial, antifungal and antiviral potential of essential oils and their chemical constituents against drug-resistant microbial pathogens. *Microb. Pathog.* **2019**, *134*, 103580. [CrossRef]
213. Guttierrez, J.; Rodriguez, G.; Barry-Ryan, C.; Bourke, P. Efficacy of Plant Essential Oils against Foodborne Pathogens and Spoilage Bacteria Associated with Ready-to-Eat Vegetables: Antimicrobial and Sensory Screening. *J. Food Prot.* **2008**, *71*, 1846–1851. [CrossRef]

214. Dušan, F.; Marián, S.; Katarína, D.; Dobroslava, B. Essential oils—Their antimicrobial activity against Escherichia coli and effect on intestinal cell viability. *Toxicol. Vitro* **2006**, *20*, 1435–1445. [CrossRef]
215. Fernández-López, J.; Viuda-Martos, M. Introduction to the Special Issue: Application of Essential Oils in Food Systems. *Foods* **2018**, *7*, 56. [CrossRef]
216. Zhu, Y.; Li, C.; Cui, H.; Lin, L. Encapsulation strategies to enhance the antibacterial properties of essential oils in food system. *Food Control* **2021**, *123*, 107856. [CrossRef]
217. Rolim, H.M.L.; Ramalho, T.C. Chapter 7—Biopolymer essential oil nanocomposite for antimicrobial packaging. In *Biopolymer-Based Nano Films*; Rai, M., dos Santos, C.A., Eds.; Elsevier: Amsterdam, The Netherlands, 2021; pp. 115–131. ISBN 978-0-12-823381-8. [CrossRef]
218. Priyadarshi, R.; Sauraj; Kumar, B.; Deeba, F.; Kulshreshtha, A.; Negi, Y.S. Chitosan films incorporated with Apricot (Prunus armeniaca) kernel essential oil as active food packaging material. *Food Hydrocoll.* **2018**, *85*, 158–166. [CrossRef]
219. Pisoschi, A.M.; Pop, A.; Georgescu, C.; Turcuş, V.; Olah, N.K.; Mathe, E. An overview of natural antimicrobials role in food. *Eur. J. Med. Chem.* **2018**, *143*, 922–935. [CrossRef] [PubMed]
220. Hoskin, D.W.; Ramamoorthy, A. Studies on anticancer activities of antimicrobial peptides. *Biochim. Biophys. Acta BBA-Biomembr.* **2008**, *1778*, 357–375. [CrossRef]
221. Burrowes, O.J.; Hadjicharalambous, C.; Diamond, G.; Lee, T.-C. Evaluation of Antimicrobial Spectrum and Cytotoxic Activity of Pleurocidin for Food Applications. *J. Food Sci.* **2004**, *69*, FMS66–FMS71. [CrossRef]
222. Wang, X.; Yue, T.; Lee, T. Development of Pleurocidin-poly(vinyl alcohol) electrospun antimicrobial nanofibers to retain antimicrobial activity in food system application. *Food Control* **2015**, *54*, 150–157. [CrossRef]
223. Antoshin, A.A.; Shpichka, A.I.; Huang, G.; Chen, K.; Lu, P.; Svistunov, A.A.; Lychagin, A.V.; Lipina, M.M.; Sinelnikov, M.Y.; Reshetov, I.V.; et al. Lactoferrin as a regenerative agent: The old-new panacea? *Pharmacol. Res.* **2021**, *167*, 105564. [CrossRef]
224. Siqueiros-Cendón, T.; Arévalo-Gallegos, S.; Iglesias-Figueroa, B.F.; García-Montoya, I.A.; Salazar-Martínez, J.; Rascón-Cruz, Q. Immunomodulatory effects of lactoferrin. *Acta Pharmacol. Sin.* **2014**, *35*, 557–566. [CrossRef]
225. Niaz, B.; Saeed, F.; Ahmed, A.; Imran, M.; Maan, A.A.; Khan, M.K.I.; Tufail, T.; Anjum, F.M.; Hussain, S.; Suleria, H.A.R. Lactoferrin (LF): A natural antimicrobial protein. *Int. J. Food Prop.* **2019**, *22*, 1626–1641. [CrossRef]
226. Krupińska, A.M.; Bogucki, Z. Clinical aspects of the use of lactoferrin in dentistry. *J. Oral Biosci.* **2021**, *63*, 129–133. [CrossRef]
227. Biernbaum, E.N.; Gnezda, A.; Akbar, S.; Franklin, R.; Venturelli, P.A.; McKillip, J.L. Lactoferrin as an antimicrobial against Salmonella enterica and Escherichia coli O157:H7 in raw milk. *JDS Commun.* **2021**, *2*, 92–97. [CrossRef]
228. Boots, J.-W.; Floris, R. Lactoperoxidase: From catalytic mechanism to practical applications. *Int. Dairy J.* **2006**, *16*, 1272–1276. [CrossRef]
229. Saeed, F.; Afzaal, M.; Tufail, T.; Ahmad, A. Use of Natural Antimicrobial Agents: A Safe Preservation Approach. In *Active Antimicrobial Food Packaging*; IntechOpen: London, UK, 2019; ISBN 978-1-78985-003-1. [CrossRef]
230. Armenteros, M.; Dalvit, P.; Leyva, V.; Ponce, P.; Alfonso, P. Risk analysis of the exacerbation of foodborne pathogens in raw milk activated with the lactoperoxidase system. *Rev. Salud Anim.* **2007**, *29*, 176–181.
231. Jooyandeh, H.; Aberoumand, A.; Nasehi, B. Application of Lactoperoxidase System in Fish and Food Products: A Review. *Am. Eurasian J. Agric. Environ. Sci.* **2011**, *10*, 89–96.
232. Seifu, E.; Buys, E.M.; Donkin, E.F. Significance of the lactoperoxidase system in the dairy industry and its potential applications: A review. *Trends Food Sci. Technol.* **2005**, *16*, 137–154. [CrossRef]
233. Munsch-Alatossava, P.; Gursoy, O.; Lorilla, P.M.; Gauchi, J.-P.; Alatossava, T. Chapter 15—Antibacterial Effects and Modes of Action of the Activated Lactoperoxidase System (LPS), of CO_2 and N_2 Gas as Food-Grade Approaches to Control Bovine Raw Milk–Associated Bacteria. In *Food Control and Biosecurity*; Holban, A.M., Grumezescu, A.M., Eds.; Handbook of Food Bioengineering; Academic Press: Cambridge, MA, USA, 2018; pp. 519–541. ISBN 978-0-12-811445-2. [CrossRef]
234. Leśnierowski, G.; Yang, T. Lysozyme and its modified forms: A critical appraisal of selected properties and potential. *Trends Food Sci. Technol.* **2021**, *107*, 333–342. [CrossRef]
235. Wu, T.; Jiang, Q.; Wu, D.; Hu, Y.; Chen, S.; Ding, T.; Ye, X.; Liu, D.; Chen, J. What is new in lysozyme research and its application in food industry? A review. *Food Chem.* **2019**, *274*, 698–709. [CrossRef]
236. Khan, M.I.; Dowarha, D.; Katte, R.; Chou, R.-H.; Filipek, A.; Yu, C. Lysozyme as the anti-proliferative agent to block the interaction between S100A6 and the RAGE V domain. *PLoS ONE* **2019**, *14*, e0216427. [CrossRef]
237. Ragland, S.A.; Criss, A.K. From bacterial killing to immune modulation: Recent insights into the functions of lysozyme. *PLoS Pathog.* **2017**, *13*, e1006512. [CrossRef]
238. Zhou, X.; Deng, J.; Fang, C.; Yu, R.; Lei, W.; He, X.; Zhang, C. Preparation and characterization of lysozyme@carbon nanotubes/waterborne polyurethane composite and the potential application in printing inks. *Prog. Org. Coat.* **2020**, *142*, 105600. [CrossRef]
239. Huang, W.; Li, X.; Xue, Y.; Huang, R.; Deng, H.; Ma, Z. Antibacterial multilayer films fabricated by LBL immobilizing lysozyme and HTCC on nanofibrous mats. *Int. J. Biol. Macromol.* **2013**, *53*, 26–31. [CrossRef]
240. Carrillo, W.; Spindola, H.; Ramos, M.; Recio, I.; Carvalho, J.E. Anti-Inflammatory and Anti-Nociceptive Activities of Native and Modified Hen Egg White Lysozyme. *J. Med. Food* **2016**, *19*, 978–982. [CrossRef]
241. Zhang, Z.; Zhou, X.; Wang, D.; Fang, C.; Zhang, W.; Wang, C.; Huang, Z. Lysozyme-based composite membranes and their potential application for active packaging. *Food Biosci.* **2021**, 101078. [CrossRef]

242. Niu, X.; Zhu, L.; Xi, L.; Guo, L.; Wang, H. An antimicrobial agent prepared by N-succinyl chitosan immobilized lysozyme and its application in strawberry preservation. *Food Control* **2020**, *108*, 106829. [CrossRef]
243. Silva, N.H.C.S.; Vilela, C.; Almeida, A.; Marrucho, I.M.; Freire, C.S.R. Pullulan-based nanocomposite films for functional food packaging: Exploiting lysozyme nanofibers as antibacterial and antioxidant reinforcing additives. *Food Hydrocoll.* **2018**, *77*, 921–930. [CrossRef]
244. Glicerina, V.; Siroli, L.; Canali, G.; Chinnici, F.; Capelli, F.; Lanciotti, R.; Colombo, V.; Romani, S. Efficacy of biodegradable, antimicrobial packaging on safety and quality parameters maintenance of a pear juice and rice milk-based smoothie product. *Food Control* **2021**, *128*, 108170. [CrossRef]
245. Wang, Y.; Xue, Y.; Bi, Q.; Qin, D.; Du, Q.; Jin, P. Enhanced antibacterial activity of eugenol-entrapped casein nanoparticles amended with lysozyme against gram-positive pathogens. *Food Chem.* **2021**, *360*, 130036. [CrossRef]
246. Quinto, E.J.; Caro, I.; Villalobos-Delgado, L.H.; Mateo, J.; De-Mateo-Silleras, B.; Redondo-Del-Río, M.P. Food Safety through Natural Antimicrobials. *Antibiotics* **2019**, *8*, 208. [CrossRef]
247. Daba, G.M.; Elkhateeb, W.A. Bacteriocins of lactic acid bacteria as biotechnological tools in food and pharmaceuticals: Current applications and future prospects. *Biocatal. Agric. Biotechnol.* **2020**, *28*, 101750. [CrossRef]
248. Agriopoulou, S.; Stamatelopoulou, E.; Sachadyn-Król, M.; Varzakas, T. Lactic Acid Bacteria as Antibacterial Agents to Extend the Shelf Life of Fresh and Minimally Processed Fruits and Vegetables: Quality and Safety Aspects. *Microorganisms* **2020**, *8*, 952. [CrossRef]
249. Woraprayote, W.; Malila, Y.; Sorapukdee, S.; Swetwiwathana, A.; Benjakul, S.; Visessanguan, W. Bacteriocins from lactic acid bacteria and their applications in meat and meat products. *Meat Sci.* **2016**, *120*, 118–132. [CrossRef] [PubMed]
250. Settier-Ramírez, L.; López-Carballo, G.; Gavara, R.; Hernández-Muñoz, P. Broadening the antimicrobial spectrum of nisin-producing Lactococcus lactis subsp. Lactis to Gram-negative bacteria by means of active packaging. *Int. J. Food Microbiol.* **2021**, *339*, 109007. [CrossRef] [PubMed]
251. Bungenstock, L.; Abdulmawjood, A.; Reich, F. Suitability of lactic acid bacteria and deriving antibacterial preparations to enhance shelf-life and consumer safety of emulsion type sausages. *Food Microbiol.* **2021**, *94*, 103673. [CrossRef] [PubMed]
252. Haakensen, M.; Dobson, C.M.; Hill, J.E.; Ziola, B. 2009 Reclassification of Pediococcus dextrinicus (Coster and White 1964) Back 1978 (Approved Lists 1980) as Lactobacillus dextrinicus comb. nov., and emended description of the genus Lactobacillus. *Int. J. Syst. Evol. Microbiol.* **2021**, *59*, 615–621. [CrossRef]
253. Gabrielsen, C.; Brede, D.A.; Nes, I.F.; Diep, D.B. Circular bacteriocins: Biosynthesis and mode of action. *Appl. Environ. Microbiol.* **2014**, *80*, 6854–6862. [CrossRef]
254. Niamah, A.K. Structure, mode of action and application of pediocin natural antimicrobial food preservative: A review. *Basrah J Agric Sci* **2018**, *31*, 59–69. [CrossRef]
255. Verma, S.K.; Sood, S.K.; Saini, R.K.; Saini, N. Pediocin PA-1 containing fermented cheese whey reduces total viable count of raw buffalo (*Bubalis bubalus*) milk. *LWT-Food Sci. Technol.* **2017**, *83*, 193–200. [CrossRef]
256. Espitia, P.J.P.; Otoni, C.G.; Soares, N.F.F. Chapter 36—Pediocin Applications in Antimicrobial Food Packaging Systems. In *Antimicrobial Food Packaging*; Barros-Velázquez, J., Ed.; Academic Press: San Diego, CA, USA, 2016; pp. 445–454. ISBN 978-0-12-800723-5. [CrossRef]
257. Meira, S.M.M.; Zehetmeyer, G.; Werner, J.O.; Brandelli, A. A novel active packaging material based on starch-halloysite nanocomposites incorporating antimicrobial peptides. *Food Hydrocoll.* **2017**, *63*, 561–570. [CrossRef]
258. Vimont, A.; Fernandez, B.; Ahmed, G.; Fortin, H.-P.; Fliss, I. Quantitative antifungal activity of reuterin against food isolates of yeasts and moulds and its potential application in yogurt. *Int. J. Food Microbiol.* **2019**, *289*, 182–188. [CrossRef]
259. Bal, C. Benefits and Uses of Mushroom. *J. Bacteriol. Mycol. Open Access* **2018**, *6*, 1. [CrossRef]
260. El-Saber Batiha, G.; Hussein, D.E.; Algammal, A.M.; George, T.T.; Jeandet, P.; Al-Snafi, A.E.; Tiwari, A.; Pagnossa, J.P.; Lima, C.M.; Thorat, N.D.; et al. Application of natural antimicrobials in food preservation: Recent views. *Food Control* **2021**, *126*, 108066. [CrossRef]
261. Shameem, N.; Kamili, A.N.; Ahmad, M.; Masoodi, F.A.; Parray, J.A. Antimicrobial activity of crude fractions and morel compounds from wild edible mushrooms of North western Himalaya. *Microb. Pathog.* **2017**, *105*, 356–360. [CrossRef]
262. Oli, A.N.; Edeh, P.A.; Al-Mosawi, R.M.; Mbachu, N.A.; Al-Dahmoshi, H.O.M.; Al-Khafaji, N.S.K.; Ekuma, U.O.; Okezie, U.M.; Saki, M. Evaluation of the phytoconstituents of Auricularia auricula-judae mushroom and antimicrobial activity of its protein extract. *Eur. J. Integr. Med.* **2020**, *38*, 101176. [CrossRef]
263. Devi, K.P.; Suganthy, N.; Kesika, P.; Pandian, S.K. Bioprotective properties of seaweeds: In vitro evaluation of antioxidant activity and antimicrobial activity against food borne bacteria in relation to polyphenolic content. *BMC Complement. Altern. Med.* **2008**, *8*, 38. [CrossRef]
264. Şen, F.; Uzunsoy, İ.; Baştürk, E.; Kahraman, M.V. Antimicrobial agent-free hybrid cationic starch/sodium alginate polyelectrolyte films for food packaging materials. *Carbohydr. Polym.* **2017**, *170*, 264–270. [CrossRef]
265. Chand, K.; Rajeshwari; Hiremathad, A.; Singh, M.; Santos, M.A.; Keri, R.S. A review on antioxidant potential of bioactive heterocycle benzofuran: Natural and synthetic derivatives. *Pharmacol. Rep.* **2017**, *69*, 281–295. [CrossRef]
266. Choe, E.; Min, D.B. Mechanisms of Antioxidants in the Oxidation of Foods. *Compr. Rev. Food Sci. Food Saf.* **2009**, *8*, 345–358. [CrossRef]

267. Fritz, K.L.; Seppanen, C.M.; Kurzer, M.S.; Saari Csallany, A. The in vivo antioxidant activity of soybean isoflavones in human subjects. *Nutr. Res.* **2003**, *23*, 479–487. [CrossRef]
268. Bera, D.; Lahiri, D.; Nag, A. Studies on a natural antioxidant for stabilization of edible oil and comparison with synthetic antioxidants. *J. Food Eng.* **2006**, *74*, 542–545. [CrossRef]
269. Jamshidian, M.; Tehrany, E.A.; Desobry, S. Release of synthetic phenolic antioxidants from extruded poly lactic acid (PLA) film. *Food Control* **2012**, *28*, 445–455. [CrossRef]
270. Bodoira, R.M.; Penci, M.C.; Ribotta, P.D.; Martínez, M.L. Chia (*Salvia hispanica* L.) oil stability: Study of the effect of natural antioxidants. *LWT* **2017**, *75*, 107–113. [CrossRef]
271. Jiang, J.; Xiong, Y.L. Natural antioxidants as food and feed additives to promote health benefits and quality of meat products: A review. *Meat Sci.* **2016**, *120*, 107–117. [CrossRef]
272. Pandey, A.K.; Kumar, P.; Singh, P.; Tripathi, N.N.; Bajpai, V.K. Essential Oils: Sources of Antimicrobials and Food Preservatives. *Front. Microbiol.* **2017**, *7*, 2161. [CrossRef]
273. Baştürk, A.; Ceylan, M.M.; Çavuş, M.; Boran, G.; Javidipour, I. Effects of some herbal extracts on oxidative stability of corn oil under accelerated oxidation conditions in comparison with some commonly used antioxidants. *LWT* **2018**, *89*, 358–364. [CrossRef]
274. Celano, R.; Piccinelli, A.L.; Pagano, I.; Roscigno, G.; Campone, L.; De Falco, E.; Russo, M.; Rastrelli, L. Oil distillation wastewaters from aromatic herbs as new natural source of antioxidant compounds. *Food Res. Int.* **2017**, *99*, 298–307. [CrossRef]
275. Dong, Z.; Xu, F.; Ahmed, I.; Li, Z.; Lin, H. Characterization and preservation performance of active polyethylene films containing rosemary and cinnamon essential oils for Pacific white shrimp packaging. *Food Control* **2018**, *92*, 37–46. [CrossRef]
276. Nerin, C.; Silva, F.; Manso, S.; Becerril, R. Chapter 6—The Downside of Antimicrobial Packaging: Migration of Packaging Elements into Food. In *Antimicrobial Food Packaging*; Barros-Velázquez, J., Ed.; Academic Press: San Diego, CA, USA, 2016; pp. 81–93. ISBN 978-0-12-800723-5. [CrossRef]
277. Mills, A. Oxygen indicators and intelligent inks for packaging food. *Chem. Soc. Rev.* **2005**, *34*, 1003–1011. [CrossRef]
278. Song, X.-C.; Canellas, E.; Wrona, M.; Becerril, R.; Nerin, C. Comparison of two antioxidant packaging based on rosemary oleoresin and green tea extract coated on polyethylene terephthalate for extending the shelf life of minced pork meat. *Food Packag. Shelf Life* **2020**, *26*, 100588. [CrossRef]
279. Zhang, D.; Chen, L.; Cai, J.; Dong, Q.; Din, Z.; Hu, Z.-Z.; Wang, G.-Z.; Ding, W.-P.; He, J.-R.; Cheng, S.-Y. Starch/tea polyphenols nanofibrous films for food packaging application: From facile construction to enhance mechanical, antioxidant and hydrophobic properties. *Food Chem.* **2021**, *360*, 129922. [CrossRef]
280. Moalla, S.; Ammar, I.; Fauconnier, M.-L.; Danthine, S.; Blecker, C.; Besbes, S.; Attia, H. Development and characterization of chitosan films carrying *Artemisia campestris* antioxidants for potential use as active food packaging materials. *Int. J. Biol. Macromol.* **2021**, *183*, 254–266. [CrossRef]
281. Abdin, M.; El-Beltagy, A.E.; El-sayed, M.E.; Naeem, M.A. Production and Characterization of Sodium Alginate/Gum Arabic Based Films Enriched with Syzygium cumini Seeds Extracts for Food Application. *J. Polym. Environ.* **2021**, *29*(10), 1–12. [CrossRef]
282. Eltabakh, M.; Kassab, H.; Badawy, W.; Abdin, M.; Abdelhady, S. Active Bio-composite Sodium Alginate/Maltodextrin Packaging Films for Food Containing Azolla pinnata Leaves Extract as Natural Antioxidant. *J. Polym. Environ.* **2021**, *29*(9), 1–11. [CrossRef]
283. Riaz, A.; Lagnika, C.; Luo, H.; Nie, M.; Dai, Z.; Liu, C.; Abdin, M.; Hashim, M.M.; Li, D.; Song, J. Effect of Chinese chives (*Allium tuberosum*) addition to carboxymethyl cellulose based food packaging films. *Carbohydr. Polym.* **2020**, *235*, 115944. [CrossRef] [PubMed]
284. Kaur, R.; Kaur, L. Encapsulated natural antimicrobials: A promising way to reduce microbial growth in different food systems. *Food Control* **2021**, *123*, 107678. [CrossRef]
285. Delshadi, R.; Bahrami, A.; Assadpour, E.; Williams, L.; Jafari, S.M. Nano/microencapsulated natural antimicrobials to control the spoilage microorganisms and pathogens in different food products. *Food Control* **2021**, *128*, 108180. [CrossRef]
286. Dima, C.; Assadpour, E.; Dima, S.; Jafari, S.M. Bioactive-loaded nanocarriers for functional foods: From designing to bioavailability. *Curr. Opin. Food Sci.* **2020**, *33*, 21–29. [CrossRef]
287. Yousefi, M.; Ehsani, A.; Jafari, S.M. Lipid-based nano delivery of antimicrobials to control food-borne bacteria. *Adv. Colloid Interface Sci.* **2019**, *270*, 263–277. [CrossRef]
288. Burgain, J.; Gaiani, C.; Linder, M.; Scher, J. Encapsulation of probiotic living cells: From laboratory scale to industrial applications. *J. Food Eng.* **2011**, *104*, 467–483. [CrossRef]
289. Jafari, S.M.; Assadpoor, E.; He, Y.; Bhandari, B. Encapsulation Efficiency of Food Flavours and Oils during Spray Drying. *Dry. Technol.* **2008**, *26*, 816–835. [CrossRef]
290. Mahdavi, S.A.; Jafari, S.M.; Ghorbani, M.; Assadpoor, E. Spray-Drying Microencapsulation of Anthocyanins by Natural Biopolymers: A Review. *Dry. Technol.* **2014**, *32*, 509–518. [CrossRef]
291. Jafari, S.M. 1—An overview of nanoencapsulation techniques and their classification. In *Nanoencapsulation Technologies for the Food and Nutraceutical Industries*; Jafari, S.M., Ed.; Academic Press: Cambridge, MA, USA, 2017; pp. 1–34. ISBN 978-0-12-809436-5. [CrossRef]
292. Aloui, H.; Khwaldia, K.; Licciardello, F.; Mazzaglia, A.; Muratore, G.; Hamdi, M.; Restuccia, C. Efficacy of the combined application of chitosan and Locust Bean Gum with different citrus essential oils to control postharvest spoilage caused by Aspergillus flavus in dates. *Int. J. Food Microbiol.* **2014**, *170*, 21–28. [CrossRef]

293. Huang, Q.; Given, P. *1952- Micro/Nano Encapsulation of Active Food Ingredients*; American Chemical Society: Washington, USA, 2009; ISBN 978-0-8412-6964-4. Available online: https://agris.fao.org/agris-search/search.do?recordID=US201300135205 (accessed on 5 July 2021).
294. Akhavan Mahdavi, S.; Jafari, S.M.; Assadpoor, E.; Dehnad, D. Microencapsulation optimization of natural anthocyanins with maltodextrin, gum Arabic and gelatin. *Int. J. Biol. Macromol.* **2016**, *85*, 379–385. [CrossRef]
295. Suganya, V.; Anuradha, V. Microencapsulation and Nanoencapsulation: A Review. *Int. J. Pharm. Clin. Res.* **2017**, *9*, 233–239. [CrossRef]
296. ANSES-French Agency for Food, Environmental and Occupational Health & Safety, France; Anastasi, E.; Riviere, G.; Teste, B. Nanomaterials in Food-Prioritisation & Assessment. *EFSA J.* **2019**, *17*, e170909. [CrossRef]
297. Chen, L.; Remondetto, G.E.; Subirade, M. Food protein-based materials as nutraceutical delivery systems. *Trends Food Sci. Technol.* **2006**, *17*, 272–283. [CrossRef]
298. Joye, I.J.; McClements, D.J. Production of nanoparticles by anti-solvent precipitation for use in food systems. *Trends Food Sci. Technol.* **2013**, *34*, 109–123. [CrossRef]
299. Khare, A.R.; Vasisht, N. Chapter 14—Nanoencapsulation in the Food Industry: Technology of the Future. In *Microencapsulation in the Food Industry*; Gaonkar, A.G., Vasisht, N., Khare, A.R., Sobel, R., Eds.; Academic Press: San Diego, CA, USA, 2014; pp. 151–155. ISBN 978-0-12-404568-2. [CrossRef]
300. Zanetti, M.; Carniel, T.K.; Dalcanton, F.; dos Anjos, R.S.; Gracher Riella, H.; de Araújo, P.H.H.; de Oliveira, D.; Antônio Fiori, M. Use of encapsulated natural compounds as antimicrobial additives in food packaging: A brief review. *Trends Food Sci. Technol.* **2018**, *81*, 51–60. [CrossRef]
301. Heckert Bastos, L.P.; Vicente, J.; Corrêa dos Santos, C.H.; Geraldo de Carvalho, M.; Garcia-Rojas, E.E. Encapsulation of black pepper (*Piper nigrum* L.) essential oil with gelatin and sodium alginate by complex coacervation. *Food Hydrocoll.* **2020**, *102*, 105605. [CrossRef]
302. Misra, S.; Trinavee, K.; Gunda, N.S.K.; Mitra, S.K. Encapsulation with an interfacial liquid layer: Robust and efficient liquid-liquid wrapping. *J. Colloid Interface Sci.* **2020**, *558*, 334–344. [CrossRef]
303. LaCoste, A.; Schaich, K.M.; Zumbrunnen, D.; Yam, K.L. Advancing controlled release packaging through smart blending. *Packag. Technol. Sci.* **2005**, *18*, 77–87. [CrossRef]
304. Buonocore, G.G.; Nobile, M.A.D.; Panizza, A.; Bove, S.; Battaglia, G.; Nicolais, L. Modeling the Lysozyme Release Kinetics from Antimicrobial Films Intended for Food Packaging Applications. *J. Food Sci.* **2003**, *68*, 1365–1370. [CrossRef]
305. Chen, X.; Chen, M.; Xu, C.; Yam, K.L. Critical review of controlled release packaging to improve food safety and quality. *Crit. Rev. Food Sci. Nutr.* **2019**, *59*, 2386–2399. [CrossRef]
306. Yam, K.L.; Zhu, X. Controlled release food and beverage packaging. In *Emerging Food Packaging Technologies*; Elsevier: Amsterdam, The Netherlands, 2012; pp. 13–26. [CrossRef]
307. Uz, M.; Altınkaya, S.A. Development of mono and multilayer antimicrobial food packaging materials for controlled release of potassium sorbate. *LWT-Food Sci. Technol.* **2011**, *44*, 2302–2309. [CrossRef]
308. Beikzadeh, S.; Akbarinejad, A.; Swift, S.; Perera, J.; Kilmartin, P.A.; Travas-Sejdic, J. Cellulose acetate electrospun nanofibers encapsulating Lemon Myrtle essential oil as active agent with potent and sustainable antimicrobial activity. *React. Funct. Polym.* **2020**, *157*, 104769. [CrossRef]
309. Bi, F.; Qin, Y.; Chen, D.; Kan, J.; Liu, J. Development of active packaging films based on chitosan and nano-encapsulated luteolin. *Int. J. Biol. Macromol.* **2021**, *182*, 545–553. [CrossRef]
310. Lei, K.; Wang, X.; Li, X.; Wang, L. The innovative fabrication and applications of carvacrol nanoemulsions, carboxymethyl chitosan microgels and their composite films. *Colloids Surf. B Biointerfaces* **2019**, *175*, 688–696. [CrossRef]
311. Alves, V.L.C.D.; Rico, B.P.M.; Cruz, R.M.S.; Vicente, A.A.; Khmelinskii, I.; Vieira, M.C. Preparation and characterization of a chitosan film with grape seed extract-carvacrol microcapsules and its effect on the shelf-life of refrigerated Salmon (Salmo salar). *LWT* **2018**, *89*, 525–534. [CrossRef]
312. Martiñon, M.E.; Moreira, R.G.; Castell-Perez, M.E.; Gomes, C. Development of a multilayered antimicrobial edible coating for shelf-life extension of fresh-cut cantaloupe (*Cucumis melo* L.) stored at 4 °C. *LWT-Food Sci. Technol.* **2014**, *56*, 341–350. [CrossRef]
313. Sadeghi, K.; Seo, J. Photografting Coating: An Innovative Approach to "Non-Migratory" Active Packaging. *Adv. Funct. Mater.* **2021**, *31*, 2010759. [CrossRef]
314. Gundewadi, G.; Rudra, S.G.; Sarkar, D.J.; Singh, D. Nanoemulsion based alginate organic coating for shelf life extension of okra. *Food Packag. Shelf Life* **2018**, *18*, 1–12. [CrossRef]
315. Lee, J.Y.; Garcia, C.V.; Shin, G.H.; Kim, J.T. Antibacterial and antioxidant properties of hydroxypropyl methylcellulose-based active composite films incorporating oregano essential oil nanoemulsions. *LWT* **2019**, *106*, 164–171. [CrossRef]
316. Mohsenabadi, N.; Rajaei, A.; Tabatabaei, M.; Mohsenifar, A. Physical and antimicrobial properties of starch-carboxy methyl cellulose film containing rosemary essential oils encapsulated in chitosan nanogel. *Int. J. Biol. Macromol.* **2018**, *112*, 148–155. [CrossRef] [PubMed]
317. Nazari, M.; Majdi, H.; Milani, M.; Abbaspour-Ravasjani, S.; Hamishehkar, H.; Lim, L.-T. Cinnamon nanophytosomes embedded electrospun nanofiber: Its effects on microbial quality and shelf-life of shrimp as a novel packaging. *Food Packag. Shelf Life* **2019**, *21*, 100349. [CrossRef]

318. Hossain, F.; Follett, P.; Salmieri, S.; Vu, K.D.; Fraschini, C.; Lacroix, M. Antifungal activities of combined treatments of irradiation and essential oils (EOs) encapsulated chitosan nanocomposite films in in vitro and in situ conditions. *Int. J. Food Microbiol.* **2019**, *295*, 33–40. [CrossRef] [PubMed]
319. Aydogdu, A.; Yildiz, E.; Aydogdu, Y.; Sumnu, G.; Sahin, S.; Ayhan, Z. Enhancing oxidative stability of walnuts by using gallic acid loaded lentil flour based electrospun nanofibers as active packaging material. *Food Hydrocoll.* **2019**, *95*, 245–255. [CrossRef]
320. EFSA Scientific Committee; Hardy, A.; Benford, D.; Halldorsson, T.; Jeger, M.J.; Knutsen, H.K.; More, S.; Naegeli, H.; Noteborn, H.; Ockleford, C.; et al. Guidance on risk assessment of the application of nanoscience and nanotechnologies in the food and feed chain: Part 1, human and animal health. *EFSA J.* **2018**, *16*, e05327. [CrossRef]

foods

Article

Thymol@activated Carbon Nanohybrid for Low-Density Polyethylene-Based Active Packaging Films for Pork Fillets' Shelf-Life Extension

Aris E. Giannakas [1,*], Vassilios K. Karabagias [1], Dimitrios Moschovas [2], Areti Leontiou [1], Ioannis K. Karabagias [1], Stavros Georgopoulos [1], Andreas Karydis-Messinis [2], Konstantinos Zaharioudakis [1], Nikolaos Andritsos [1], George Kehayias [1], Apostolos Avgeropoulos [2], Charalampos Proestos [3] and Constantinos E. Salmas [2,*]

1 Department of Food Science and Technology, University of Patras, 30100 Agrinio, Greece; vkarampagias@upatras.gr (V.K.K.); aleontiu@upatras.gr (A.L.); ikaraba@upatras.gr (I.K.K.); sgeorgop@upatras.gr (S.G.); zacharioudakis.k@upatras.gr (K.Z.); nandritsos@upatras.gr (N.A.); gkechagi@upatras.gr (G.K.)

2 Department of Material Science and Engineering, University of Ioannina, 45110 Ioannina, Greece; dmoschov@uoi.gr (D.M.); karydis.and@gmail.com (A.K.-M.); aavger@uoi.gr (A.A.)

3 Laboratory of Food Chemistry, Department of Chemistry, National and Kapodistrian University of Athens Zografou, 15771 Athens, Greece; harpro@chem.uoa.gr

* Correspondence: agiannakas@upatras.gr (A.E.G.); ksalmas@uoi.gr (C.E.S.)

Citation: Giannakas, A.E.; Karabagias, V.K.; Moschovas, D.; Leontiou, A.; Karabagias, I.K.; Georgopoulos, S.; Karydis-Messinis, A.; Zaharioudakis, K.; Andritsos, N.; Kehayias, G.; et al. Thymol@activated Carbon Nanohybrid for Low-Density Polyethylene-Based Active Packaging Films for Pork Fillets' Shelf-Life Extension. *Foods* **2023**, *12*, 2590. https://doi.org/10.3390/foods12132590

Academic Editors: Sergio Torres-Giner, Amparo Chiralt and Chelo Gonzalez-Martinez

Received: 4 June 2023
Revised: 29 June 2023
Accepted: 30 June 2023
Published: 3 July 2023

Abstract: The replacement of food packaging additives and preservatives with bio-based antioxidant/antibacterial compounds has been a common practice in recent years following the trend of bioeconomy and nanotechnology. Such bio-additives are often enclosed in nanocarriers for a controlled release process. Following this trend in this work, a thymol (TO)-rich activated carbon (AC) nanohybrid was prepared and characterized physicochemically with various techniques. This TO@AC nanohybrid, along with the pure activated carbon, was extruded with low-density polyethylene (LDPE) to develop novel active packaging films. The codenames used in this paper were LDPE/xTO@AC and LDPE/xAC for the nanohybrid and the pure activated carbon, respectively. X-ray diffractometry, Fourier-transform infrared spectroscopy, and scanning electron microscopy measurements showed high dispersity of both the TO@AC nanohybrid and the pure AC in the LDPE matrix, resulting in enhanced mechanical properties. The active film with 15 wt.% of the TO@AC nanohybrid (LDPE/15TO@AC) exhibited a 230% higher water/vapor barrier and 1928% lower oxygen permeability than the pure LDPE film. For this active film, the highest antioxidant activity referred to the DPPH assay (44.4%), the lowest thymol release rate ($k_2 \approx 1.5\ s^{-1}$), and the highest antibacterial activity were recorded, resulting in a 2-day extension of fresh pork fillets' shelf-life.

Keywords: activated carbon; thymol; low-density polyethylene; active packaging; control release; kinetics; pork fillets; heme iron; shelf-life

1. Introduction

Today, the need to enhance the safety and shelf-life of foods is driving research and development (R&D) in the food industry towards the investigation of novel packaging techniques. Such techniques aim not only to protect food, but also to react with it in order to enhance the preservation and the safety of the product, and to extend its shelf-life [1]. This type of packaging is called "active food packaging" [1–3]. Moreover, one of the most recent trends in active food packaging is the removal of the chemical additives from the food and the active packaging, or at least to control the release of such food additives into the food [4,5]. In this spirit, it is also suggested to replace chemical food preservatives with natural, abundant compounds that have demonstrated antioxidant and antibacterial activity [4]. Such bioactive antioxidant/antibacterial compounds include essential oils (EOs) and their components, along with other phytochemicals like monoterpenes, sesquiterpenes, alcohols, phenols, aldehydes, ketones, and esters [6–8]. Over the past few years, considerable effort has been devoted to the development of polymers or biopolymer-based active packaging films incorporated with various EOs, such as oregano EO [9,10], thyme EO [9–11], basil EO [12,13], orange EO [14], lemongrass EO [15], or their derivatives, such as thymol [16], carvacrol [17], cinnamaldehyde [18], and limonene [19]. The main disadvantage of directly incorporating EOs into the polymer or biopolymer matrix is that their volatile nature leads to rapid loss and, therefore, the weakening of the active properties of the packaging. One famous proposed technique to balance this drawback is to adsorb these EOs or their derivatives into inexpensive, naturally abundant porous materials such as nanoclays, and afterwards to incorporate the resulting EO/nanoclay hybrid structure into the polymer or biopolymer matrix [20]. This approach enables the development of polymeric or biopolymer/EO/nanoclay active packaging films that exhibit enhanced mechanical, tensile, and water/oxygen barrier properties, as well as controlled release of antioxidant and antibacterial substances [21–23]. Nanoclays such as montmorillonite (MMT) and halloysite nanotubes (HNTs) have been extensively studied as EO nanocarriers in the development of polymeric or biopolymer/EO/nanoclay active packaging films [24,25]. Recently, well-known multifunctional materials, such as natural zeolite (NZ) [26–28], silica-based materials (MCM and SBA-15) [29,30], and carbon-based materials (activated carbon-AC) [31–33], have been suggested as nanocarriers for EOs and their derivatives because of their advanced adsorption capacity. The advantages of such materials compared to nanoclays are due to their larger specific surface areas and their ability to adsorb and load higher amounts of EOs or their derivatives. Among various carbon-based nanomaterials, the food-biomass-derived activated carbon (AC) has the advantage of low cost, non-toxicity, and degradability [31,34]. The effectiveness of AC is strongly dependent on its pore structure and surface chemistry [31]. AC is used in packaging films because of its adsorption and release capabilities. Thus, it can be used as a powerful adsorbent to remove odors and food spoilage substances from packaged foods, or as a nanocarrier for bioactive compounds to be released into food [33,35].

Low-density polyethylene (LDPE) is one of the most widely used materials for flexible food packaging. Although it is a polymer produced via petrochemical processes, in the future it could be produced via the bioethanol process as a bio-based material. It exhibits satisfactory strength-at-break properties up to −60 °C and good water barrier properties, but its high oxygen permeability makes it unsuitable as a food packaging material and vulnerable to oxidation deterioration phenomena of foods such as meat products. Oxidative and microbial deterioration is the most critical factor for pork meat's degradation and the end of its shelf-life [36]. Therefore, in order to transform the bio-based LDPE into a suitable food packaging material, it is essential to modify it by enhancing its oxygen barrier properties through the development of an LDPE-based active packaging material. This will be more valuable if the modification process follows the spirit of the circular economy, which proposes the use of bio-based raw materials such as EOs and biomass-derived AC. Recently, Macca carbon (MC) powder biomass, which is derived from macadamia nuts, was incorporated into LDPE by melt-compounding and subsequent melt-extrusion processes [37]. The content of MC powder in the LDPE matrix was varied from 0.5 up to 5 wt.%. The film with MC powder at 0.5 wt.% exhibited improved characteristics of antibacterial activity and storage performance, allowing lettuce and strawberries to remain fresh outside the refrigerator for more than 7 and 5 days, respectively, but any further addition of such carbon material led to a distortion of the film. Based on this paper, our research group started to study the use of AC-based powders as novel nanoreinforcements and antimicrobial/antioxidant agent carriers, as well as the incorporation of the produced novel materials with LDPE for the development of active packaging films.

In recent studies, our group developed a "green" method for the extraction of a thymol-rich (TO) fraction from thyme oil. This oil was subsequently adsorbed on halloysite nanotubes (HNTs) and natural zeolite (NZ) to create innovative TO@HNT and TO@NZ nanohybrids [38,39]. These nanohybrids were successfully incorporated into low-density polyethylene (LDPE) to develop promising LDPE/TO@HNT and LDPE/TO@NZ active packaging films [40,41]. These active films were successful at preserving low-fat pork fillets by preventing their lipid oxidation process and heme iron loss [40,41]. In this study, we applied the same green method to incorporate TO in AC to obtain a novel TO@AC nanohybrid, with the prospect that the TO@AC nanohybrid would act as a novel TO nanocarrier by absorbing higher amounts of TO than the recently used HNT and NZ nanocarriers [40,41]. The obtained TO@AC nanohybrid was characterized physicochemically through XRD analysis and FTIR spectroscopy, as well as TG and DSC experiments. Next, the TO@AC nanohybrid was incorporated into the LDPE matrix in various contents via an extrusion-molding process to develop novel LDPE/TO@AC active packaging films. The LDPE/TO@AC film with the best packaging properties, as well as LDPE/AC and pure LDPE films, was utilized to wrap pork fillets, and their shelf-life was evaluated by measuring their lipid oxidation through the thiobarbituric-acid-reactive substances (TBARS) method, their heme iron values, and the total bacterial count values.

The overall novelty of the present study is that, in the spirit of the circular economy, it proposes one more successful application for the valorization of a part of the 6 Mtons per year of globally landfilled spent coffee [42], combined with the production of natural preservation additives such as EOs. The final novel product will be an improved, environmentally friendly food packaging film capable of food preservation via "green methods" and the shelf-life extension of packaged products. The innovations of this work can be summarized as follows: (i) the development and characterization of novel TO@AC nanohybrids for the first time, to the best of our knowledge; (ii) the incorporation of pure AC and modified TO@AC nanohybrids into LDPE at high contents of 5, 10, and 15 wt.%; (iii) the development, characterization, and application of such novel LDPE/TO@AC active packaging films for pork fillet preservation for the first time, to the best of our knowledge; and (iv) in alignment with recent works [40,41], the identification of a direct linear relationship between the TBARS lipid oxidation values and heme iron values of pork fillets.

2. Materials and Methods

2.1. Materials

Thyme oil used was supplied by Chemco (Via Achille Grandi, 13—13/A, 42030 Vezzano sul Crostolo, Reggio Emilia, Italy). LDPE (CAS No. 9002-88-4) with a density of 0.925 g cm^{-3} and a melting point of 116 °C was supplied by Sigma-Aldrich (Sigma-Aldrich, St. Louis, MO, USA). Fresh pork fillets of the "skalopini" type were a kind offering from Aifantis Company (Aifantis Group—Head Quarters, Acheloos Bridge, Agrinio, Greece 30100). Three fresh and deboned pork fillets of the "skalopini type" with an approximate weight of 700 g each were provided by a local meat processing plant (Aifantis Company) within one hour after the slaughter. Activated carbon (AC) from spent coffee of the students' café at the University of Ioannina was prepared via a pyrolysis process and a treatment with KOH, as described in a previous report [42]. The Brunauer–Emmett–Teller surface area of the AC was 1372 m^2/g, and the micropore volume was 84.6% [42].

2.2. Preparation of TO@AC Nanohybrid

The process followed for the modification of the pure AC with a TO-rich fraction was the same as the process that was recently used for the modification of HNTs and NZ [38,39]. As illustrated in Figure 1, the process was carried out in two steps: In the first step, a distillation took place using a distillation apparatus on 20 mL of pure thyme oil at 200 °C to obtain a TO-rich fraction (14 mL) and separate the most volatile limonene–cymene fraction (6 mL). In the second step, TO's adsorption on AC occurred by placing 3 g of an AC bed above the spherical glass container containing the TO-rich fraction and evaporating the TO-rich fraction by heating it to 300 °C. Thus, the TO@AC nanohybrid was obtained and kept for further physicochemical characterization.

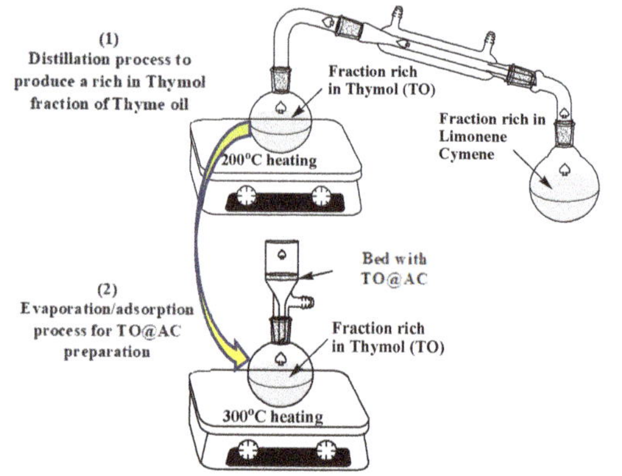

Figure 1. Schematic presentation of the process used for the TO@AC nanohybrid's preparation.

2.3. Preparation of LDPE/AC and LDPE/TO@AC Films

The preparation of LDPE/AC and LDPE/TO@AC films was performed by using a Mini Lab twin-screw extruder (Haake Mini Lab II, Thermo Scientific, ANTISEL, S.A., Athens, Greece) operated at 140 °C and 100 rpm for 3 min [21]. LDPE pellets were mixed with pure AC and the previously prepared powder of the TO@AC nanohybrid to achieve nominal AC and TO@AC weight percentages of 5, 10, and 15% (see Figure 2). The resultant samples were labeled as LDPE/5AC, LDPE/10AC, LDPE/15AC, LDPE/5TO@AC, LDPE/10TO@AC, and LDPE/15TO@AC. A pure LDPE sample was also prepared for comparison. By following the extrusion process, the LDPE/AC and LDPE/TO@AC samples, as well as the pure LDPE, were thermomechanical transformed into films (see Figure 2). This transformation was achieved by operating a hydraulic press with heated platens at 110 °Cand a constant pressure of 2 MPa for 2 min [12,43].

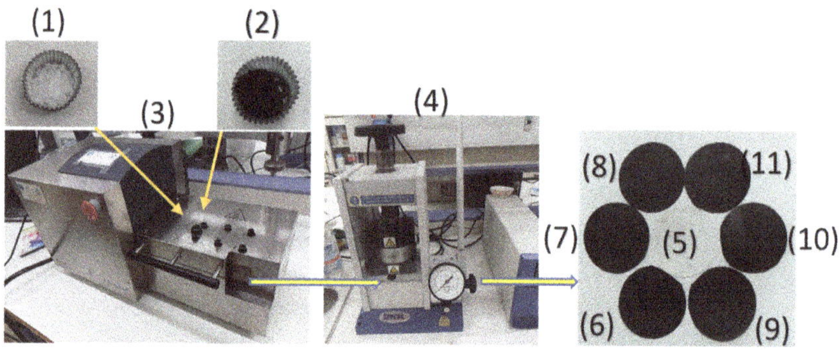

Figure 2. Schematic representation of the extrusion process for the preparation of LDPE/AC and LDPE/TO@AC films: (1) LDPE pellets, (2) TO@AC powder, (3) Mini Lab twin-screw extruder, (4) lab hydraulic press with heated platens, (5) LDPE film, (6) LDPE/5AC film, (7) LDPE/10AC film, (8) LDPE/15AC film, (9) LDPE/5TO@AC film, (10) LDPE/10TO@AC film, and (11) LDPE/15TO@AC film.

2.4. Physicochemical Characterization of AC, TO@AC Powders, and LDPE/AC and LDPE/TO@AC Films

The methods used to carry out the physicochemical characterization of AC, TO@AC powders, and LDPE/AC and LDPE/TO@AC films were as described in [38–41]. To study possible crystallinity changes in the AC during the process of TO@AC nanohybrid development, XRD analyses were performed on both pure AC and TO@AC. The crystal structures of the resulting LDPE/AC and LDPE/TO@AC films, as well as the pure LDPE film, were also examined via XRD analyses. All XRD measurements were conducted using a Bruker XRD D8 Advance diffractometer (Bruker, Analytical Instruments, S.A., Athens, Greece) with a LINXEYE XE high-resolution energy-dispersive detector, under experimental conditions set in accordance with previous reports. The possible interactions between AC and the adsorbed TO molecules were examined with FTIR spectroscopy measurements on pure AC and the modified TO@AC nanohybrid. Additionally, the interactions between AC, TO@AC, and the LDPE matrix were investigated via FTIR spectroscopy measurements on the pure LDPE film and all of the obtained LDPE/AC films and LDPE/TO@AC active films. The FTIR measurements employed an FT/IR-6000 JASCO Fourier-transform spectrometer (JASCO, Interlab, S.A., Athens, Greece). Differential scanning calorimetry (DSC) measurements were conducted on pure AC and on the modified TO@AC nanohybrid by using a DSC214 Polyma Differential Scanning Calorimeter (NETZSCH manufacturer, Selb, Germany). The total amount of the TO fraction adsorbed onto AC was determined through thermogravimetric analysis (TGA) experiments performed on pure AC and on the modified TO@AC nanohybrid, utilizing a PerkinElmer Pyris Diamond TGA/DTA instrument (Interlab, S.A., Athens, Greece). Lastly, the surface and cross-sectional morphologies of all resulting LDPE/AC and LDPE/TO@AC films, as well as the pure LDPE film, were scrutinized via SEM analysis with an accompanying EDX analysis. For the SEM analysis, a JEOL JSM-6510 LV SEM Microscope (Ltd., Tokyo, Japan) equipped with an X-Act EDS-detector (Oxford Instruments, Abingdon, Oxfordshire, UK) with an acceleration voltage of 20 kV was employed.

2.5. Tensile Measurements of LDPE/AC and LDPE/TO@AC Films

For all the obtained LDPE/AC and LDPE/TO@AC films, as well as for the pure LDPE film, tensile properties were measured according to according to the ASTM D638 method using a Simantzü AX-G 5kNt instrument (Simantzu. Asteriadis, S.A., Athens, Greece), following the methodology described previously in [21,40,44].

2.6. Water Vapor Transmission Rate (WVTR) and Water Diffusion Coefficient (D_{wv})

The WVTR values for all of the obtained LDPE/AC and LDPE/TO@AC films, as well as for the pure LDPE film, were measured with a handmade apparatus according to the ASTM E96/E 96M-05 method [43,45]. From the WVTR values, the water vapor diffusion coefficient values (D_{wv}) were calculated according to the methodology described in detail recently [44,46].

2.7. Oxygen Transmission Rate Values (O.T.R.) and Oxygen Diffusion Coefficient (P_{O2})

For all of the obtained LDPE/AC and LDPE/TO@AC films, as well as for the pure LDPE film, O.T.R. values were measured by using an oxygen permeation analyzer (O.P.A., 8001, Systech Illinois Instruments Co., Johnsburg, IL, USA) according to the ASTM D 3985 method (23 °C and 0% RH). From these values, oxygen diffusion coefficient (P_{O2}) values were obtained by following the methodology described in detail recently [44,46].

2.8. Total Antioxidant Activity of Films

The antioxidant activity of all obtained films was measured using the 2,2-diphenyl-1-picrylhydrazyl radical (DPPH) assay, as described by Brand-Williams et al. [47] and modified recently [48,49]. The DPPH assay method has the advantages of a cheap, simple, and trustworthy method but also exhibits some limitations. The evaluation of the antioxidant capacity of a material by the change in DPPH• UV–vis absorbance at a wavelength of 517 nm must be carefully interpreted, since the light absorbance of the DPPH• at this wavelength may be diminished because of the influence of some other factors, e.g., the reaction of the DPPH• with the analyzed sample, pH, O_2, light, type of solvent, etc. This is why, as mentioned, the method added a buffer solution in all samples to keep the pH stable. Also, dark-colored and hermetically sealed vials were used to avoid light sensitivity and oxidation processes caused by O_2. In 4 mL of 30 ppm DPPH ethanolic solution, approximately 5 mg of LDPE/5TO@AC, LDPE/10TO@AC, and LDPE/15TO@AC film was added, and the absorbance at 517 nm was recorded as a function of time for 1 h (60 min), 2 h (120 min), 3 h (180 min), 1 day (1440 min), and 2 days (2880 min). As a blank sample, the absorbance of 4 mL of 30ppm DPPH ethanolic solution without the addition of any film was also recorded as a function of time. The % antioxidant activity was calculated by using Equation (1):

$$\% \text{ film antioxidant activity} = [(A_{t,film} - A_{t,blank})/A_0] \times 100 \quad (1)$$

where $A_{t,film}$ is the absorbance of each film as a function of time, $A_{t,blank}$ is the absorbance of the blank sample as a function of time, and A_0 is the initial absorbance of each sample solution.

2.9. TO Release Test—Calculation of Released TO wt.% Content and TO's Released Diffusion Coefficient (D_{TO})

The experiments of TO's release from all of the obtained LDPE/TO@AC active films were conducted using an AXIS AS-60 moisture analyzer (AXIS Sp. z o.o. ul. Kartuska 375b, 80-125 Gdańsk) according to the methodology described previously [41]. Approximately 300 to 500 mg of each film was used for the controlled release experiments. The diffusion coefficient (D_{TO}) for TO's release from the obtained LDPE/xTO@AC films was calculated according to Equation (2):

$$\frac{m_t}{m_\infty} = \sqrt{4\frac{D.t}{\pi.l^2}} \quad (2)$$

where m_t and m_∞ are the amount of TO released form the film after time t and after the equilibrium time $t_{eq}\rightarrow\infty$, respectively, D is the diffusion coefficient, and l is the average film thickness.

The linearization of Equation (2) leads to the slightly modified Equation (3):

$$\left(\frac{m_t}{m_\infty}\right)^2 = 4\frac{D \times t}{\pi \times \ell^2} \quad (3)$$

By employing the pseudo-second-order sorption mechanism model [42], we calculated the desorption rate constant k_2 and the maximum TO desorbed q_e according to the following equation:

$$q_t = \frac{q_e^2 \times k_2 \times t}{q_e \times k_2 \times t + 1} \quad (4)$$

where $q_t = m_t/m_0$ and $q_e = 1$.

2.10. Packaging Preservation Test of "Skaloppini Type" Fresh Pork Meat Fillets

The pork fillets were wrapped aseptically between two LDPE@15TO@AC active films with a diameter of 11 cm (see Figure 3). For comparison, pork fillets were also wrapped with two LDPE/15AC films and two pure LDPE films (control samples) (see Figure 3). All of the as-wrapped pork fillets were folded inside the packaging paper of the local meat processing plant Aifantis, without the inner membrane used by the company for packaging. After packaging, the fillets were placed in a preservation chamber at 4 ± 1 °C, and then the lipid oxidation, the heme iron content, the total variable counts (TVCs), and the sensory analysis were measured.

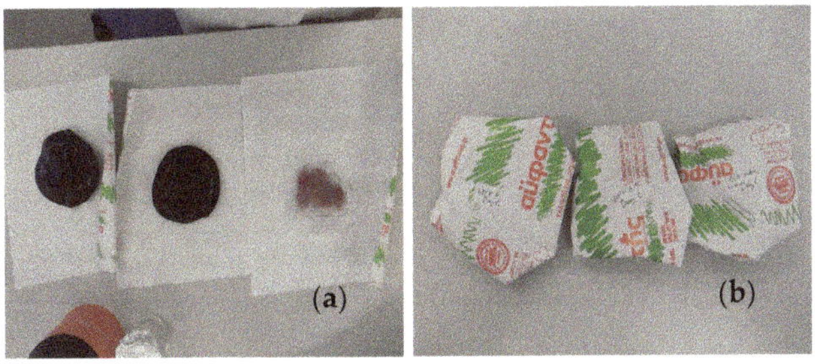

Figure 3. (**a**) Wrapping of fresh pork fillets inside two LDPE@15TO@AC, LDPE/15AC, and pure LDPE films; (**b**) folding of wrapped pork fillets inside the packaging paper of the Aifantis company.

2.11. Packaging Test of Fresh Pork Fillets

2.11.1. Lipid Oxidation of Pork Fillets with Thiobarbituric-Acid-Reactive Substances

The thiobarbituric-acid-reactive substances (TBARS) values of the packed pork fillets after 2, 4, 6, 8, 10, and 12 days were determined using the method of Tarladgis et al. [50], as recently modified by Karabagias et al. [51]. The methodology for determination of the TBARS values of packaged fresh pork fillets was as described in detail recently [40,41].

2.11.2. Heme Iron Content

The heme iron content of the packaged fresh pork fillets was determined according to the method reported by Clark et al. [52], as described in detail recently [40,41]. Heme iron content analyses were carried out every 2 days up to 12 days of storage.

2.11.3. Total Variable Counts (TVCs) of Pork Fillets

The TVCs were monitored every 2 days up to 8 days of storage at 4 ± 1 °C. Ten grams of pork fillet was removed aseptically from each packaging treatment using a spoon, transferred to a stomacher bag (Seward Medical, Worthing, West Sussex, UK) containing 90 mL of sterile buffered peptone water (BPW, NCM0015A, Heywood, BL97JJ, UK) (0.1 g/100 mL of distilled water), and homogenized using a stomacher (LAB Blender 400, Seward Medical, UK) for 90 s at room temperature. For the microbial enumeration, 0.1 mL of serial dilutions (1:10 diluents, buffered peptone water) of pork meat homogenates was spread on the surface of plate count agar (PCA, NCM0010A, Heywood UK). The TVCs were determined every 2 days up to 8 days of storage at 4 ± 1 °C after incubation of each plate for 2 days at 30 °C [53].

2.11.4. Sensory Analysis Testing of Pork Fillets

The sensory properties of pork fillets were scaled from 0 (for the least liked sample) to 5 (most liked sample) points by seven experienced members of the Food Science and Technology Department. At each sampling day, color, odor, cohesion, and taste were evaluated. The test samples were cooked, cut to 1.50 cm × 1.50 cm, and then served at 60 °C. Palate cleansers—room-temperature distilled water and unsalted crackers—were provided between samples. The panelists did not taste those samples that exceeded the limit of acceptability for TVCs (higher than 7 log CFU/g) [54].

2.12. Statistical Analysis

In this study, an extensive array of tests was conducted on a minimum of three separate samples. These tests encompassed an assessment of structural, mechanical, and barrier properties, alongside a range of others, including antioxidant activities, controlled release, TBARS, heme iron, and TVC tests. The results presented in the tables throughout this work are mean values derived from these measurements, with each table also displaying the standard deviation. Equal-means hypothesis testing was performed to evaluate the equality of the means. IBM's SPSS Statistics software, version 20, was used to perform the necessary statistical analyses. Hypothesis tests were carried out via the ANOVA method, using the Tukey criteria for equal-variance assumptions, setting confidence intervals at 95% and significance levels at $p < 0.05$. In addition, the assurance of mean equality (EA%) or inequality (IA%) was examined based on empirical equations previously reported in the literature [55]. Across all tests, it was observed that the mean values of the various sample properties that were statistically different exhibited an inequality assurance factor (IA%) of over 56%, while the mean values that were statistically equal exhibited an equality assurance factor (EA%) of over 61%. More details about the ANOVA results are available in the relevant tables.

3. Results and Discussion

3.1. Physicochemical Characterization of AC and TO@AC Nanohybrid

The XRD plot of pure AC powder (depicted as line (1) in Figure 4a) is consistent with a typical amorphous material. After the adsorption of TO onto AC to develop the TO@AC nanohybrid, a broad peak appears at around 18° 2theta in the XRD plot of the TO@AC nanohybrid (see line (2) in Figure 4a), suggesting that the adsorption of TO has subtly influenced the amorphous structure of pure AC. Line (1) in the FTIR graph (Figure 4b) corresponds to the FTIR plot of pure thyme oil. The bands observable at around 3530 to 3433 cm^{-1} are attributed to the stretching vibration of O-H groups. The bands at around 3100–3000 cm^{-1} are ascribed to the stretching vibrations of aromatic and alkenic groups of TO molecules. The bands at 2958 and at 2868 cm^{-1} are assigned to the stretching mode of C-H groups. The bands between 1500 cm^{-1} and 1300 cm^{-1} are assigned to the bending of C-H on the C-O-H group and the bending of aliphatic CH_2 groups [38–40]. The FTIR spectra of pure AC (see line (2) in Figure 4b) prominently display bands at 1590 cm^{-1} and 1260 cm^{-1}. The band at 1590 cm^{-1} corresponds to the stretching vibration modes of carbonylic (-COO-) groups, while the band at 1260 cm^{-1} is assigned to the stretching vibration of C-O groups [42]. The band with a maximum at approximately 3420–3440 cm^{-1} is attributed to the O–H stretching mode of hydroxyl groups of the adsorbed water molecules [56]. In the FTIR plot of the modified TO@AC nanohybrid (line (3) in Figure 4b), it is evident that in addition to the bands of the pure AC, the characteristic bands of TO also exist in the ranges of 2800–3100 cm^{-1}, 1300–1500 cm^{-1}, and 500–1000 cm^{-1} [38–40]. This observation signifies the adsorption of TO molecules onto the internal surface area of pure AC, without causing a shift in the characteristic peaks of AC. This suggests that the TO molecules are more likely physiosorbed than chemisorbed on the surface of AC, favoring the physisorption of TO molecules on the AC surface for controlled-release applications of such TO@AC nanohybrids.

The thermogravimetric analysis (TG) plot of pure AC, shown as line (1) in Figure 4c, reveals a mass-loss step below 200 °C, which is consistent with the desorption of AC moisture [57]. The TG plot of AC remains almost unchanged until 600 °C, where a second mass-loss step commences [58]. This step correlates with the decomposition of cellulose, lignin, and hemicellulose matter, along with the expulsion of volatile matter that occurs during the carbonization process [59]. The TG plot of the TO@AC nanohybrid (see line (2) in Figure 4c) displays a first mass-loss step below 300 °C and a second one beginning at around 350–400 °C and ending around 500 °C. This pattern suggests that a fraction of TO desorbs from the AC surface in the temperature range from 150 to 300 °C, while another TO fraction desorbs in a higher temperature range from 350 to 500 °C. Considering the previously studied textural features of the AC [42], the Brunauer–Emmett–Teller surface area (S_{gBET}) of the AC was 1372 m^2/g, with the micropore volume being 84.6% and exhibiting micropores at D_{micro1} = 1.28 and D_{micro2} = 1.6 nm [41]. Given that the size of a TO molecule is roughly equal to that of a phenol molecule (about 0.6–0.8 nm), a fraction of TO could be adsorbed within the micropore structure of AC. Consequently, the TO fraction residing in the micropores of AC requires more energy to desorb (second mass-loss step). It is therefore suggested that the first mass-loss step from 150 to 300 °C corresponds to the desorption of TO molecules adsorbed in the mesoporous AC structure, while the second mass-loss step from 350 to 500 °C corresponds to the TO molecules desorbed from the microporous AC structure. In Figure 3c, the % mass loss of the first step and the % total mass loss are calculated by subtracting the mass values of TO@AC from the mass values of pure AC at 300 °C and 500 °C, respectively. The total TO mass adsorbed was found to be equal to 50.7 wt%—a value considerably higher than those recently obtained for TO's adsorption on natural zeolite (35.5 wt%) [41] and HNTs (31.4 wt%) [40]. This outcome reveals that AC is a promising bio-based material for application as a nanocarrier in controlled-release applications within the active food packaging sector. The DSC plot of pure AC (see line (1) in Figure 4d) exhibits an exothermic peak at approximately 157 °C, corresponding to the enthalpy of water molecules' desorption. The DSC plot of the modified TO@AC displays a peak at around 217 °C. This peak aligns more closely with the boiling point of TO (232 °C) than the boiling point of limonene (176 °C), indicating that the fraction of molecules adsorbed in AC is rich in TO molecules, which is consistent with previous publications [40,41].

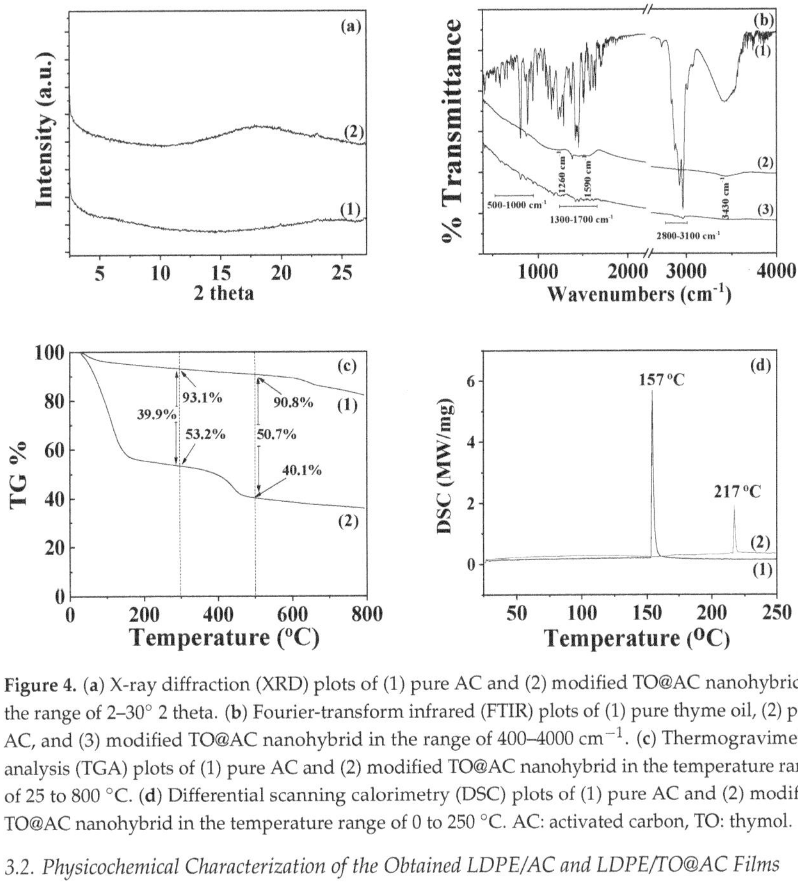

Figure 4. (**a**) X-ray diffraction (XRD) plots of (1) pure AC and (2) modified TO@AC nanohybrid in the range of 2–30° 2 theta. (**b**) Fourier-transform infrared (FTIR) plots of (1) pure thyme oil, (2) pure AC, and (3) modified TO@AC nanohybrid in the range of 400–4000 cm^{-1}. (**c**) Thermogravimetric analysis (TGA) plots of (1) pure AC and (2) modified TO@AC nanohybrid in the temperature range of 25 to 800 °C. (**d**) Differential scanning calorimetry (DSC) plots of (1) pure AC and (2) modified TO@AC nanohybrid in the temperature range of 0 to 250 °C. AC: activated carbon, TO: thymol.

3.2. Physicochemical Characterization of the Obtained LDPE/AC and LDPE/TO@AC Films

3.2.1. XRD Analysis

In Figure 5a, the XRD plots of all LDPE/AC and LDPE/TO@AC films, as well as the pure LDPE film, are shown for comparison.

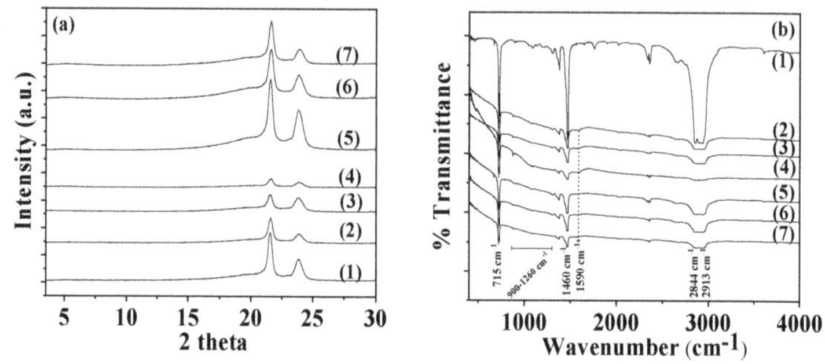

Figure 5. (**a**) X-ray diffraction (XRD) plots of pure LDPE and all LDPE/AC and LDPE/TO@AC films. (**b**) Fourier-transform infrared (FTIR) plots of pure LDPE and all LDPE/AC and LDPE/TO@AC films. LDPE: low-density polyethylene, AC: activated carbon, TO: thymol.

In all plots, the characteristic peaks of LDPE's crystal phase at Bragg angles of $2\theta 1 = 21.5°$ and $23.75°$ are observed. It can also be observed, that with the addition of both AC and TO@AC, LDPE's peaks decreased. In the case of LDPE/AC, films the decrease in LDPE's characteristic peaks is higher than in the case of LDPE/TO@AC. This is an indication of higher dispersion of the hydrophobic modified TO@AC hybrid in the LDPE matrix compared with pure AC.

3.2.2. FTIR Spectroscopy

In Figure 5b, the FTIR plots of all obtained LDPE/AC and LDPE/TO@AC films, as well as the pure LDPE film, are shown for comparison. In all cases, the characteristic peaks of LDPE are obtained. More specifically, the bands at 1460 and 715 cm^{-1} are assigned to the asymmetric stretching of the CH_3 group, the group wagging of the CH_2 group, and the group rocking of the CH_2 group of the LDPE. In both LDPE/AC and LDPE/TO@AC films, the characteristic peaks of LDPE are decreased and the characteristic peak of AC at 1590 cm^{-1} is obtained, along with the broad peaks of AC in the range of 900–1260 cm^{-1}. No shift in LDPE's characteristic peaks is observed, implying no chemical bonding between the LDPE matrix and AC or TO@AC's chemical groups [60]. With a more careful glance, it can be observed that the characteristic peak of AC at 1590 cm^{-1} is more prominent in the case of all LDPE/AC films than in the case of LDPE/TO@AC films. This could be an indication of higher dispersion of the modified and more hydrophobic TO@AC nanohybrid in the LDPE matrix compared with the relevant of pure AC, which is in accordance with the aforementioned XRD observations. Finally, no TO peaks are observed in the case of TO@AC-containing films, and this is an indication that TO molecules are blended inside the LDPE polymer matrix.

3.2.3. SEM Images

The surface/cross-section morphologies of the pure polymer matrix LDPE, pure AC, and all of the obtained LDPE/xAC and LDPE/xTO@AC films were investigated using an SEM instrument, and the results confirmed that the AC and the hybrid TO@AC nanostructure were homogeneously dispersed in the polymer matrix. The SEM images (surface and cross-section) in Figure 6a,b exhibit the expected homogeneous structure of the pristine polymer matrix. In Figure 6c, the surface morphology of activated carbon (AC) with highly porous characteristics is shown, resembling a honeycomb structure. The micro- and mesoporous surface morphology in AC acts as an absorbent from the packaged food or as a nanocarrier for bioactive compounds released in the packaged food [61].

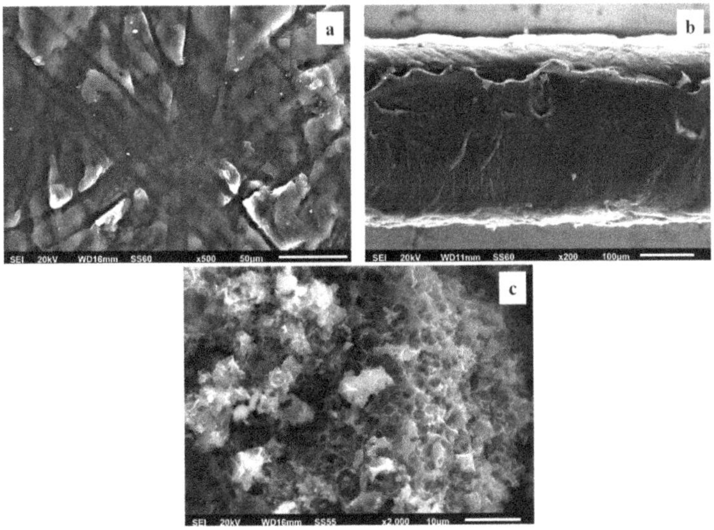

Figure 6. Scanning electron microscopy (SEM) images of the (**a**) surface and (**b**) cross-section for the pure LDPE film, and (**c**) surface morphology of AC. LDPE: low-density polyethylene, AC: activated carbon.

Surface and relative cross-section images of LDPE/xAC and LDPE/xTO@AC with different ratios (5, 10, 15%) of AC and TO@AC are presented in Figures 7–9.

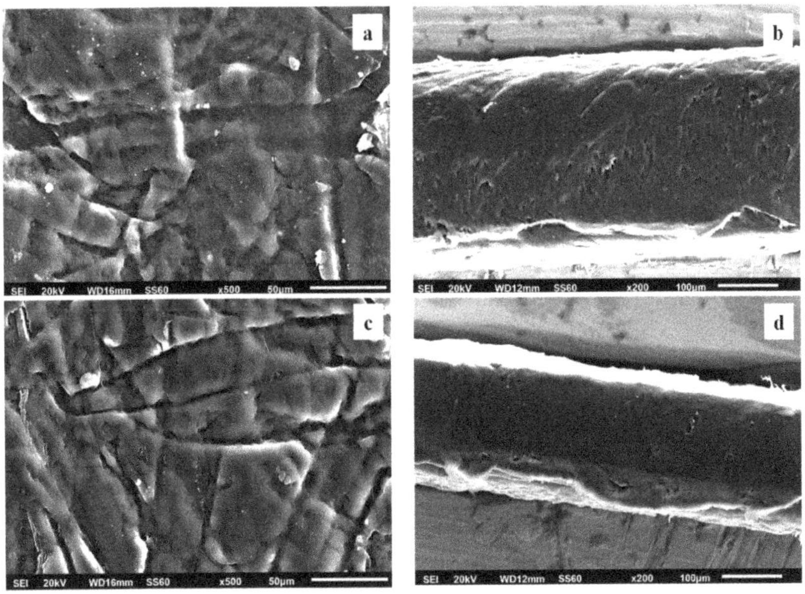

Figure 7. Scanning electron microscopy (SEM) images of the surface (**a**,**c**) and cross-section (**b**,**d**) of the nanocomposite films of (**a**,**b**) LDPE/5AC and (**c**,**d**) LDPE/5TO@AC. LDPE: low-density polyethylene, AC: activated carbon, TO: thymol.

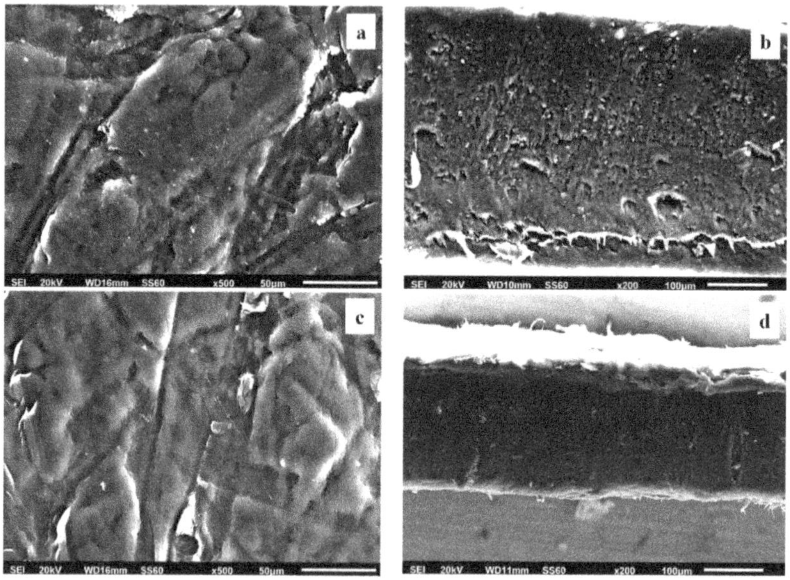

Figure 8. Scanning electron microscopy (SEM) images of the (**a**,**c**) surface and (**b**,**d**) cross-section of the nanocomposite films of (**a**,**b**) LDPE/10AC and (**c**,**d**) LDPE/10TO@AC. LDPE: low-density polyethylene, AC: activated carbon, TO: thymol.

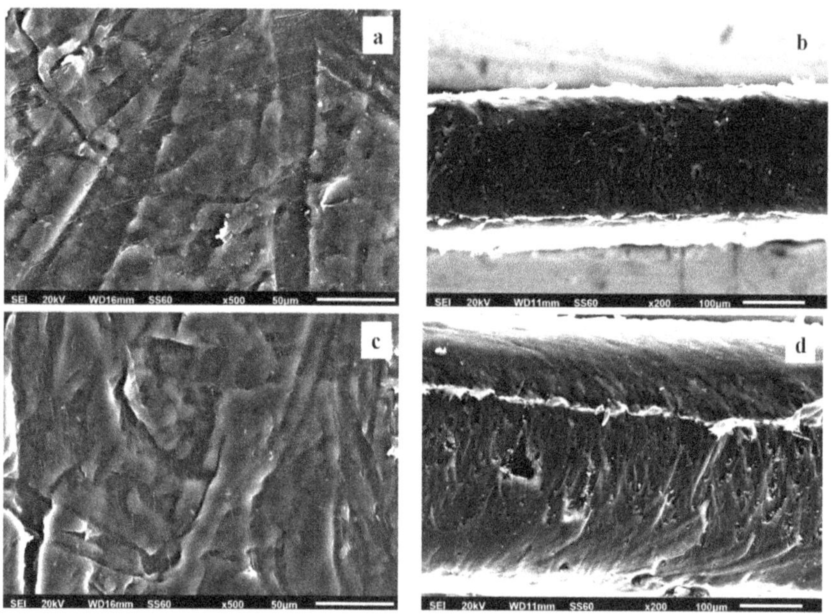

Figure 9. Scanning electron microscopy (SEM) images of the (**a**,**c**) surface and (**b**,**d**) cross-section of the nanocomposite films of (**a**,**b**) LDPE/15AC and (**c**,**d**) LDPE/15TO@AC. LDPE: low-density polyethylene, AC: activated carbon, TO: thymol.

Based on the SEM studies, it should be mentioned that in all cases of LDPE/xAC and LDPE/xTO@AC films, both AC and TO@AC were homogeneously dispersed, indicating their enhanced compatibility and their interfacial adhesion with the polymer matrix of LDPE. This result agreed with the XRD and FTIR results discussed above. It is obvious that pure AC derived from spent coffee has a high dispersibility in the LDPE matrix due to its hydrophobic nature [42]. Its hydrophobic nature is enhanced by the adsorbed TO molecules, further enhancing its compatibility with LDPE.

3.3. Tensile Properties of LDPE/AC and LDPE/TO@AC Films

The calculated elastic modulus (E), ultimate strength (σ_{uts}), and elongation at break (%ε) values for all obtained LDPE/AC and LDPE/TO@AC films, as well as for the pure LDPE film, are listed in Table 1 for comparison.

Table 1. Elastic modulus (E), ultimate strength (σ_{uts}), and elongation at break (%ε) values for all obtained LDPE/AC and LDPE/TO@AC films, as well as for pure LDPE film. LDPE: low-density polyethylene, AC: activated carbon, TO: thymol.

	E (MPa)	σ_{uts} (MPa)	%ε
LDPE	183.3 ± 28.8	12.6 ± 0.5	29.3 ± 8.1
LDPE/5AC	282.7 ± 17.1 ^{a}	10.9 ± 1.5	33.4 ± 2.7
LDPE/10AC	312.7 ± 21.4 ^{b}	11.8 ± 1.9	63.4 ± 14.0 ^{d}
LDPE/15AC	327.3 ± 21.8 ^{b}	12.7 ± 1.9	51.0 ± 12.7
LDPE/5TO@AC	286.0 ± 14.5 ^{a,c}	11.2 ± 1.7	43.3 ± 17.5
LDPE/10TO@AC	363.5 ± 19.4	12.1 ± 2.4	75.8 ± 15.1
LDPE/15TO@AC	291.0 ± 19.0 ^{c}	11.8 ± 1.0	64.8 ± 18.6 ^{d}

^{a,b,c,d} Indices for statistically equal mean values according to the ANOVA comparison method and Tukey's criteria for the assumption of equal variances. Significance level: $p < 0.05$.

As shown in Table 1, the incorporation of pure AC powder and the modified TO@AC nanohybrid in the LDPE matrix did not significantly affect the ultimate strength values of LDPE/AC and LDPE/TO@AC films as compared to the ultimate strength value of the pure LDPE film. On the other hand, the incorporation of pure AC powder and the modified TO@AC nanohybrid in the LDPE increased the elastic modulus and % elongation at break values of both LDPE/AC and LDPE/TO@AC films. In advance, the highest % elongation at break value was obtained for LDPE/TO@AC, as compared to the % elongation at break value of the LDPE/AC film. The higher % elongation at break values of the LDPE/AC films as compared to the % elongation at break value of the pure LDPE film suggest the high dispersity and compatibility of AC powder in the LDPE matrix. In contrast to our previous reports, where inorganic nanostructures such as HNTs and NZ were incorporated in the LDPE matrix and led to fragile LDPE/HNT and LDPE/NZ films, the AC powder seems to be more hydrophobic, less inorganic and, thus, more compatible with the LDPE matrix [40,41]. Recently, Sadakpipanich et al. [37] showed that the incorporation of MC activated carbon into the LDPE increased the tensile strength and decreased the elongation at break of the obtained LDPE/MC films. These results are in contrast with the results presented here. However, Sadakpipanich et al. [37] have incorporated MC contents lower than 5 wt.%. into LDPE films. Thus, herein, it is reported for first time that the incorporation of high contents of AC (varying from 5 to 15%) in LDPE drives the development of LDPE/AC films with similar or higher elongation at break values in comparison to pure LDPE. It seems that the higher amounts of AC incorporated into LDPE somehow react as plasticizer. In advance, the plasticization capacity of AC increased for TO@AC nanohybrids because of the presence of TO molecules [12,62,63]. Overall, the highest elastic modulus and % elongation at break values were found for the LDPE/10TO@AC film. The addition of AC or TO@AC in the LDPE matrix imposes a limit. Above this amount, a worse dispersion and homogeneity start to exist. This is why the mechanical properties of the LDPE/15TO@AC became worse in comparison to the relevant properties of the LDPE/10TO@AC film. So, this film had 98.3% and 158.7% higher elastic modulus, and % elongation at break values, respectively, than the pure LDPE film. The abovementioned conclusions were tested via the ANOVA statistical tool, and the results showed that for a significance level of $p = 0.05$ the mean value of the Young's modulus of LDPE/5AC was statistically equal to the relevant value of the LDPE/5TO@AC. Moreover, the mean value of Young's modulus of the LDPE/5TO@AC film was statistically equal to the respective value of the LDPE/15TO@AC film, but not equal to the relevant value of the LDPE/5AC film. Equality of the mean values of the elongation at break property was also observed for the LDPE/10AC and LDPE/15TO@AC films. Finally, more details for the statistical procedure comparing the mean values are available in Section 2.12.

3.4. Water and Oxygen Barrier Properties of the LDPE/AC and LDPE/TO@AC Films

In Table 2, the obtained water vapor transmission rate (WVTR) and oxygen transmission rate (OTR) values of all LDPE/AC and LDPE/TO@AC films, as well as for the pure LDPE film, are listed. From these values, the water vapor diffusion coefficient (D_{wv}) values and the oxygen permeability (Pe_{O2}) values for all LDPE/AC and LDPE/TO@AC films, as well as for the pure LDPE film, were calculated, as listed in Table 2 for comparison.

Table 2. WVTR, D_{wv}, OTR, and Pe_{O2} values for all LDPE/AC and LDPE/TO@AC films, as well as for the pure LDPE film. LDPE: low-density polyethylene, AC: activated carbon, TO: thymol.

	Film Thickness (mm)	WVTR (10^{-7} gr cm^{-2} s^{1})	D_{wv} (10^{-4} cm^2/s)	Film Thickness (mm)	OTR (mL m^{-2} day^{-1})	Pe_{O2} (10^{-8} cm^2/s)
LDPE	0.270 ± 0.012	5.89 ± 0.13	3.77 ± 0.17 ^{a}	0.270 ± 0.010	6407 ± 120	20.02 ± 2.41
LDPE/5AC	0.256 ± 0.011	6.42 ± 0.56	3.73 ± 0.07 ^{a}	0.266 ± 0.004	1365 ± 62	4.19 ± 0.2
LDPE/10AC	0.141 ± 0.014	7.31 ± 0.20	2.35 ± 0.02	0.378 ± 0.010	818 ± 15	3.57 ± 0.21
LDPE/15AC	0.176 ± 0.010	5.02 ± 0.73	1.99 ± 0.31 ^{b}	0.410 ± 0.015	675 ± 34	3.20 ± 0.19 ^{c}
LDPE/5TO@AC	0.332 ± 0.012	3.96 ± 0.53	2.84 ± 0.23	0.170 ± 0.015	1593 ± 67	3.12 ± 0.17 ^{c}
LDPE/10TO@AC	0.186 ± 0.016	4.55 ± 0.13	1.87 ± 0.21 ^{b}	0.311 ± 0.005	642 ± 17	2.11 ± 0.16
LDPE/15TO@AC	0.241 ± 0.013	2.62 ± 0.23	1.47 ± 0.04	0.351 ± 0.010	228 ± 6	0.92 ± 0.04

^{a,b,c} Indices for statistically equal mean values according to the ANOVA comparison method and Tukey's criteria for the assumption of equal variances. Significance level: $p < 0.05$.

As shown in Table 2, both the AC powder and the TO@AC nanohybrid successfully increased both the water and oxygen barriers. The higher the AC and TO@AC wt.% contents, the stronger the obtained water and oxygen barriers. A higher increase in the water and oxygen barriers was observed for the TO@AC-containing films, in accordance with the higher dispersion of LDPE/TO@AC as compared to LDPE/AC films, as shown in the SEM images section above. So, the lowest water vapor diffusion coefficient (D_{wv}) and oxygen permeability (Pe_{O2}) values were obtained for the LDPE/15TO@AC film. For this film, the obtained water vapor diffusion coefficient (D_{wv}) value was 55% lower than the water vapor diffusion coefficient (D_{wv}) value of the pure LDPE film, while its oxygen permeability (Pe_{O2}) value was 95.4% lower than the oxygen permeability (Pe_{O2}) value of the pure LDPE film. Sadakpipanich et al. [37] developed LDPE-based films by incorporating AC produced from biomass derived from macadamia nut cultivation. Their findings showed lower water and oxygen barrier properties by varying the AC's wt%. content from 1 to 5%. Therefore, herein, we have shown for the first time that AC derived from spent coffee is a promising material to be applied as a nanoreinforcement to reduce water/oxygen permeability in active packaging films when incorporated into LDPE films in high wt% contents, from 5 to 15%. Its capacity to increase the water/oxygen barrier is maximized when modified with TO to obtain TO@AC nanohybrids. The abovementioned conclusions were tested via the ANOVA statistical tool, and the results showed that for a significance level of $p = 0.05$, the water vapor barriers of pure LDPE and LDPE/5AC are equal. Equality of the mean values of this property was also observed for the LDPE/15AC and LDPE/10TO@AC films. Finally, the LDPE/15AC film exhibited the same mean value for oxygen barrier as the LDPE/5TO@AC film. More details for the mean values comparing statistical procedures are available in Section 2.12.

3.5. Antioxidant Activity of LDPE/TO@AC Films

The plots of % antioxidant activity of all LDPE/TO@AC films as a function of time are shown in Figure 10.

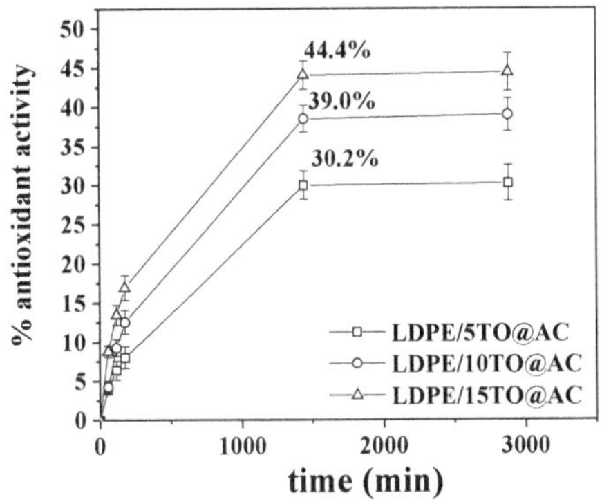

Figure 10. Plots of calculated % antioxidant activity of LDPE/5TO@AC, LDPE/10TO@AC, and LDPE/15TO@AC films as a function of time. DPPH assay method as described by Brand-Williams et al. [47] and modified recently [48,49]. LDPE: low-density polyethylene, AC: activated carbon, TO: thymol.

As shown in Figure 10, the total antioxidant activity of the films increased rapidly during the first day and then remained constant. As expected, the antioxidant activity increased as the wt%. TO content in the films increased. So, after 1 day of incubation, the LDPE/5TO@AC, LDPE/10TO@AC, and LDPE/15TO@AC films reached a total antioxidant activity of 30.2%, 39.0%, and 44.4%, respectively.

3.6. TO Release Test—Calculation of Released TO wt.% Content and TO's Released Diffusion Coefficient (D_{TO}) of LDPE/TO@AC Films

By using the data given in Tables S1–S3 and the Equations (2)–(4), the diffusion coefficient of TO released, the total amount of TO released (q_e), and the desorption rate constant value (K_2) for all studied LDPE/xTO@AC films were calculated, as listed in Table 3 for comparison.

Table 3. Calculated values of the diffusion coefficient of TO molecule, total desorbed amount of TO (q_e), and desorption rate constant (K_2) for all obtained LDPE/xTO@AC active films. LDPE: low-density polyethylene, AC: activated carbon, TO: thymol.

	$D_{TO} \times 10^{-7}$ (cm²/s)	q_e	K_2 (s^{-1})
LDPE/5TO@AC	1.39 ± 0.45	0.023 ± 0.002	2.425 ± 0.419
LDPE/10TO@AC	2.00 ± 0.33	0.032 ± 0.006	2.444 ± 0.355
LDPE/15@TOAC	1.47 ± 0.86	0.044 ± 0.009	1.488 ± 0.142

As shown in Table 3, as the wt.% content of TO@AC increased, the value of q_e increased, and the value of the constant K_2 decreased. This means that by increasing the content of TO@AC nanohybrid in the LDPE/xTO@AC active films, the total amount of TO released increases and its release rate is reduced. On the other hand, the calculated values of the TO diffusion coefficient were 1.39×10^{-7} cm²/s for the LDPE/5TO@AC film, 2.00×10^{-7} cm²/s for the LDPE/10TO@AC film, and 1.47×10^{-7} cm²/s for the LDPE/15TO@AC film. The values of the TO diffusion coefficient and K_2 release rate constant reported here were an order of magnitude higher than the TO diffusion coefficient values reported recently for the TO diffusion coefficient and K_2 release rate constant from LDPE/xTO@NZ films [41].

This means that AC releases higher amounts of TO and at higher rates than natural zeolite (NZ) [41].

3.7. Lipid Oxidation of Pork Fillets

The calculated TBARS values of the low-fat pork fillets wrapped with pure LDPE, LDPE/15AC, and LDPE/15TO@AC films are shown in Table 4 for comparison.

Table 4. Calculated TBARS and heme iron content values of pork fillets wrapped with pure LDPE, LDPE/15AC, and LDPE/15TOAC films, with respect to storage time. LDPE: low-density polyethylene, AC: activated carbon, TO: thymol.

TBARS	Day 0	Day 2	Day 4	Day 6 AVG ± SD (mg/kg)	Day 8	Day 10	Day 12
Control	0.23 [a] ± 0.01	0.33 [b] ± 0.02	0.46 [d] ± 0.01	0.63 [g] ± 0.03	0.98 [j] ± 0.02	1.21 [m] ± 0.02	1.36 [p] ± 0.02
LDPE/15AC	-	0.28 [b] ± 0.01	0.40 [e] ± 0.01	0.53 [h] ± 0.01	0.83 [k] ± 0.03	1.10 [n] ± 0.04	1.28 [q] ± 0.01
LDPE/15TO@AC	-	0.23 [c] ± 0.01	0.33 [f] ± 0.02	0.44 [i] ± 0.02	0.75 [l] ± 0.03	0.95 [o] ± 0.03	1.13 [r] ± 0.02
				ANOVA			
F	-	39.577	63.033	82.765	91.031	67.336	149.763
p	-	0.000	0.000	0.000	0.000	0.000	0.000
Fe	**Day 0**	**Day 2**	**Day 4**	**Day 6 AVG ± SD (μg/g)**	**Day 8**	**Day 10**	**Day 12**
Control	10.18 [a] ± 0.12	8.00 [b] ± 0.12	6.16 [e] ± 0.12	5.12 [h] ± 0.19	3.54 [k] ± 0.16	2.22 [n] ± 0.27	1.28 [p] ± 0.15
LDPE/15AC	-	8.96 [c] ± 0.09	6.80 [f] ± 0.09	5.72 [i] ± 0.12	4.14 [l] ± 0.18	2.48 [n] ± 0.15	1.82 [q] ± 0.09
LDPE/15TO@AC	-	9.66 [d] ± 0.12	7.92 [g] ± 0.12	7.40 [j] ± 0.27	5.70 [m] ± 0.16	3.84 [o] ± 0.18	2.48 [r] ± 0.15
				ANOVA			
F	-	162.781	186.000	99.771	135.130	52.083	60.200
p	-	0.000	0.000	0.000	0.000	0.000	0.000

Different letters in the same column indicate statistically significant differences at the confidence level $p < 0.05$ (see also Tables S4 and S5).

The calculated TBARS values from day 0 to the 12th day were similar to those recently reported in [40,41]. From Table 4, it can be seen that both the LDPE/15AC and LDPE/15TO@AC films were successful in reducing the obtained TBARS values during the 12 days of storage in comparison to the relevant TBARS values of the pure LDPE film. In addition, the lowest TBARS values during the 12 days of storage were obtained for the LDPE/15TO@AC active film.

So, in accordance with recent reports [40,41], both the LDPE/15AC and LDPE/15TO@AC active films succeeded in significantly decreasing the deterioration of pork fillets due to the lipid oxidation process.

3.8. Heme Iron of Pork Fillets

Table 4 also shows the calculated heme iron values. These values, as expected, were decreased over the 12 days of storage for all tested films. The greatest decrease in the heme iron values was obtained for the pork fillets wrapped with the LDPE/15TO@AC active film. For the pork fillets wrapped with the LDPE/15AC active film, the heme iron values were higher than the relevant heme iron values of pork fillets wrapped with the LDPE/15TO@AC film but lower than the relevant heme iron values of pork filets wrapped with the pure LDPE film. Thus, both LDPE/15AC and LDPE/15TO@AC films succeeded in preserving the pork fillets with higher heme iron contents, which is beneficial from a nutritional point of view. In advance, in accordance with previous recent reports, the heme iron contents of pork fillets have a linear correlation with the obtained TBARS values [40,41].

3.9. Correlation of TBARS and Heme Iron

Throughout storage, the bivariate Pearson's correlation analysis was performed on the obtained TBARS and heme iron content values of the pork fillets. The results of the analysis indicated that significant and positive correlations were observed between the two methods throughout the storage period. The Supplementary Materials provide the corresponding correlations (see Table S6), which were determined in relation to the packaging treatment and storage time.

3.10. Microbiological Changes of Pork Fillets

The TVC gives a quantitative estimate of the population of microorganisms such as bacteria, yeasts, and molds in a food sample capable of forming visible colonies. The microorganisms present in pork fillets, either as a part of their natural microflora or as the result of cross-contamination from other sources, are mostly aerobic microorganisms, and their population is an indicator of the product's microbiological quality [53]. Table 5 shows the changes in the TVC of pork fillets as a function of the film used and the storage time.

Table 5. Calculated TVC values of pork fillets wrapped with pure LDPE, LDPE/15AC, and LDPE/15TOAC with respect to storage time. LDPE: low-density polyethylene, AC: activated carbon, TO: thymol. PCA used: (PCA, NCM0010A, Heywood UK). The plates seeded by the spread plate method were incubated for 2 days at 30 °C.

	Days				
Sample Name	0	2	4	6	8
			logCFU/g (Avg ± SD)		
LDPE	3.15 ± 0.06 [a]	4.56 ± 0.32 [b]	5.65 ± 0.24 [d]	6.98 ± 0.41 [f]	7.88 ± 0.20 [h]
LDPE/15AC	3.15 ± 0.06 [a]	4.15 ± 0.12 [c]	5.52 ± 0.18 [d]	6.68 ± 0.32 [f]	7.53 ± 0.21 [h]
LDPE/15TO@AC	3.15 ± 0.06 [a]	3.86 ± 0.09 [c]	4.53 ± 0.08 [e]	5.77 ± 0.17 [g]	6.84 ± 0.14 [i]
			ANOVA		
F *	-	8.914	35.032	11.935	28.478
P *	-	0.0160	0.0000	0.0080	0.0010

Identical letters in each column indicate no significant statistical differences. * ANOVA results, confidence level $p < 0.05$ (see also Table S7).

As shown in Table 5, the initial value of TVC was 3.15 log cfu/g, indicating a very good microbiological quality of the pork meat. According to the ICMSF [64], the upper microbiological limit for acceptable quality of foods' TVC is 7 log cfu/g. This limit value (7 log cfu/g) was almost reached for pork fillets wrapped with pure LDPE and LDPE/15AC films on the 6th day of storage. For pork fillets wrapped with the LDPE/15TO@AC active film, the TVC limit value (7 log cfu/g) was reached on the 8th day of storage. Thus, it is obvious that the TVC values for pork fillets wrapped with the pure LDPE and LDPE/15AC films were significantly higher ($p < 0.05$) in comparison to the TVC values of pork fillets wrapped with the LDPE/15TO@AC active film. In advance, no statistically significant differences ($p > 0.05$) were observed for pork fillets wrapped with the pure LDPE and LDPE/15AC films. For the pork fillets wrapped with the LDPE/15AC film, we observed significantly lower TVC values from pork fillets wrapped with the pure LDPE film only on the 2nd day of storage. So, it seems that the LDPE/15AC film, which has a higher oxygen barrier than the pure LDPE film, succeeded in preventing microbiological growth in pork fillets during the first 2 days of storage. From day 2 to day 8, the active LDPE/15TO@AC film succeeded in preventing microbiological growth in pork fillets in comparison to the pure LDPE and LDPE/15AC films, due to its ability to release TO in the fillets. Overall, we can conclude that the LDPE/15TO@AC active film succeeded in extending the microbiological shelf-life of pork fillets by 2 days. Considering that both the LDPE/15AC film and the LDPE/15TO@AC film resulted in higher water/oxygen barrier properties than the pure LDPE film, we can conclude that the release of TO is the key factor for the shelf-life extension of wrapped pork fillets.

3.11. Sensory Evaluation of Pork Fillets

Sensory properties such as color, odor, taste, and cohesion are major factors for consumers to accept a food product. Off odors in spoiled pork meat could be related to compounds that originate from the growth of certain microorganisms, e.g., dimethyl disulfide, dimethyl sulfide, and propylene sulfide generated by *Pseudomonas* spp. or acetoin, diacetyl, and 3-methylbutanol produced by *homofermentative LAB*, *Enterobacteriaceae*, *or Brochothrix thermosphacta* [54]. Chemical compounds, such as ammonia or amines resulting from protein breakdown, as well as ketones and aldehydes resulting from lipid oxidation, could be responsible for the development of an off odor in meat [54]. The sensory evaluation results of the present study are displayed in Table 6.

Table 6. Taste and odor of wrapped pork fillets during storage at 4 ± 1 °C.

Sample Name	Taste 0 Day	Taste 2nd Day	Taste 4th Day	Taste 6th Day	Odor 0 Day	Odor 2nd Day	Odor 4th Day	Odor 6th Day
LDPE	5.00 ± 0.00 ^{a}	4.70 ± 0.29 ^{b}	4.46 ± 0.49 ^{c}	-	5.00 ± 0.00 ^{e}	4.70 ± 0.19 ^{f}	4.16 ± 0.51 ^{g}	3.22 ± 0.36 ^{h}
LDPE/15AC	5.00 ± 0.00 ^{a}	4.68 ± 0.22 ^{b}	4.42 ± 0.35 ^{c}	-	5.00 ± 0.00 ^{e}	4.70 ± 0.41 ^{f}	4.16 ± 0.44 ^{g}	3.44 ± 0.22 ^{h}
LDPE/15TO@AC	5.00 ± 0.00 ^{a}	4.88 ± 0.22 ^{b}	4.70 ± 0.30 ^{c}	4.10 ± 0.16 ^{d}	5.00 ± 0.00 ^{e}	4.86 ± 0.13 ^{f}	4.5 ± 0.31 ^{g}	3.92 ± 0.16 ^{i}
	Color 0 Day	Color 2nd Day	Color 4th Day	Color 6th Day	Cohesion 0 Day	Cohesion 2nd Day	Cohesion 4th Day	Cohesion 6th Day
LDPE	5.00 ± 0.00 ^{j}	4.58 ± 0.39 ^{k}	4.08 ± 0.57 ^{l}	2.96 ± 0.15 ^{m}	5.00 ± 0.00 ^{p}	4.48 ± 0.29 ^{q}	3.78 ± 0.30 ^{r}	2.84 ± 0.11 ^{s}
LDPE/15AC	5.00 ± 0.00 ^{j}	4.32 ± 0.30 ^{k}	4.00 ± 0.53 ^{l}	3.20 ± 0.16 ^{n}	5.00 ± 0.00 ^{p}	4.38 ± 0.40 ^{q}	4.12 ± 0.51 ^{r}	3.44 ± 0.30 ^{t}
LDPE/15TO@AC	5.00 ± 0.00 ^{j}	4.58 ± 0.32 ^{k}	4.30 ± 0.34 ^{l}	4.02 ± 0.19 ^{o}	5.00 ± 0.00 ^{p}	4.46 ± 0.27 ^{q}	4.26 ± 0.43 ^{r}	3.94 ± 0.18 ^{u}

Identical letters in each column indicate no significant statistical differences ($p > 0.05$) (see also Tables S8–S11).

The results showed that the LDPE, LDPE/15AC, and LDPE/15TO@AC films had a similar impact on the color, odor, taste, and cohesion attributes of the pork fillet samples throughout the first 4 days of storage time. At day 6, the pork meat samples packaged in the LDPE, LDPE/15AC, and LDPE/15TO@AC films showed statistically significant differences ($p < 0.05$) in all sensory properties, except for the attribute of odor, which was similar for the LDPE and LDPE/15AC films. Also, on day 6, the samples packaged in LDPE and LDPE/15AC exceeded 7 log CFU/g. On the other hand, the pork meat samples packaged in LDPE/15TO@AC, although no statistically significant differences ($p < 0.05$) were observed in the first 4 days for the sensory analysis parameters (Table 6), had a characteristic and pleasant mild spicy flavor throughout the 6 days of storage according to the panelists' scores, whereas the TVC did not exceed 7 logCFU/g given the antimicrobial properties of thyme essential oil [51]. This finding is probably related to the low fat content of the "scaloppini" type pork fillets studied and the impact of TO [40]. In addition, this packaging technology may be beneficial for the meat industry in terms of "marinated" pork meat under active packaging.

4. Conclusions

In the current study, the preparation and characterization of a novel TO@AC nanohybrid, the development and characterization of innovative LDPE/xTO@AC active packaging films, and their application for the preservation of pork fillets were studied. It is obvious from the results that the hydrophobic AC derived from spent coffee proved to be an excellent nanocarrier for essential oil components such as TO, yielding a TO@AC nanohybrid with a loading capacity of 100% (g_{TO}/g_{AC}). This high quantity of the physiosorbed TO molecules in the AC pore structure resulted in LDPE/xTO@AC films with TO release rates higher than the TO release rates recently reported for similar LDPE/xTO@NZ active films where TO molecules were adsorbed on a natural zeolite (NZ) nanofiller. XRD, FTIR results, and SEM images demonstrated that both pure AC and TO@AC materials exhibited excellent dispersity and compatibility with the LDPE matrix. The high dispersity of both pure AC and the TO@AC nanohybrid in LDPE resulted in LDPE/xAC and LDPE/xTO@AC films having superior mechanical properties compared to the relevant properties of the pure LDPE film. However, it seems that an upper limit to the weight concentration for the

TO@AC material in the LDPE matrix exists and is equal to 10%. The additional loading of such material to 15% started to cause inhomogeneities and the mechanical properties started to become worse compared with the relevant properties of the pure LDPE.

Nevertheless, because such mechanical behavior still remains far improved compared with the behavior of the pure LDPE film, it makes sense to characterize the LDPE/15TO@AC film as the most promising candidate for food preservation (Young's modulus 1.5 times higher and elongation at break 2 times higher than the relevant values of pure LDPE), because the water/vapor diffusion coefficient is 230% lower than that of pure LDPE, the oxygen permeability coefficient is 1928% lower than that of pure LDPE, the TO controlled release rate constant is the lowest ($k_2 = 1.488\ s^{-1}$), the antioxidant activity according to the DPPH assay is the highest (44.4%), and the antibacterial activity according to TVC is the best.

All of the abovementioned parameters strongly suggest that the LDPE/15TO@AC active film could potentially extend the shelf-life of pork fillets by at least 2 days.

Supplementary Materials: The following supporting information can be downloaded at: https://www.mdpi.com/article/10.3390/foods12132590/s1. Table S1: Data from thermogravimetric experiments used control release experiments of thymol from LDPE/5TO@AC film sample; Table S2. Data from thermogravimetric experiments used for control release experiments of thymol from LDPE/10TO@AC film sample; Table S3. Data from thermogravimetric experiments used for control release experiments of thymol from LDPE/15TO@AC film sample; Table S4: Multiple comparison tests of TBARS values of pork meat during storage with respect to packaging treatments; Table S5: Multiple comparison tests of heme iron values of pork meat during storage with respect to packaging treatment; Table S6: Pearson's correlation of heme iron with TBARS values of pork meat during storage within each packaging treatments; Table S7: ANOVA Results of TVC; Table S8: Multiple Comparisons tests; Table S9: ANOVA ODOR; Table S10: ANOVA COLOR; Table S11: ANOVA COHESION.

Author Contributions: Synthesis experiment design, A.E.G. and C.E.S.; characterization measurements and interpretation, A.E.G., V.K.K., A.L., I.K.K., D.M., A.K.-M., N.A., A.A., G.K., S.G. and C.E.S.; paper writing, A.E.G., C.E.S., V.K.K. and A.L.; overall evaluation of this work, A.E.G. and C.E.S.; experimental data analysis and interpretation, A.E.G., V.K.K., A.A., I.K.K., D.M., A.A., S.G., G.K., C.P. and C.E.S.; XRD, OTR, tensile measurements, antioxidant activity, WVTR experimental measurements, and release kinetics test, V.K.K., A.E.G. and C.E.S.; pork fillets' shelf-life packaging test, V.K.K., A.L., K.Z., I.K.K., N.A., S.G., G.K. and A.E.G.; SEM images D.M., A.K.-M. and A.A.; FTIR analysis, A.E.G. and C.E.S. All authors have read and agreed to the published version of the manuscript.

Funding: This research received no external funding.

Data Availability Statement: The datasets generated for this study are available upon request to the corresponding author.

Acknowledgments: The authors want to thank the meat processing company Aifantis Company (Aifantis Group—Head Quarters, Acheloos Bridge, Agrinio, Greece 30100, e-Mail: info@aifantis.gr, website: www.aifantis-group.com/) for the kind offer of "skalopini type" fresh pork fillets.

Conflicts of Interest: The authors declare no conflict of interest.

References

1. Ahmed, W.; Haque, A.; Mohibbullah, M.; Khan, S.I.; Islam, M.A.; Mondal, H.T.; Ahmmed, R. A Review on Active Packaging for Quality and Safety of Foods: Current Trends, Applications, Prospects and Challenges. *Food Packag. Shelf Life* **2022**, *33*, 100913. [CrossRef]
2. Soltani Firouz, M.; Mohi-Alden, K.; Omid, M. A Critical Review on Intelligent and Active Packaging in the Food Industry: Research and Development. *Food Res. Int.* **2021**, *141*, 110113. [CrossRef] [PubMed]
3. Yildirim, S.; Röcker, B.; Pettersen, M.K.; Nilsen-Nygaard, J.; Ayhan, Z.; Rutkaite, R.; Radusin, T.; Suminska, P.; Marcos, B.; Coma, V. Active Packaging Applications for Food. *Compr. Rev. Food Sci. Food Saf.* **2018**, *17*, 165–199. [CrossRef] [PubMed]
4. Carocho, M.; Morales, P.; Ferreira, I.C.F.R. Natural Food Additives: Quo Vadis? *Trends Food Sci. Technol.* **2015**, *45*, 284–295. [CrossRef]
5. Saltmarsh, M. Recent Trends in the Use of Food Additives in the United Kingdom. *J. Sci. Food Agric.* **2015**, *95*, 649–652. [CrossRef]

6. Almeida-Souza, F.; Magalhães, I.F.B.; Guedes, A.C.; Santana, V.M.; Teles, A.M.; Mouchrek, A.N.; Calabrese, K.S.; Abreu-Silva, A.L. Safety Assessment of Essential Oil as a Food Ingredient. In *Essential Oils: Applications and Trends in Food Science and Technology*; Santana de Oliveira, M., Ed.; Springer International Publishing: Cham, Switzerland, 2022; pp. 123–171. ISBN 978-3-030-99476-1.
7. Carpena, M.; Nuñez-Estevez, B.; Soria-Lopez, A.; Garcia-Oliveira, P.; Prieto, M.A. Essential Oils and Their Application on Active Packaging Systems: A Review. *Resources* **2021**, *10*, 7. [CrossRef]
8. Essential Oils in Food Preservation, Flavor and Safety—1st Edition. Available online: https://www.elsevier.com/books/essential-oils-in-food-preservation-flavor-and-safety/preedy/978-0-12-416641-7 (accessed on 11 January 2023).
9. Coşkun, B.K.; Çalikoğlu, E.; Emiroğlu, Z.K.; Candoğan, K. Antioxidant Active Packaging with Soy Edible Films and Oregano or Thyme Essential Oils for Oxidative Stability of Ground Beef Patties. *J. Food Qual.* **2014**, *37*, 203–212. [CrossRef]
10. Hosseini, S.F.; Zandi, M.; Rezaei, M.; Farahmandghavi, F. Two-Step Method for Encapsulation of Oregano Essential Oil in Chitosan Nanoparticles: Preparation, Characterization and in Vitro Release Study. *Carbohydr. Polym.* **2013**, *95*, 50–56. [CrossRef]
11. Wang, L.; Liu, T.; Liu, L.; Liu, Y.; Wu, X. Impacts of Chitosan Nanoemulsions with Thymol or Thyme Essential Oil on Volatile Compounds and Microbial Diversity of Refrigerated Pork Meat. *Meat Sci.* **2022**, *185*, 108706. [CrossRef]
12. Giannakas, A.E.; Salmas, C.E.; Leontiou, A.; Baikousi, M.; Moschovas, D.; Asimakopoulos, G.; Zafeiropoulos, N.E.; Avgeropoulos, A. Synthesis of a Novel Chitosan/Basil Oil Blend and Development of Novel Low Density Poly Ethylene/Chitosan/Basil Oil Active Packaging Films Following a Melt-Extrusion Process for Enhancing Chicken Breast Fillets Shelf-Life. *Molecules* **2021**, *26*, 1585. [CrossRef]
13. Salmas, C.E.; Giannakas, A.E.; Baikousi, M.; Leontiou, A.; Siasou, Z.; Karakassides, M.A. Development of Poly(L-Lactic Acid)/Chitosan/Basil Oil Active Packaging Films via a Melt-Extrusion Process Using Novel Chitosan/Basil Oil Blends. *Processes* **2021**, *9*, 88. [CrossRef]
14. Chaiwarit, T.; Ruksiriwanich, W.; Jantanasakulwong, K.; Jantrawut, P. Use of Orange Oil Loaded Pectin Films as Antibacterial Material for Food Packaging. *Polymers* **2018**, *10*, 1144. [CrossRef] [PubMed]
15. Yang, H.-J.; Song, K.B. Application of Lemongrass Oil-Containing Polylactic Acid Films to the Packaging of Pork Sausages. *Korean J. Food Sci. Anim. Resour.* **2016**, *36*, 421–426. [CrossRef] [PubMed]
16. Sivaram, S.; Somanathan, H.; Kumaresan, S.M.; Muthuraman, M.S. The Beneficial Role of Plant Based Thymol in Food Packaging Application: A Comprehensive Review. *Appl. Food Res.* **2022**, *2*, 100214. [CrossRef]
17. Ochoa-Velasco, C.E.; Pérez-Pérez, J.C.; Varillas-Torres, J.M.; Navarro-Cruz, A.R.; Hernández-Carranza, P.; Munguía-Pérez, R.; Cid-Pérez, T.S.; Avila-Sosa, R. Starch Edible Films/Coatings Added with Carvacrol and Thymol: In Vitro and In Vivo Evaluation against Colletotrichum Gloeosporioides. *Foods* **2021**, *10*, 175. [CrossRef] [PubMed]
18. Srisa, A.; Harnkarnsujarit, N. Antifungal Films from Trans-Cinnamaldehyde Incorporated Poly(Lactic Acid) and Poly(Butylene Adipate-Co-Terephthalate) for Bread Packaging. *Food Chem.* **2020**, *333*, 127537. [CrossRef] [PubMed]
19. Arrieta, M.P.; López, J.; Ferrándiz, S.; Peltzer, M.A. Characterization of PLA-Limonene Blends for Food Packaging Applications. *Polym. Test.* **2013**, *32*, 760–768. [CrossRef]
20. Giannakas, E.A.; Leontion, A. Montmorillonite Composite Materials and Food Packaging. In *Packaging Materials*; Wiley: Hoboken, NJ, USA, 2018; pp. 1–71.
21. Giannakas, A. Na-Montmorillonite Vs. Organically Modified Montmorillonite as Essential Oil Nanocarriers for Melt-Extruded Low-Density Poly-Ethylene Nanocomposite Active Packaging Films with a Controllable and Long-Life Antioxidant Activity. *Nanomaterials* **2020**, *10*, 1027. [CrossRef]
22. Giannakas, A.E.; Salmas, C.E.; Karydis-Messinis, A.; Moschovas, D.; Kollia, E.; Tsigkou, V.; Proestos, C.; Avgeropoulos, A.; Zafeiropoulos, N.E. Nanoclay and Polystyrene Type Efficiency on the Development of Polystyrene/Montmorillonite/Oregano Oil Antioxidant Active Packaging Nanocomposite Films. *Appl. Sci.* **2021**, *11*, 9364. [CrossRef]
23. Giannakas, A.E.; Salmas, C.E.; Leontiou, A.; Moschovas, D.; Baikousi, M.; Kollia, E.; Tsigkou, V.; Karakassides, A.; Avgeropoulos, A.; Proestos, C. Performance of Thyme Oil@Na-Montmorillonite and Thyme Oil@Organo-Modified Montmorillonite Nanostructures on the Development of Melt-Extruded Poly-L-Lactic Acid Antioxidant Active Packaging Films. *Molecules* **2022**, *27*, 1231. [CrossRef]
24. Saucedo-Zuñiga, J.N.; Sánchez-Valdes, S.; Ramírez-Vargas, E.; Guillen, L.; Ramos-deValle, L.F.; Graciano-Verdugo, A.; Uribe-Calderón, J.A.; Valera-Zaragoza, M.; Lozano-Ramírez, T.; Rodríguez-González, J.A.; et al. Controlled Release of Essential Oils Using Laminar Nanoclay and Porous Halloysite/Essential Oil Composites in a Multilayer Film Reservoir. *Microporous Mesoporous Mater.* **2021**, *316*, 110882. [CrossRef]
25. de Oliveira, L.H.; Trigueiro, P.; Souza, J.S.N.; de Carvalho, M.S.; Osajima, J.A.; da Silva-Filho, E.C.; Fonseca, M.G. Montmorillonite with Essential Oils as Antimicrobial Agents, Packaging, Repellents, and Insecticides: An Overview. *Colloids Surf. B Biointerfaces* **2022**, *209*, 112186. [CrossRef] [PubMed]
26. Blinka, T.A.; Edwards, F.B.; Miranda, N.R.; Speer, D.V.; Thomas, J.A. Zeolite in Packaging Film. U.S. Patent 5834079A, 10 November 1998.
27. de Araújo, L.O.; Anaya, K.; Pergher, S.B.C. Synthesis of Antimicrobial Films Based on Low-Density Polyethylene (LDPE) and Zeolite A Containing Silver. *Coatings* **2019**, *9*, 786. [CrossRef]
28. Dogan, H.; Koral, M.; İnan, T.Y. Ag/Zn Zeolite Containing Antibacterial Coating for Food-Packaging Substrates. *J. Plast. Film. Sheeting* **2009**, *25*, 207–220. [CrossRef]

29. Gargiulo, N.; Attianese, I.; Buonocore, G.G.; Caputo, D.; Lavorgna, M.; Mensitieri, G.; Lavorgna, M. α-Tocopherol Release from Active Polymer Films Loaded with Functionalized SBA-15 Mesoporous Silica. *Microporous Mesoporous Mater.* **2013**, *167*, 10–15. [CrossRef]
30. Khezri, K.; Roghani-Mamaqani, H. Effect of MCM-41 Nanoparticles on ARGET ATRP of Styrene: Investigating Thermal Properties. Available online: https://journals.sagepub.com/doi/abs/10.1177/0021998314535961?journalCode=jcma (accessed on 20 November 2019).
31. Chaemsanit, S.; Matan, N.; Matan, N. Activated Carbon for Food Packaging Application: Review. *Walailak J. Sci. Technol. (WJST)* **2018**, *15*, 255–271. [CrossRef]
32. Sobhan, A.; Muthukumarappan, K.; Wei, L.; Van Den Top, T.; Zhou, R. Development of an Activated Carbon-Based Nanocomposite Film with Antibacterial Property for Smart Food Packaging. *Mater. Today Commun.* **2020**, *23*, 101124. [CrossRef]
33. Mitura, K.; Kornacka, J.; Kopczyńska, E.; Kalisz, J.; Czerwińska, E.; Affeltowicz, M.; Kaczorowski, W.; Kolesińska, B.; Frączyk, J.; Bakalova, T.; et al. Active Carbon-Based Nanomaterials in Food Packaging. *Coatings* **2021**, *11*, 161. [CrossRef]
34. Scientific Opinion on the Safety Evaluation of the Active Substances, Activated Carbon, Water, Iron Powder, Kaolin Calcined, Sulphur and Sodium Chloride for Use as Active Component in Food Contact Materials | EFSA. Available online: https://www.efsa.europa.eu/en/efsajournal/pub/2643 (accessed on 14 June 2023).
35. Idrees, M.; Rangari, V.; Jeelani, S. Sustainable Packaging Waste-Derived Activated Carbon for Carbon Dioxide Capture. *J. CO2 Util.* **2018**, *26*, 380–387. [CrossRef]
36. Pomponio, L.; Ruiz-Carrascal, J. Oxidative Deterioration of Pork during Superchilling Storage. *J. Sci. Food Agric.* **2017**, *97*, 5211–5215. [CrossRef]
37. Sakdapipanich, J.; Rodgerd, P.; Sakdapipanich, N. A Low-Density Polyethylene (LDPE)/Macca Carbon Advanced Composite Film with Functional Properties for Packaging Materials. *Polymers* **2022**, *14*, 1794. [CrossRef] [PubMed]
38. Giannakas, A.E.; Salmas, C.E.; Moschovas, D.; Zaharioudakis, K.; Georgopoulos, S.; Asimakopoulos, G.; Aktypis, A.; Proestos, C.; Karakassides, A.; Avgeropoulos, A.; et al. The Increase of Soft Cheese Shelf-Life Packaged with Edible Films Based on Novel Hybrid Nanostructures. *Gels* **2022**, *8*, 539. [CrossRef] [PubMed]
39. Salmas, C.E.; Giannakas, A.E.; Moschovas, D.; Kollia, E.; Georgopoulos, S.; Gioti, C.; Leontiou, A.; Avgeropoulos, A.; Kopsacheili, A.; Avdulai, L.; et al. Kiwi Fruits Preservation Using Novel Edible Active Coatings Based on Rich in Thymol Halloysite Nanostructures and Chitosan/Polyvinyl Alcohol Gels. *Gels* **2022**, *8*, 823. [CrossRef]
40. Giannakas, A.E.; Salmas, C.E.; Moschovas, D.; Karabagias, V.K.; Karabagias, I.K.; Baikousi, M.; Georgopoulos, S.; Leontiou, A.; Katerinopoulou, K.; Zafeiropoulos, N.E.; et al. Development, Characterization, and Evaluation as Food Active Packaging of Low-Density-Polyethylene-Based Films Incorporated with Rich in Thymol Halloysite Nanohybrid for Fresh "Scaloppini" Type Pork Meat Fillets Preservation. *Polymers* **2023**, *15*, 282. [CrossRef] [PubMed]
41. Salmas, C.E.; Giannakas, A.E.; Karabagias, V.K.; Moschovas, D.; Karabagias, I.K.; Gioti, C.; Georgopoulos, S.; Leontiou, A.; Kehayias, G.; Avgeropoulos, A.; et al. Development and Evaluation of a Novel-Thymol@Natural-Zeolite/Low-Density-Polyethylene Active Packaging Film: Applications for Pork Fillets Preservation. *Antioxidants* **2023**, *12*, 523. [CrossRef] [PubMed]
42. Asimakopoulos, G.; Baikousi, M.; Kostas, V.; Papantoniou, M.; Bourlinos, A.B.; Zbořil, R.; Karakassides, M.A.; Salmas, C.E. Nanoporous Activated Carbon Derived via Pyrolysis Process of Spent Coffee: Structural Characterization. Investigation of Its Use for Hexavalent Chromium Removal. *Appl. Sci.* **2020**, *10*, 8812. [CrossRef]
43. Giannakas, A.; Giannakas, A.; Ladavos, A. Preparation and Characterization of Polystyrene/Organolaponite Nanocomposites. *Polym. Plast. Technol. Eng.* **2012**, *51*, 1411–1415. [CrossRef]
44. Giannakas, A.E.; Salmas, C.E.; Moschovas, D.; Baikousi, M.; Kollia, E.; Tsigkou, V.; Karakassides, A.; Leontiou, A.; Kehayias, G.; Avgeropoulos, A.; et al. Nanocomposite Film Development Based on Chitosan/Polyvinyl Alcohol Using ZnO@Montmorillonite and ZnO@Halloysite Hybrid Nanostructures for Active Food Packaging Applications. *Nanomaterials* **2022**, *12*, 1843. [CrossRef]
45. Giannakas, A.; Stathopoulou, P.; Tsiamis, G.; Salmas, C. The Effect of Different Preparation Methods on the Development of Chitosan/Thyme Oil/Montmorillonite Nanocomposite Active Packaging Films. *J. Food Process. Preserv.* **2019**, *44*, e14327. [CrossRef]
46. Salmas, C.E.; Giannakas, A.E.; Baikousi, M.; Kollia, E.; Tsigkou, V.; Proestos, C. Effect of Copper and Titanium-Exchanged Montmorillonite Nanostructures on the Packaging Performance of Chitosan/Poly-Vinyl-Alcohol-Based Active Packaging Nanocomposite Films. *Foods* **2021**, *10*, 3038. [CrossRef]
47. Brand-Williams, W.; Cuvelier, M.E.; Berset, C. Use of a Free Radical Method to Evaluate Antioxidant Activity. *LWT Food Sci. Technol.* **1995**, *28*, 25–30. [CrossRef]
48. Karabagias, I.K.; Dimitriou, E.; Kontakos, S.; Kontominas, M.G. Phenolic Profile, Colour Intensity, and Radical Scavenging Activity of Greek Unifloral Honeys. *Eur. Food Res. Technol.* **2016**, *242*, 1201–1210. [CrossRef]
49. Karabagias, I.K.; Maia, M.; Karabagias, V.K.; Gatzias, I.; Badeka, A.V. Characterization of Eucalyptus, Chestnut and Heather Honeys from Portugal Using Multi-Parameter Analysis and Chemo-Calculus. *Foods* **2018**, *7*, 194. [CrossRef]
50. Tarladgis, B.G.; Watts, B.M.; Younathan, M.T.; Dugan, L. A Distillation Method for the Quantitative Determination of Malonaldehyde in Rancid Foods. *J. Am. Oil Chem. Soc.* **1960**, *37*, 44–48. [CrossRef]
51. Karabagias, I.; Badeka, A.; Kontominas, M.G. Shelf Life Extension of Lamb Meat Using Thyme or Oregano Essential Oils and Modified Atmosphere Packaging. *Meat Sci.* **2011**, *88*, 109–116. [CrossRef] [PubMed]

52. Clark, E.M.; Mahoney, A.W.; Carpenter, C.E. Heme and Total Iron in Ready-to-Eat Chicken. *J. Agric. Food Chem.* **1997**, *45*, 124–126. [CrossRef]
53. Assanti, E.; Karabagias, V.K.; Karabagias, I.K.; Badeka, A.; Kontominas, M.G. Shelf Life Evaluation of Fresh Chicken Burgers Based on the Combination of Chitosan Dip and Vacuum Packaging under Refrigerated Storage. *J. Food Sci. Technol.* **2021**, *58*, 870–883. [CrossRef]
54. Hematizad, I.; Khanjari, A.; Basti, A.A.; Karabagias, I.K.; Noori, N.; Ghadami, F.; Gholami, F.; Teimourifard, R. In Vitro Antibacterial Activity of Gelatin-Nanochitosan Films Incorporated with Zataria Multiflora Boiss Essential Oil and Its Influence on Microbial, Chemical, and Sensorial Properties of Chicken Breast Meat during Refrigerated Storage. *Food Packag. Shelf Life* **2021**, *30*, 100751. [CrossRef]
55. Darvish, M.; Ajji, A. Synergistic Antimicrobial Activities of Limonene with Mineral Carriers in LDPE Films for Active Packaging Application. *Sci. J. Chem.* **2022**, *10*, 32. [CrossRef]
56. Yakout, S.M.; Sharaf El-Deen, G. Characterization of Activated Carbon Prepared by Phosphoric Acid Activation of Olive Stones. *Arab. J. Chem.* **2016**, *9*, S1155–S1162. [CrossRef]
57. Pathak, U.; Jhunjhunwala, A.; Roy, A.; Das, P.; Kumar, T.; Mandal, T. Efficacy of Spent Tea Waste as Chemically Impregnated Adsorbent Involving Ortho-Phosphoric and Sulphuric Acid for Abatement of Aqueous Phenol—Isotherm, Kinetics and Artificial Neural Network Modelling. *Environ. Sci. Pollut. Res.* **2020**, *27*, 20629–20647. [CrossRef] [PubMed]
58. Bansal, R.C.; Goyal, M. *Activated Carbon Adsorption*; CRC Press: Boca Raton, FL, USA, 2005; ISBN 978-0-429-11418-2.
59. Hadoun, H.; Sadaoui, Z.; Souami, N.; Sahel, D.; Toumert, I. Characterization of Mesoporous Carbon Prepared from Date Stems by H3PO4 Chemical Activation. *Appl. Surf. Sci.* **2013**, *280*, 1–7. [CrossRef]
60. Giannakas, A.; Salmas, C.; Leontiou, A.; Tsimogiannis, D.; Oreopoulou, A.; Braouhli, J. Novel LDPE/Chitosan Rosemary and Melissa Extract Nanostructured Active Packaging Films. *Nanomaterials* **2019**, *9*, 1105. [CrossRef] [PubMed]
61. Moosavi, S.; Lai, C.W.; Gan, S.; Zamiri, G.; Akbarzadeh Pivehzhani, O.; Johan, M.R. Application of Efficient Magnetic Particles and Activated Carbon for Dye Removal from Wastewater. *ACS Omega* **2020**, *5*, 20684–20697. [CrossRef]
62. Zubair, M.; Shahzad, S.; Hussain, A.; Pradhan, R.A.; Arshad, M.; Ullah, A. Current Trends in the Utilization of Essential Oils for Polysaccharide- and Protein-Derived Food Packaging Materials. *Polymers* **2022**, *14*, 1146. [CrossRef]
63. Aitboulahsen, M.; El Galiou, O.; Laglaoui, A.; Bakkali, M.; Hassani Zerrouk, M. Effect of Plasticizer Type and Essential Oils on Mechanical, Physicochemical, and Antimicrobial Characteristics of Gelatin, Starch, and Pectin-Based Films. *J. Food Process. Preserv.* **2020**, *44*, e14480. [CrossRef]
64. Stewart, G.S.A.B. Micro-Organisms in Food—2. Sampling for Microbiological Analysis: Principles and Specific Applications: ICMSF, Blackwell Scientific Publications, Oxford, 1986. 310 Pp. Price: £19·50 (Cloth). *Meat Sci.* **1987**, *19*, 315. [CrossRef]

foods

MDPI

Article

Application of Ultrasound Pre-Treatment for Enhancing Extraction of Bioactive Compounds from Rice Straw

Pedro A. V. Freitas *, Chelo González-Martínez and Amparo Chiralt

Institute of Food Engineering for Development, Universitat Politècnica de València, 46022 Valencia, Spain; cgonza@tal.upv.es (C.G.-M.); dchiralt@tal.upv.es (A.C.)

* Correspondence: pedroafreitas3@gmail.com; Tel.: +34-600-761-730

Received: 22 October 2020; Accepted: 10 November 2020; Published: 12 November 2020

check for updates

Abstract: The extraction of water-soluble bioactive compounds using different green methods is an eco-friendly alternative for valorizing agricultural wastes such as rice straw (RS). In this study, aqueous extracts of RS (particles < 500 μm) were obtained using ultrasound (US), reflux heating (HT), stirring (ST) and a combination of US and ST (USST) or HT (USHT). The extraction kinetics was well fitted to a pseudo-second order model. As regards phenolic compound yield, the US method (342 mg gallic acid (GAE). 100 g^{-1} RS) was more effective than the ST treatment (256 mg GAE·100 g^{-1} RS), reaching an asymptotic value after 30 min of process. When combined with HT (USHT), the US pre-treatment led to the highest extraction of phenolic compounds from RS (486 mg GAE·100 g^{-1} RS) while the extract exhibited the greatest antioxidant activity. Furthermore, the USHT extract reduced the initial counts of *Listeria innocua* by 1.7 logarithmic cycles. Therefore, the thermal aqueous extraction of RS applying the 30 min US pre-treatment, represents a green and efficient approach to obtain bioactive extracts for food applications.

Keywords: bioactive compounds; water extraction; antioxidant activity; pseudo-second order law; antimicrobial activity; heating; combined methods

1. Introduction

A rising global population and limited natural resources have encouraged the food industries to seek new environmentally-friendly alternatives for agro-industrial waste management [1,2]. One of the primary agro-industrial wastes, rice straw (RS) is a by-product obtained from the rice grain (*Oryza sativa* L.) harvesting, which is considered one of the most important global food crops [3]. Indeed, the Food and Agriculture Organization of the United Nations reported that the annual world rice production was approximately 782 million tons, with 90% production from the Asian continent [4]. Considering that one kilogram of rice grain provides 1.5 kg RS and the largest portion of RS is commonly burnt in the field, there is a continuous increase of air pollution associated with the release of dioxins, affecting human health and reducing the soil feasibility [3,5]. In this way, efforts have recently been made to find new alternatives for the reuse of RS, such as bioethanol production [6], silica extraction [7], paper production [8], cellulose isolation [9] or bioactive compound extraction [10,11].

RS contains about 35% cellulose, 20% lignin, 18% hemicellulose and 15% ashes (dry weight basis) [12]. Likewise, RS has been considered an excellent source of bioactive compounds, including phenolic compounds with antioxidant and antimicrobial potential. Menzel et al. [10] identified bioactive compounds from RS aqueous and alcoholic extracts using ultra-high-performance liquid chromatography coupled to high-resolution mass spectrometry. They found several types of compounds with excellent antioxidant properties, such as ferulic acid, protocatechuic acid, *p*-coumaric acid, caffeic acid, vanillic acid, tricin and vanillin. Some of these phenolic compounds, such as *p*-coumaric, and ferulic acid, have not been found in significant quantities in other vegetable and fruit matrices [13].

In recent years, new technologies have been applied in extraction methodologies in order to overcome problems related to traditional techniques, such as limitations in the extract yield and the fact that they consume a great deal of energy and time [14]. Ultrasound-assisted extraction (UAE) is an eco-friendly and efficient process that applies acoustic cavitation using an ultrasonic probe device in a closed flow-through mode in a determined matrix to extract target compounds. The ultrasound waves, ranging from 20 to 1000 kHz, have been applied to improve the extraction by the solvent acoustic cavitation. The shear forces promoted from the rupture of the bubbles disrupt the cell matrices, extracting higher amounts of compounds by means of different physical phenomena [14–16]. Many studies have shown the efficiency of UAE coupled with the greater antioxidant and antimicrobial activity of extracts from different vegetal matrices [17–21]. To the best of our knowledge, UAE has not been previously applied to extract phenolic compounds from the RS matrix to obtain bioactive extracts.

The aim of this study was to improve the water extraction effectiveness of bioactive compounds from rice straw by applying probe-based UAE. The performance of the UAE treatments and their combination with conventional heating and stirring extraction methods was evaluated in terms of the extraction kinetics of phenolic compounds, as well as the antioxidant and antimicrobial activity of the obtained extracts. Likewise, the mass transfer phenomenon was correlated with the possible physical changes that occurred in the ground RS plant matrix.

2. Materials and Methods

2.1. Chemicals

Gallic acid, Folin-Ciocalteau reagent (2N), methanol, and 2,2-Diphenyl-1-picrylhydrazyl (DPPH) were purchased from Sigma-Aldrich (St. Louis, MO, USA). Sodium carbonate was obtained from PanReac Quimica S.L.U. (Castellar del Vallés, Barcelona, Spain). Tryptone soy broth, tryptone soy agar, phosphate-buffered saline, and peptone water were purchased from Scharlab (Barcelona, Spain). Strains of *L. innocua* (CECT 910) and *Escherichia coli* (CECT 101) were purchased from the Spanish Type Collection (CECT, University of Valencia, Valencia, Spain).

2.2. Plant Material Preparation

RS (*Oryza sativa* L.), *J. Sendra var.*, was obtained from L'Albufera rice fields, Valencia, Spain, and supply by the "Banco de Paja" (Valencia, Spain), from the harvest 2019. Samples were processed by drying the RS at 50 ± 2 °C under vacuum (0.5 mm bar) for 16 h. After that, the RS was milled using a milling machine (IKA, model M20, Germany) operating in 3 cycles of 90 s each. The ground RS was sieved to obtain particles of under 0.5 mm (representing 98% wt. of the milled RS) (Figure 1) and stored in a desiccator at 20 ± 2 °C until further use.

2.3. Aqueous Extraction of RS Powder

The aqueous extracts of phenolic compounds were obtained using an RS: water ratio of 1:20 (m/v). As shown in Figure 1, different extraction approaches were used based on ultrasound (US), stirring (ST), heating under reflux condition (HT), and a combination. The US method was carried out using an ultrasonic homogenizer (Vibra CellTM VCX750, Sonics & Material, Inc., Newtown, CT, USA) equipped with a high-intensity probe operating in a continuous mode. RS and water were mixed and sonicated for 60 min at 25 °C (maintaining the sample in an ice bath to prevent heating), using 750 W power, 20 kHz frequency and 40% sonication amplitude. The ST extract was obtained by stirring the dispersion for 60 min at 350 rpm and 25 °C. The HT method consisted of heating the RS suspension at 100 °C for 60 min using a typical reflux device. Two combined extraction methods were carried out by successively applying 30 min US, as a pre-treatment, and ST for 60 min (USST) or HT for 60 min (USHT). The US pretreatment time was established based on the results obtained for the US treatment.

In every process, the supernatant obtained after each treatment was filtered using a qualitative filter (Filter-lab S.A., Cataluña, Spain) and stored in a dark bottle at 4 ± 2 °C until used.

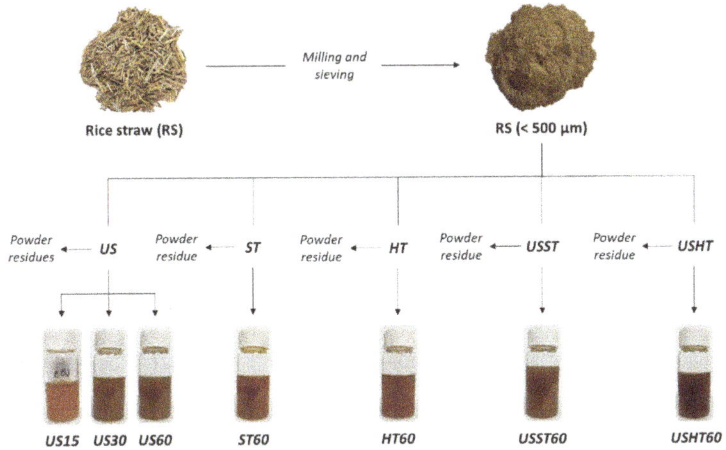

Figure 1. Schematic illustration and appearance of RS extracts obtained from different extraction processes: ultrasound (US), stirring (ST), heating under reflux condition (HT), and a combined treatment (USST and USHT). The number in the sample code indicates the processing time without considering the pre-treatment time.

The extraction kinetics was evaluated in terms of the total phenolic content by sampling and analyzing the extract after different process times: 2, 5, 10, 15, 20, 25, 30, 40, 50, and 60 min. To perform the other analyses, extracts obtained from the US method were characterized after 15, 30, and 60 min (US15, US30, and US60), and the ST and HT extracts were analyzed after 60 min extraction time (ST60 and HT60). Likewise, combined USST60 and USHT60 treatments were also characterized. Each treatment was carried out in triplicate and analyzed in terms of total solid yield (TSY), total phenolic content (TPC), antioxidant, and antibacterial activities. TSY was determined by drying the RS extracts at 70 ± 2 °C under vacuum for 6 h until constant weight. After extraction, the residues were collected, and their particle size distribution and microstructure were compared with those of the initial RS powder.

2.4. Analysis of Total Phenolic Content

TPC for the different extraction methods over time were determined according to the modified Folin-Ciocalteu method by Menzel et al. [10]. The TPC was determined from the absorbance values, using the linear equation fitted to the stand curve using gallic acid ($Abs_{725\ nm}$ = 0.099 (gallic acid) + 0.043; $R^2 = 0.9991$). The results were expressed as mg gallic acid equivalents (mg GAE) per 100 g of dried RS. All measurements were taken in triplicate.

2.5. Modelling Extraction Kinetics

A pseudo-second order rate law was applied to model the extraction kinetics, according to Ho et al. [22]. The dissolution rate of RS compounds, contained within the RS matrix, into water solution can be written as:

$$\frac{d(C_t)}{dt} = k(C_e - C_t)^2 \quad (1)$$

where k is the second-order extraction rate constant ($L{\cdot}mg^{-1}{\cdot}min^{-1}$), C_e ($mg{\cdot}L^{-1}$) and C_t ($mg{\cdot}L^{-1}$) are the total phenol concentrations in the solvent phase at equilibrium and at any time t (min), respectively.

Equation (1) was integrated under the boundary conditions $t = 0$ to t and $C_t = C_0$ to C_t, thus obtaining Equation (2):

$$C_t = \frac{ktC_e(C_e - C_0) + C_0}{kt(C_e - C_0) + 1} \quad (2)$$

when $C_0 = 0$, Equation (2) can be linearized as Equation (3) and the initial extraction rate, h (mg·L^{-1}·min^{-1}), can be obtained when t approaches 0 (Equation (4)):

$$\frac{t}{C_t} = \frac{1}{k(C_e)^2} + \frac{t}{C_e} \quad (3)$$

$$h = k(C_e)^2 \quad (4)$$

C_e, k, and h were determined from the linear fitting of the data in the plot t/C_t versus t, as inferred from Equations (3) and (4). In the case of treatments applied after 30 min US pre-treatment, when $C_o \neq 0$, a non-linear fitting was applied, using the SOLVER tool from EXCEL to obtain k and C_e values.

The quality of fit for each extraction method was statistically evaluated by the determination coefficient (R^2) (Equation (4)), and the average absolute relative deviation ($AARD\%$) (Equation (5)), which measure the model's predictive effectiveness.

$$R^2 = \frac{\sum_{i=1}^{N}\left(y_i^{cal} - y_i^{exp}\right)^2}{\sum_{i=1}^{N}\left(y_i^{cal} - y_m\right)^2} \quad (5)$$

$$AARD\% = 100 \times \frac{1}{N}\sum_{i=1}^{N}\left|\frac{y_i^{cal} - y_i^{exp}}{y_i^{exp}}\right| \quad (6)$$

where N is the number of the data set; y_i^{cal} is the predictive value of the model; y_i^{exp} is the experimental value; and y_m is the mean value.

2.6. Antioxidant Activity by DPPH Radical Scavenging Method

The antiradical activities of the different RS extracts were determined using the free radical 2,2-Diphenyl-1-pikryl-hydrazyl (DPPH) method, with some modifications [23]. For each RS extract, different concentrations were mixed with the 6.22×10^{-2} mM DPPH methanolic solution ($Abs_{515nm} = 0.7 \pm 0.2$) to a final volume of 4 mL. The initial DPPH concentration in the reaction medium was determined from a calibration curve ($Abs_{515nm} = 11.324\ [DPPH] - 0.038$; $R^2 = 0.9992$). The antiradical activity was evaluated by EC_{50}, which is defined as the amount of antioxidant necessary to reduce the initial DPPH concentration by 50% when reaction stability has been reached. The reaction stability times between the extract and DPPH solutions were 180 min for the US (15, 30, and 60 min), ST60, and their combined treatment (USST60), and 45 min for the HT60 and USHT60 treatments. To obtain the EC_{50} values, the plots % $[DPPH]_{remaining}$ vs. mg solid extract/mg DPPH were obtained, where:

$$\%[DPPH]_{remaining} = \left(\frac{[DPPH]_t}{[DPHH]_{t=0}}\right) \times 100 \quad (7)$$

$[DPPH]_t$ is the DDPH concentration value when the reaction was stable, and $[DPPH]_{t=0}$ is the DPPH initial concentration.

2.7. Antibacterial Bioactivity

For both *L. innocua* and *E. coli* strains (stored at −20 °C), a stock solution was obtained by transferring amounts of bacteria strain into 10 mL tryptic soy broth (TSB) using an inoculation loop twice. After incubation at 37 °C for 24 h, the working solution was prepared by pipetting 10 µL of the stock solution in 10 mL TSB and incubating at 37 °C for 24 h. The working solution was serially diluted in TSB to obtain 10^5 CFU·mL^{-1} (concentration was validated by incubation of 100 µL in tryptic soy agar at 37 °C for 24 h and counting).

The antimicrobial activity of the RS extract against *L. innocua* and *E. coli* strains were analyzed using previously described methods [24,25]. The analysis was carried out in standard 96-well microtiter plates (well volume of 200 µL). For each microorganism, 100 µL of 10^5 CFU·mL^{-1} bacteria solution was pipetted in each well together with different RS extract solution volumes. Then, the final volume in each well was completed to 200 µL with TSB and incubated at 37 °C for 24 h. The tested concentrations of freeze-dried RS extract dissolved in TSB were 96, 104, 112, 120, 128, 136, 144, 152, and 160 mg/mL^{-1}. Typically, the analysis controls were wells containing only bacterial suspension, extract solution and TSB solution. The final counts of each bacteria after incubation were determined in tryptic soy agar media for each well. Thus, 100 µL of each well was transferred in tryptic soy agar plates and incubated at 37 °C for 24 h to obtain the counts. The analysis was performed in duplicate.

2.8. Particle Size Distribution in RS Powder

The particle size distributions of RS powders before and after the extraction processes were determined in triplicate using a laser-diffraction particle size analyzer (Malvern Instruments, Malvern, UK) coupled with a Scirocco 2000 dry dispersion unit. A refractive index of 1.520 and an absorption of 0.1 was considered. Samples were fed into the system at a feed rate of 60% and a pressure of 2.5 bar until reaching an obscuration rate of 1.26%. The parameter D_{43} and the mean particle size distribution curves were obtained.

2.9. High-Resolution Field Emission Scanning Electron Microscopy (FESEM)

The morphologies of the RS particles after the extraction processes were characterized using a High-Resolution Field Emission Scanning Electron Microscope (GeminiSEM 500, Zeiis, Oxford Instruments, Oxford, UK). Before the microscopy observations, the samples were coated with platinum using an EM MED020 sputter coater (Leica Biosystems, Barcelona, Spain). The images were taken under vacuum and 2.0 kV acceleration voltage.

2.10. Statistical Analysis

The data were submitted to analysis of variance (ANOVA), and Tukey's studentized range HSD (honestly significant difference) test using a Minitab statistical program (version 17). This was performed to determine whether there were significant differences among the extraction methods, using the least significant difference (α) of 5%.

3. Results and Discussion

3.1. Extraction Kinetics of Phenolic Compounds

The performance of the different aqueous extraction methods was compared in terms of the total phenolic content (TPC) reached after different process times (Figure 2a, points). As previously observed by other authors [14,17,26,27] for extraction processes from plant matrices, the compound extraction rate was higher in the first stage of the process (10–15 min), progressively decreasing till the concentration reached an asymptotic value, considered as the equilibrium value (C_e). This is coherent with the greater driving force for mass transfer after short process times [28] while also reflecting the faster initial desorption of compounds from the surface of solid particles and the slower diffusion of compounds extracted from their more internal parts. For the combined USST treatment, the extraction rate was very low throughout the time due to the small difference between the initial TPC concentration (obtained during the 30 min US pretreatment) and that reached at equilibrium, which indicates the limited capacity of simple stirring to improve phenol extraction after the US pretreatment. However, in the USHT treatment, with the same initial TPC value, the equilibrium concentration increased with respect to that obtained in the USST process. Thus, different treatments led to distinct asymptotic values of C_e, depending on the temperature and the application of the US step. Higher temperatures and the US pre-treatment promoted the extraction of the phenolic compounds, giving rise to higher TPC values at equilibrium (C_e).

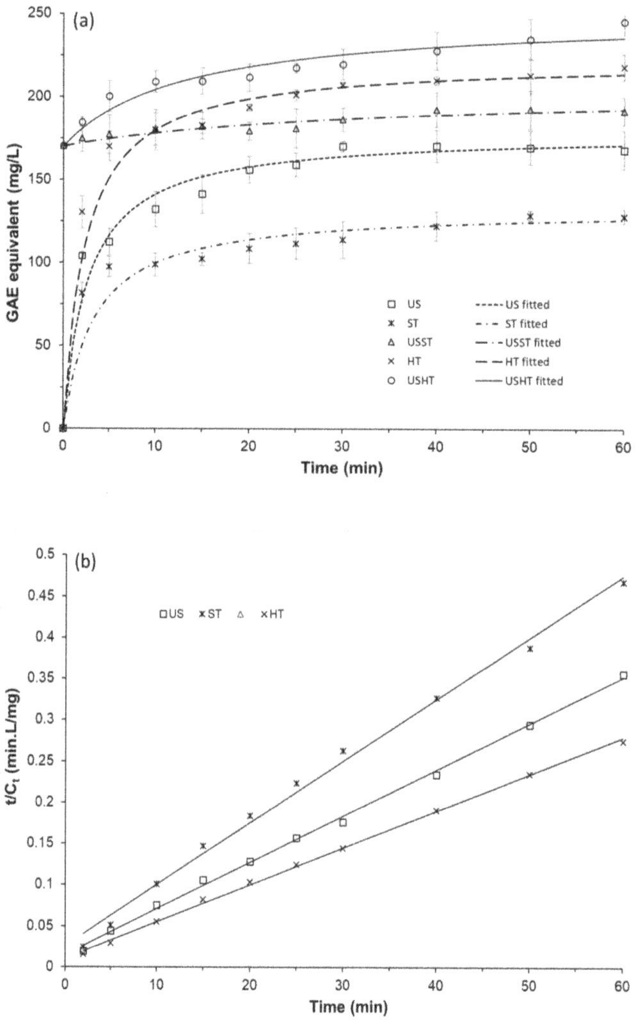

Figure 2. (**a**) Phenolic content of aqueous extracts obtained in the different processes as a function of time (points) and fitted kinetic model (lines). (**b**) Linearization of experimental points from the ST, US, and HT methods according to the pseudo-second order rate law. (US: ultrasound, ST: stirring, HT: heating, USST: stirring with 30 min of US pre-treatment, USHT: heating with 30 min of US pre-treatment).

Therefore, the US method was notably more efficient than the ST treatment (both at 25 °C), the former reaching a higher asymptotic value from practically 30 min extraction time. Thus, 30 min was chosen as the US pretreatment time for combined processes. The combined USST method yielded higher final TPC values than the US, although these values were lower than those reached with a higher extraction temperature (100 °C) in the HT process. The HT extraction behavior suggested that the high temperature significantly favors the release of phenolic compounds from the plant matrix into the extraction solvent, as previously reported by other authors [29,30]. The combination of 30 min US and 60 min HT (USHT process) gave rise to the highest yield in total phenolic content, which indicates the efficiency of the combination of US and thermal treatments as a means of favoring the extraction of phenolic compounds from the RS matrix. Phenolic compounds are linked through acetal, ether or ester covalent bonds with the lignocellulosic fraction of the plant matrix [31,32], and the debonding of phenolic compounds is promoted at high temperatures [33,34]. Thus, the combination of the structural effects provoked by sonication in the solid particles, and the cleavage of phenolic covalent bonds promoted by high temperatures, gave rise to the most efficient method with which to enhance the phenolic extraction yield. Other authors also found an increase in the extraction of phenolic compounds by applying heating or ultrasound to different plant matrices [17,30,33].

3.2. Extraction Kinetics Modeling

A pseudo-second order rate equation was adjusted to model the extraction kinetics. The kinetic parameters, the average absolute relative deviation (AARD%), and the determination coefficients (R^2) of the fitted model, are shown in (Table 1). The goodness of model prediction can be seen in Figure 2a, where the experimental points and fitted curves can be observed. Figure 2b shows the fitting of the linearized model when $C_0 = 0$ (cases without US pre-treatment). Likewise, the values of statistical parameters, R^2 (>0.98) and AARD% (<10%), also revealed the goodness of the fitted model, in agreement with other studies [27]. The pseudo-second order rate model is coherent with a two-step mass transfer process for the phenolic compounds: an initial and intensive washing-out of the solute on the RS particle surface, followed by a slow diffusion stage from the interior of the solid particles to the liquid solvent [14,22,28].

Table 1. Kinetic parameters obtained for the different extraction processes.

Extraction Method [1]	k ($\times 10^{-3}$) ($L \cdot mg^{-1} \cdot min^{-1}$)	C_e ($mg \cdot L^{-1}$)	h ($mg \cdot L^{-1} \cdot min^{-1}$)	R^2	AARD%
US	2.12	178.6	67.60	0.998	4.89
ST	2.24	133.3	39.82	0.994	7.76
HT	1.91	222.2	94.32	0.992	4.08
USST	0.84	205.3	-	0.980	1.14
USHT	0.93	250.3	-	0.985	2.14

[1] ST: stirring, US: ultrasound, HT: heating, USST and UHHT: combination of 30 min US and ST or HT, respectively.

For treatments without US pre-treatment ($C_0 = 0$), the initial extraction rates were calculated (h values) and are shown in Table 1. The HT treatment presented the highest h value, followed by the US and ST treatments. This finding agrees with the phenol debonding action of high temperatures that promoted phenol extraction from the exposed surface of solid particles. Ultrasonic cavitation at a lower temperature also promoted the fast extraction of phenolic compounds due to the US mechanical action that implied a greater solid surface exposure to the extraction process. However, this effect had a lower impact on the initial extraction rate than the high temperature. The ST treatment exhibited the lowest h value, due to the worse ability of low temperatures to debond phenols and the lack of enough mechanical energy to disrupt the particles and increase the extraction surface. The equilibrium concentration values were also positively affected by high temperatures and US application, coherently with their effects on the initial extraction rate (h). Thus, the US method increased the phenolic extraction by around 34% with respect to the ST, but it was 24% less extractive than the HT process.

The US pre-treatment (30 min) promoted the phenol extraction in both USST and USHT processes carried out at 25 and 100 °C, respectively. The phenolic equilibrium concentration was increased by around 54% and 13% with respect to that obtained in the ST and HT processes, respectively. Likewise, the C_e value obtained in the USHT treatment increased by around 40% with respect to that obtained in the US treatment, but only by 12% with respect to the HT. This finding agrees with the fact that high temperatures are more effective than ultrasonic cavitation at extracting phenolic compounds.

Nevertheless, the combination of both effects in the USHT process produced the maximum extraction of phenolic compounds.

The constant rate values (*k*) were barely affected by either the temperature or the US effect but fell greatly in processes with a US pretreatment (USST and USHT); this was due to the substantial reduction in the process driving force, since the initial concentration of the extract (170 mg·L^{-1}) was nearer the equilibrium value.

3.3. Changes in the Plant Tissue Produced by Different Extraction Processes

Figure 3 shows the particle size distribution of the RS powder and the solid extraction residues after the different treatments: US at 15 (US15), 30 (US30) and 60 (US60) min, ST at 60 min (ST60), HT at 60 min (HT60), and the combined methods (with 30 min US pretreatment) at 60 min (USST60 and USHT60). Before the extraction, the RS particles exhibited a monomodal distribution, with an average particle diameter (D_{43}) of 295 μm. As can be observed, the RS distribution showed fractions of the particles with mean diameters larger than the opening size of the used sieve (500 μm). Nevertheless, the RS particles are mostly rod-shaped, with a smaller diameter than length, as shown in Figure 4. Thus, these particles can pass through the mesh even if the length is longer than the opening size. In contrast, the laser diffraction method delivers a particle diameter corresponding to an equivalent sphere with the same diffraction pattern and the length of the particles mainly determines their mean radius of gyration and so, the equivalent sphere diameters.

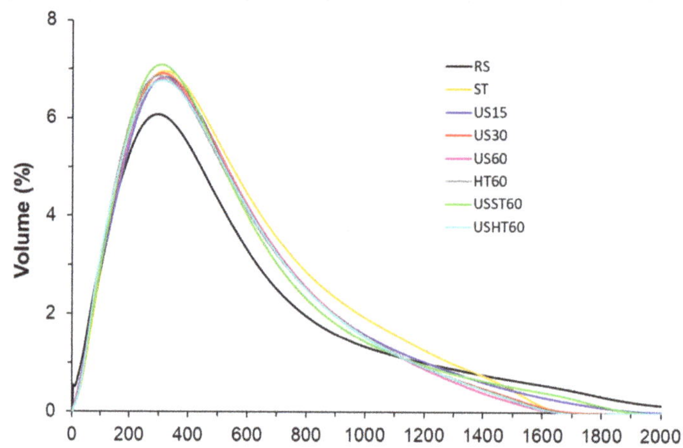

Figure 3. Particle size distribution obtained for the residual powder from the different extraction processes at a determined time (RS: rice straw, US: ultrasound, ST: stirring, HT: heating at reflux, USST, and USHT: combined treatments). The number in the sample code indicates the processing time.

The extraction processes provoked changes in the particle size distribution to a similar extent in every treatment. The distributions maintained the monomodal characteristic, exhibiting D_{43} values of 308 μm, but were slightly narrower, with a smaller presence of the biggest particles (>1400 μm) and a greater one of particles with an intermediate diameter (300–1000 μm). This behavior was more marked in treatments where the US or high temperatures were applied, suggesting the particles become rounder during extraction. Compared with the other processes, however, it was not especially notable for the US treatments, in which more particle fragmentation would be expected due to the implosion of cavitation bubbles on the particle surface and the inter-particle collisions. Particle fragmentation was found by Machado et al. (2019) in the US extraction of trace metals from globe artichoke leaves and soybean seeds [35]. Nevertheless, US treatments in RS extraction processes could promote other physical changes than particle fragmentation, such as erosion, detexturation, sonocapillarity, or sonoporation, as described by other authors [36,37], which favored the leaching of soluble compounds. Likewise, considering the limitations of the experimental method as a means of characterizing rod-shaped particles, the particle sizes obtained and their potential changes during extraction are not conclusive.

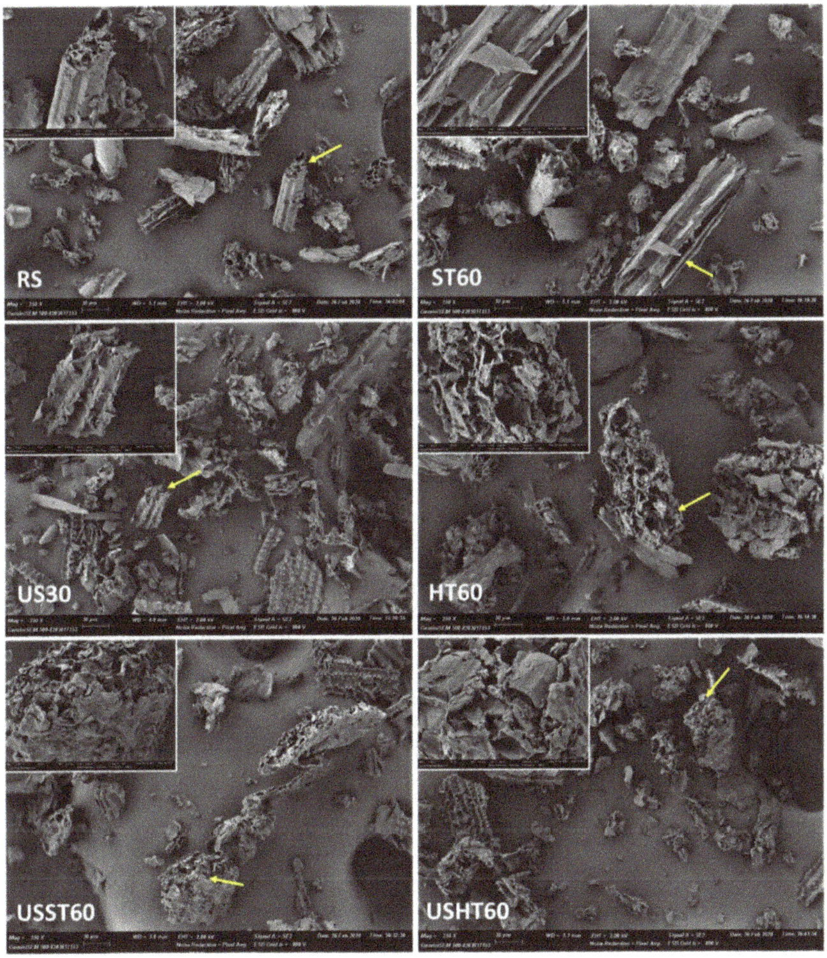

Figure 4. High-resolution field emission scanning electron microscopy (FESEM) images of the particles from the residual powder and untreated RS (1300× magnification). (RS: rice straw, US: ultrasound, ST: stirring, HT: heating at reflux, USST, and USHT: combined treatments). The number in the sample code indicates the processing time. Micrographs of different particles (arrow marked) at higher magnification (×1300) were also included in each case.

Microstructural changes in RS powders were also analyzed by FESEM (Figure 4) to identify the potential effects of the different treatments on the particles that would explain the different mass transfer patterns. The untreated sample (Figure 4) showed the RS fragments with different dimensions that reveals the original structure of the RS tissue. This has numerous vascular bundles in different layers, consisting of a waxy cuticle, silicate epidermis, cortex (collenchyma), a thick lignified layer (sclerenchyma) and ground tissue (parenchyma) that contained cellulose microfibrils covered with lignin and hemicellulose [38]. Most of the particles resulting from the milling operation exhibit the fibrillar nature of the original tissue. These fibrillar fragments are mainly constituted by the cellulose-hemicellulose-lignin complex where phenolic compounds are bonded. Micrographs obtained at higher magnification were also shown in Figure 4 to better appreciate the morphology of the different particles before and after the extraction process.

From qualitative analyses of FESEM observations (Figure 4), a greater proportion of particles more laminar than tubular in shape appeared in micrographs corresponding to treatments in which US was applied. This suggests a certain degree of plane de-bonding in the fibril structures, resulting from the cavitation phenomena. This fibril destruction would contribute to the promotion of mass transfer during the extraction process in line with the increase of the surface exposed to the solvent action. Other effects, such as sonocapillarity or sonoporation, previously described [36,37], could also occur, but they are difficult to identify in the already unstructured ground tissue. In the thermal treatment without US application (HT60), the presence of flakes or de-bonded lamellas is less noticeable, but particles appeared more distorted, thus revealing the erosive thermal effect associated with the cleavage of phenolic structures from the carbohydrate-lignin complex. Moreover, the more planar particles would have a gyration radius, and thus a mean equivalent sphere diameter, similar to the corresponding original fibrils, which would explain the lack of notable differences in the particle size distribution obtained by laser diffraction in the treatments submitted, or not, to US.

Then, the separation of different planes in the carbohydrate-lignin assemblies provoked by the US and phenol thermal de-bonding as a result of thermal treatment gave rise to the most effective extraction of phenolic compounds when combined in the USHT treatment.

3.4. Bioactive Characterization of the Extracts

The RS extracts obtained from different extraction methods were characterized in terms of total solid yield (TSY), total phenolic content (TPC), antioxidant (Table 2), and antibacterial activities. RS contains a wide variety of compounds, such as lignin, cellulose, tannins, hemicellulose, sterols, flavonoids, phenolic aldehydes, and hydroxycinnamic acids, which may or not have bioactive properties [11,12,21]. The amount and proportion of these extracted compounds depend strictly on the extraction parameters, such as the method, type and proportion of extractor solvent, time, and temperature.

Table 2. Total solid yield (TSY), total phenolic content (TPC) and antioxidant capacity (EC_{50}) of the different extracts. Phenolic content and antioxidant capacity were expressed per mass unit of RS and dry extract.

Method [1]	TSY (g Dry Extract. 100 g^{-1} RS) *	TPC_1 (mg GAE 100^{-1} g RS) *	TPC_2 (mg GAE. g^{-1} Dry Extract) *	$EC_{50(1)}$ (g RS. mg^{-1} DPPH) *	$EC_{50(2)}$ (mg Dry Extract. mg^{-1} DPPH) *
ST60	5.61 ± 0.07^{a}	256 ± 3^{a}	$45.7 \pm 0.1^{a,b}$	19.7 ± 0.6^{a}	11.0 ± 0.5^{a}
US15	5.7 ± 0.3^{a}	284 ± 17^{b}	49.2 ± 0.9^{a}	16.0 ± 1.3^{b}	9.2 ± 0.3^{b}
US30	7.48 ± 0.05^{b}	342 ± 10^{c}	$45.7 \pm 1.0^{a,b}$	12.5 ± 0.1^{c}	9.35 ± 0.18^{b}
US60	9.48 ± 0.17^{c}	354 ± 12^{c}	$37.4 \pm 1.9^{c,d}$	9.8 ± 0.3^{d}	9.3 ± 0.3^{b}
HT60	9.6 ± 0.6^{c}	459 ± 6^{d}	47.0 ± 3.0^{a}	7.8 ± 0.5^{e}	7.49 ± 0.05^{c}
USST60	9.0 ± 0.4^{c}	382 ± 6^{e}	$42.3 \pm 2.4^{b,c}$	10.4 ± 0.3^{f}	9.39 ± 0.10^{b}
USHT60	13.95 ± 0.13^{d}	486 ± 4^{f}	34.8 ± 0.5^{d}	4.6 ± 0.3^{g}	6.3 ± 0.4^{d}

[1] ST: stirring treatment, US: ultrasound treatment, HT: heating treatment, USST and UHHT: combination of 30 min US, and ST or HT, respectively. The number in the sample code indicates the process time. * Different superscript letters in the same column indicate significant differences ($\alpha = 0.05$).

As shown in (Table 2), non-statistical differences were found between the TSY values of ST60 and US15 treatments ($p > 0.05$) while the total solid extraction efficiency of the US treatment significantly increased ($p < 0.05$) when using longer sonication times, as has been previously observed by other authors [29]. No statistical differences were observed ($p > 0.05$) between the TSY of the US60 and HT60 treatments carried out at 25 and 100 °C, respectively. However, although the effects induced by ultrasonic cavitation and heating were comparable in terms of the total water solubilization capacity of the compounds present in the RS matrix, differences in the extract composition could provide distinct bioactive properties. The combination of US and heating (USHT60) yielded the highest solid leaching from RS ($p < 0.05$). Thus, as previously commented on, the US pre-treatment led to there being more points of the RS structure accessible to the solvent, thereby improving the mass transfer potential of RS compounds. The US pre-treatment increased the total solid leaching by around 60% and 45% when used before the stirring and heating processes, respectively. Heat treatment provokes the softening of the plant tissue, weakening the cell wall integrity and the bond hydrolysis of the phenolic compounds linked to the carbohydrate-lignin complex, thus enhancing the solubility of phenols and other compounds [29,36,37].

The total phenolic content of the different extracts was expressed with respect to the RS mass (TPC_1) and to the total extracted solids (TPC_2). The former permits the observation of what treatment was more effective at extracting phenolic compounds from the raw material, whereas the latter supplies information on how rich in phenolic compounds the different extracts, which also contain other soluble compounds, are. Thermal treatments at 100 °C, with and without US pre-treatment, yielded the highest extraction of phenolic compounds from RS, according to what is described by other authors when analyzing the temperature effect on the phenolic de-bonding from the cellular structure. The US pre-treatment enhanced this effect in line with the increase in the solid matrix surface available for the extraction, provoked by the cavitation phenomenon. The US pre-treatment improved the TSY of the ST treatment and the extraction of phenolic compounds from the RS matrix, but the extract's richness in phenols was slightly reduced due to the simultaneous promotion of the extraction of other components. In the same way, in the HT process, although the US pre-treatment enhanced the phenolic extraction, it led to the release of a greater quantity of other solids, thus implying an increase of 45% in the total solid extraction, but of only 6% in the phenolic compounds from the RS. The TPC values of the extracts (TPC_2) revealed that the greater total solid extraction does not guarantee a higher content of phenolic or active compounds in the extract. Thus, similar TPC_2 values were obtained for the US60 and HT60 treatments despite the different total solid yield. The poorest extract in phenols (lowest TPC_2 value) was that of the USHT60 treatment with the highest TSY. Therefore, what this treatment promoted was the extraction of non-phenolic compounds as opposed to phenolic. The extracts from the US15 and US30 treatments had similar phenolic contents (TPC_2 values), however, longer US treatment times slightly decreased the phenolic richness of the extracts, thus indicating a greater promotion of the extraction of non-phenolic compounds. In this sense, other authors [39] indicated that 20 min of sonication time was enough to extract phenolic compounds from soy beverages.

A DPPH radical scavenging assay was used to analyze the antioxidant activities of the RS extracts obtained from different extraction methods. The results were expressed in terms of the EC_{50} parameter, referred to the mass of RS and extract ($EC_{50(1)}$ and $EC_{50(2)}$, respectively). Thus, $EC_{50(2)}$ indicates the radical scavenging capacity of the extracts, whereas $EC_{50(1)}$ reflects the ability of the extraction method to extract compounds with more radical scavenging capacity from the RS. A significant correlation was observed between the capacity of the extraction method to obtain phenolic compounds from RS (TPC_1) and the radical scavenging capacity ($EC_{50(1)}$), with r values of −0.960. This indicates that the treatments that extracted a greater quantity of phenolic compounds from RS also extracted more compounds with radical scavenging capacity, which can be mainly attributed to the phenols. However, the parallel extraction of other compounds, as affected by the kind of treatment, determined the final ratio of the phenols in the extracts and their $EC_{50(2)}$ values. As can be observed in Table 2, the ST60 exhibited the lowest radical scavenging capacity (the highest EC_{50} values). Other authors [10] found a similar $EC_{50(2)}$ value (12.0 mg freeze-dried extract/mg DPPH) when extracting compounds from RS under similar conditions. Although the US treatments promoted the enrichment of the extracts in non-phenolic compounds when the extraction time lengthened, the extracts for 15, 30, and 60 min exhibited a similar radical scavenging capacity ($p > 0.05$), which was greater than that of the ST extract. The combination of the US with ST (USST60) did not imply extracts with more radical scavenging activity than that obtained by the US pre treatment, nor was the extract enriched in phenolic compounds. In contrast, the thermal treatment produced the extracts with some of the highest radical scavenging capacities;

this was enhanced by the US pre-treatment, despite the fact that US application more effectively promoted the extraction of non-phenolic compounds and, globally, the USHT60 extract had the lowest total phenolic content. This finding indicates that the thermal treatment promotes the extraction of the more active compounds (phenolic or not); this was enhanced when US pretreatment was applied in line with the higher surface exposure in the solid particles due to the cavitation phenomenon. Other authors [30] also observed that the microwave and conventional oven heating pre-treatments promoted a significant increase in the antioxidant activity of extracts from fennel seeds, which was attributed to the promotion of flavonoid extraction. Flavonoid extraction from baobab seeds was also promoted when US treatments were carried out at higher temperatures [17]. Several compounds with proven antioxidant capacity, such as ferulic, protocatechuic, *p*-coumaric, caffeic, and vanillic acids, tricin, and vanillin have been identified in RS extracts by other authors [10,40].

In brief, the obtained results suggested that the combined ultrasound-heating method promoted a greater extraction of compounds from RS with antioxidant properties, diluted with other non-phenolic components, and, from a practical point of view, the obtained USHT60 extract exhibited the best potential to act as an efficient antioxidant material.

Considering the ability of the US and HT combination to extract bioactive compounds in terms of the radical scavenging capacity, the corresponding extract (USHT60) was also evaluated as to its antibacterial activity against *L. innocua* and *E. coli* strains. The extract obtained in the corresponding US30 pretreatment (richer in phenolic compounds) was also analyzed. Thus, different concentrations of freeze-dried extracts (96–160 $mg \cdot mL^{-1}$ TSB) were applied in the well microtiter plates in order to determine the minimum inhibitory concentration. No total inhibition of bacterial growth was observed for the extract over the considered concentration range. However, the most heavily-concentrated USHT60 extract at 160 $mg \cdot mL^{-1}$ reduced the initial counts of *L. innocua* by 1.7 log CFU, although it was not effective at reducing the counts of *E. coli*. At the same concentration, the US30 extract was not effective at reducing the initial counts of *E. coli* or *L. innocua*, which suggests that the extraction of more active compounds was achieved during the thermal step of the extraction process. Although extracts obtained from RS contain antimicrobial compounds, such as ferulic and *p*-coumaric acids [10,21] with proven antibacterial activity against these strains, their concentration could not reach the minimal inhibitory concentration of the bacteria, or their interactions with other extract components could limit their effectiveness as antibacterial agents.

4. Conclusions

RS is an agro-industrial waste that contains a wide variety of bioactive compounds of technological interest for food application purposes, but the extraction conditions greatly influence the extraction efficiency of the target compounds. The application of ultrasound was notably more effective at extracting water-soluble phenolic compounds than simple stirring, as revealed by the higher extract yields and antioxidant activity of the extracts. Nevertheless, a high extraction temperature produced materials with improved antioxidant activities, in line with the promotion of the cleavage of covalent bonds between the phenolic compounds and the lignocellulosic fraction. This thermal effect was greatly enhanced when ultrasound pretreatment was applied due to the increase in the substrate surface exposed to the extraction. Therefore, a combination of 30 min US plus 60 min thermal treatment in water reflux is recommended to obtain solid extracts with great antioxidant activity and notable anti-listerial effect, which could be used in the food or pharmaceutical industries.

Author Contributions: P.A.V.F.: conceptualization, methodology, formal analysis, investigation, writing—original draft preparation, writing—review and editing; C.G.-M.: conceptualization, methodology, investigation, writing—original draft preparation, writing—review and editing; A.C.: conceptualization, methodology, investigation, writing—original draft preparation, writing—review and editing. All authors have read and agreed to the published version of the manuscript.

Funding: This research was funded by Agencia Estatal de Investigación (Spain), grant number PID2019-105207RB-I00.

Acknowledgments: Author P.A.V.F. is grateful to Generalitat Valenciana for the GrisoliaP/2019/115 grant.

Conflicts of Interest: The authors declare no conflict of interest.

References

1. Sharma, B.; Vaish, B.; Monika; Singh, U.K.; Singh, P.; Singh, R.P. Recycling of Organic Wastes in Agriculture: An Environmental Perspective. *Int. J. Environ. Res.* **2019**, *13*, 409–429. [CrossRef]
2. Ng, H.-M.; Sin, L.T.; Tee, T.-T.; Bee, S.-T.; Hui, D.; Low, C.-Y.; Rahmat, A.R. Extraction of cellulose nanocrystals from plant sources for application as reinforcing agent in polymers. *Compos. Part B: Eng.* **2015**, *75*, 176–200. [CrossRef]
3. Peanparkdee, M.; Iwamoto, S. Bioactive compounds from by-products of rice cultivation and rice processing: Extraction and application in the food and pharmaceutical industries. *Trends Food Sci. Technol.* **2019**, *86*, 109–117. [CrossRef]
4. FAOSTAT. Available online: http://www.fao.org/faostat/en/#data/QC/visualize (accessed on 4 November 2020).
5. Sarkar, N.; Ghosh, S.K.; Bannerjee, S.; Aikat, K. Bioethanol production from agricultural wastes: An overview. *Renew. Energy* **2012**, *37*, 19–27. [CrossRef]
6. Takano, M.; Hoshino, K. Bioethanol production from rice straw by simultaneous saccharification and fermentation with statistical optimized cellulase cocktail and fermenting fungus. *Bioresour. Bioprocess* **2018**, *5*, 16. [CrossRef]
7. Krishania, M.; Kumar, V.; Sangwan, R.S. Integrated approach for extraction of xylose, cellulose, lignin and silica from rice straw. *Bioresour. Technol. Rep.* **2018**, *1*, 89–93. [CrossRef]
8. Jani, S.M.; Rushdan, I. Mechanical properties of beating pulp and paper from rice straw. *J. Trop. Agric. Food Sci.* **2016**, *44*, 103–109.
9. Elhussieny, A.; Faisal, M.; D'Angelo, G.; Aboulkhair, N.T.; Everitt, N.M.; Fahim, I.S. Valorisation of shrimp and rice straw waste into food packaging applications. *Ain Shams Eng. J.* **2020**, S2090447920300101. [CrossRef]
10. Menzel, C.; González-Martínez, C.; Vilaplana, F.; Diretto, G.; Chiralt, A. Incorporation of natural antioxidants from rice straw into renewable starch films. *Int. J. Biol. Macromol.* **2020**, *146*, 976–986. [CrossRef]
11. Li, Y.; Qi, B.; Luo, J.; Khan, R.; Wan, Y. Separation and concentration of hydroxycinnamic acids in alkaline hydrolyzate from rice straw by nanofiltration. *Sep. Purif. Technol.* **2015**, *149*, 315–321. [CrossRef]
12. Barana, D.; Salanti, A.; Orlandi, M.; Ali, D.S.; Zoia, L. Biorefinery process for the simultaneous recovery of lignin, hemicelluloses, cellulose nanocrystals and silica from rice husk and Arundo donax. *Ind. Crop. Prod.* **2016**, *86*, 31–39. [CrossRef]
13. Adom, K.K.; Liu, R.H. Antioxidant Activity of Grains. *J. Agric. Food Chem.* **2002**, *50*, 6182–6187. [CrossRef] [PubMed]
14. Cheung, Y.-C.; Wu, J.-Y. Kinetic models and process parameters for ultrasound-assisted extraction of water-soluble components and polysaccharides from a medicinal fungus. *Biochem. Eng. J.* **2013**, *79*, 214–220. [CrossRef]
15. Ojha, K.S.; Aznar, R.; O'Donnell, C.; Tiwari, B.K. Ultrasound technology for the extraction of biologically active molecules from plant, animal and marine sources. *TRAC Trends Anal. Chem.* **2020**, *122*, 115663. [CrossRef]
16. Luque-García, J.L.; Luque de Castro, M.D. Ultrasound: A powerful tool for leaching. *TRAC Trends Anal. Chem.* **2003**, *22*, 41–47. [CrossRef]
17. Ismail, B.B.; Guo, M.; Pu, Y.; Wang, W.; Ye, X.; Liu, D. Valorisation of baobab (*Adansonia digitata*) seeds by ultrasound assisted extraction of polyphenolics. Optimisation and comparison with conventional methods. *Ultrason. Sonochem.* **2019**, *52*, 257–267. [CrossRef] [PubMed]
18. Sumere, B.R.; de Souza, M.C.; dos Santos, M.P.; Bezerra, R.M.N.; da Cunha, D.T.; Martinez, J.; Rostagno, M.A. Combining pressurized liquids with ultrasound to improve the extraction of phenolic compounds from pomegranate peel (*Punica granatum* L.). *Ultrason. Sonochem.* **2018**, *48*, 151–162. [CrossRef] [PubMed]
19. Wang, L.; Boussetta, N.; Lebovka, N.; Vorobiev, E. Selectivity of ultrasound-assisted aqueous extraction of valuable compounds from flesh and peel of apple tissues. *LWT* **2018**, *93*, 511–516. [CrossRef]
20. Dias, A.L.B.; Arroio Sergio, C.S.; Santos, P.; Barbero, G.F.; Rezende, C.A.; Martínez, J. Ultrasound-assisted extraction of bioactive compounds from dedo de moça pepper (*Capsicum baccatum* L.): Effects on the vegetable matrix and mathematical modeling. *J. Food Eng.* **2017**, *198*, 36–44. [CrossRef]

21. Karimi, E.; Mehrabanjoubani, P.; Keshavarzian, M.; Oskoueian, E.; Jaafar, H.Z.; Abdolzadeh, A. Identification and quantification of phenolic and flavonoid components in straw and seed husk of some rice varieties (*Oryza sativa* L.) and their antioxidant properties: Identification and quantification of phenolic and flavonoid. *J. Sci. Food Agric.* **2014**, *94*, 2324–2330. [CrossRef]
22. Ho, Y.S.; McKay, G. Sorption of dye from aqueous solution by peat. *Eng. J.* **1998**, *70*, 115–124. [CrossRef]
23. Brand-Williams, W.; Cuvelier, M.E.; Berset, C. Use of a free radical method to evaluate antioxidant activity. *LWT—Food Sci. Technol.* **1995**, *28*, 25–30. [CrossRef]
24. Abdi, R.D.; Kerro Dego, O. Antimicrobial activity of Persicaria pensylvanica extract against Staphylococcus aureus. *Eur. J. Integr. Med.* **2019**, *29*, 100921. [CrossRef]
25. Requena, R.; Jiménez-Quero, A.; Vargas, M.; Moriana, R.; Chiralt, A.; Vilaplana, F. Integral Fractionation of Rice Husks into Bioactive Arabinoxylans, Cellulose Nanocrystals, and Silica Particles. *ACS Sustain. Chem. Eng.* **2019**, *7*, 6275–6286. [CrossRef]
26. Wang, Y.; Liu, J.; Liu, X.; Zhang, X.; Xu, Y.; Leng, F.; Avwenagbiku, M.O. Kinetic modeling of the ultrasonic-assisted extraction of polysaccharide from Nostoc commune and physicochemical properties analysis. *Int. J. Biol. Macromol.* **2019**, *128*, 421–428. [CrossRef]
27. González, N.; Elissetche, J.; Pereira, M.; Fernández, K. Extraction of polyphenols from and: Experimental kinetics, modeling and evaluation of their antioxidant and antifungical activities. *Ind. Crop. Prod.* **2017**, *109*, 737–745. [CrossRef]
28. Dutta, R.; Sarkar, U.; Mukherjee, A. Pseudo-kinetics of batch extraction of *Crotalaria juncea* (Sunn hemp) seed oil using 2-propanol. *Ind. Crop. Prod.* **2016**, *87*, 9–13. [CrossRef]
29. Tabaraki, R.; Heidarizadi, E.; Benvidi, A. Optimization of ultrasonic-assisted extraction of pomegranate (*Punica granatum* L.) peel antioxidants by response surface methodology. *Sep. Purif. Technol.* **2012**, *98*, 16–23. [CrossRef]
30. Hayat, K.; Abbas, S.; Hussain, S.; Shahzad, S.A.; Tahir, M.U. Effect of microwave and conventional oven heating on phenolic constituents, fatty acids, minerals and antioxidant potential of fennel seed. *Ind. Crop. Prod.* **2019**, *140*, 111610. [CrossRef]
31. Xu, G.; Ye, X.; Chen, J.; Liu, D. Effect of Heat Treatment on the Phenolic Compounds and Antioxidant Capacity of Citrus Peel Extract. *J. Agric. Food Chem.* **2007**, *55*, 330–335. [CrossRef]
32. Purohit, A.J.; Gogate, P.R. Ultrasound-Assisted Extraction of β-Carotene from Waste Carrot Residue: Effect of Operating Parameters and Type of Ultrasonic Irradiation. *Sep. Sci. Technol.* **2015**, *50*, 1507–1517. [CrossRef]
33. Wanyo, P.; Meeso, N.; Siriamornpun, S. Effects of different treatments on the antioxidant properties and phenolic compounds of rice bran and rice husk. *Food Chem.* **2014**, *157*, 457–463. [CrossRef] [PubMed]
34. Niwa, Y.; Miyachi, Y. Antioxidant action of natural health products and Chinese herbs. *Inflammation* **1986**, *10*, 79–91. [CrossRef] [PubMed]
35. Machado, I.; Faccio, R.; Pistón, M. Characterization of the effects involved in ultrasound-assisted extraction of trace elements from artichoke leaves and soybean seeds. *Ultrason. Sonochem.* **2019**, *59*, 104752. [CrossRef]
36. Chemat, F.; Rombaut, N.; Sicaire, A.-G.; Meullemiestre, A.; Fabiano-Tixier, A.-S.; Abert-Vian, M. Ultrasound assisted extraction of food and natural products. Mechanisms, techniques, combinations, protocols and applications. A review. *Ultrason. Sonochem.* **2017**, *34*, 540–560. [CrossRef] [PubMed]
37. Cravotto, G.; Cintas, P. Power ultrasound in organic synthesis: Moving cavitational chemistry from academia to innovative and large-scale applications. *Chem. Soc. Rev.* **2006**, *35*, 180–196. [CrossRef] [PubMed]
38. Seo, D.-J.; Sakoda, A. Assessment of the structural factors controlling the enzymatic saccharification of rice straw cellulose. *Biomass Bioenergy* **2014**, *71*, 47–57. [CrossRef]
39. Rostagno, M.A.; Palma, M.; Barroso, C.G. Ultrasound-assisted extraction of isoflavones from soy beverages blended with fruit juices. *Anal. Chim. Acta* **2007**, *597*, 265–272. [CrossRef]
40. Shi, J.; Wang, Y.; Wei, H.; Hu, J.; Gao, M.-T. Structure analysis of condensed tannin from rice straw and its inhibitory effect on Staphylococcus aureus. *Ind. Crop. Prod.* **2020**, *145*, 112130. [CrossRef]

foods

Article

Films Based on Thermoplastic Starch Blended with Pine Resin Derivatives for Food Packaging

Cristina Pavon [1,*], Miguel Aldas [1,2], Juan López-Martínez [1], Joaquín Hernández-Fernández [3,4] and Marina Patricia Arrieta [5,6]

[1] Instituto de Tecnología de Materiales (ITM), Universitat Politècnica de València (UPV), 03801 Alcoy, Spain; miguel.aldas@epn.edu.ec (M.A.); jlopezm@mcm.upv.es (J.L.-M.)
[2] Departamento de Ciencia de Alimentos y Biotecnología, Facultad de Ingeniería Química y Agroindustria, Escuela Politécnica Nacional, Quito 170517, Ecuador
[3] Research Group in Polymer Science, Engineering and Sustainability, Esenttia, Mamonal Industrial Zona, km. 8, Cartagena 130013, Colombia; hernandez548@hotmail.com
[4] Department of Natural and Exact Sciences, Universidad de la Costa, Calle 58 # 55–66, Barranquilla 080002, Colombia
[5] Departamento de Ingeniería Química Industrial y del Medio Ambiente, Escuela Técnica Superior de Ingenieros Industriales, Universidad Politécnica de Madrid (ETSII-UPM), Calle José Gutiérrez Abascal 2, 28006 Madrid, Spain; m.arrieta@upm.es
[6] Grupo de Investigación: Polímeros, Caracterización y Aplicaciones (POLCA), 28006 Madrid, Spain
* Correspondence: crisppavonv@gmail.com; Tel.: +34-644343735

Abstract: Completely biobased and biodegradable thermoplastic starch (TPS) based materials with a tunable performance were prepared for food packaging applications. Five blends were prepared by blending TPS with 10 wt%. of different pine resins derivatives: gum rosin (GR), disproportionated gum rosin (RD), maleic anhydride-modified gum rosin (CM), pentaerythritol ester of gum rosin (LF), and glycerol ester of gum rosin (UG). The materials were characterized in terms of thermo-mechanical behavior, surface wettability, color performance, water absorption, X-ray diffraction pattern, and disintegration under composting conditions. It was determined that pine resin derivatives increase the hydrophobicity of TPS and also increase the elastic component of TPS which stiffen the TPS structure. The water uptake study revealed that GR and LF were able to decrease the water absorption of TPS, while the rest of the resins kept the water uptake ability. X-ray diffraction analyses revealed that GR, CM, and RD restrain the aging of TPS after 24 months of aging. Finally, all TPS-resin blends were disintegrated under composting conditions during the thermophilic incubation period (90 days). Because of the TPS-resin blend's performance, the prepared materials are suitable for biodegradable rigid food packaging applications.

Keywords: bioplastic; thermoplastic starch; pine resin; gum rosin; disintegration; packaging

Citation: Pavon, C.; Aldas, M.; López-Martínez, J.; Hernández-Fernández, J.; Arrieta, M.P. Films Based on Thermoplastic Starch Blended with Pine Resin Derivatives for Food Packaging. *Foods* **2021**, *10*, 1171. https://doi.org/10.3390/foods10061171

Academic Editor: Mario M. Martinez

Received: 27 April 2021
Accepted: 21 May 2021
Published: 23 May 2021

1. Introduction

Synthetic plastics production and applications have experienced exponential growth since the beginning of the polymer industry on a large scale back in 1940 and 1950 [1]. Despite the widespread use of plastics as materials are relatively new in history, after World War II [1], they have become essential for the world economy and modern life activities [2]. Synthetic plastics provide many benefits as membranes for water purification, food packaging to prevent spoilage, or lightweight transportation to reduce fuel consumption [3]. However, the current linear-economy model is unsustainable as the production and disposal mechanisms of polymers did not consider their end-of-life issues [2]. Thus, there is a need for the plastic industry to move towards a circular economy model, particularly in the case of short-term applications.

Polymeric waste causes worldwide environmental pollution and huge economic and material value loss [2,3]. Furthermore, the majority of synthetic plastics are not biodegradable, which has increased the disposed of polymer waste amount in the past decades [4,5]. Moreover, the annual production of plastics keeps increasing, from 15 million MT in the sixties it has reached 359 million MT in 2018 [6] and it is expected to triple by 2050 (reaching 1.12 billion tons) [2,5]. In this frame, many researchers and companies intend to produce sustainable polymers to gradually replace fossil-based polymers [2,5]. As a result, several sustainable polymers have been developed from renewable raw materials, for instance, biopolymers such as cellulose [7], lignin [8], plant-based fatty acids, or polymers from animal origin such as chitin [9], caseinates [10], etc. Among them, carbohydrates are convenient raw biopolymers due to their availability, inexpensiveness, and stereochemical diversity [11]. Moreover, defined properties of these biopolymers can be attained through their chemical modification and/or physical blending [12,13].

Starch is produced by plants such as wheat, corn, rice, bean, tapioca, and peas [14]. Starch is composed of two polysaccharides (amylose and amylopectin). Starch is widely used thanks to its low cost and high availability [15,16]. To use starch in the plastic industry, it can be plasticized in the presence of plasticizers, high temperatures, and shear stress [15]. Thermoplastic starch (TPS) has gained considerable attention during the last years for the development of biodegradable starch-based food packaging materials or edible coatings [17]. TPS is made from edible starch plasticized with food-grade plasticizers and thus allowed for food contact applications. Moreover, TPS is biobased and biodegradable, which from an environmental perspective allows closing the loop of circular economy [18]. However, TPS industrial application is somewhat limited due to its poor mechanical performance, low water resistance, and the undesirable changes in the thermomechanical characteristics of the material caused by the re-crystallization and retrogradation that its structure is subject to [18–20]. Thus, in addition to increasing its water resistance, improvements on the thermo-mechanical performance of TPS-based formulations are needed to extend its industrial applications [21]. To enable TPS industrial applications in the food packaging sector several strategies have been proven, such as chemical modifications [17], blending [19,22], and/or the development of composites and nanocomposites [23]. Melt-blending strategies seem to be an effective way to tailor TPS properties by a simple, industrially scalable, and cost-effective plastic processing method [18]. However, TPS-based formulations usually show the glass transition temperature above or below room temperature as a function of the plasticizer and the water content [23]. Above T_g, the polymeric matrix loses its rigidity leading to plastic deformation making the material unsuitable for warm or hot food applications [18].

The revalorization of food and agro-industrial wastes into environmentally friendly materials has considerably increased during the last years [14,24]. In this sense, the revalorization of pine resin and gum rosin derivatives has gained interest in the food packaging field during the last years as natural low-cost additives (i.e., stabilizers, compatibilizers, and/or plasticizers) [18,19,25–27]. Resin is exudated from Conifers as a defense mechanism in wounds of their tissues or cuts of the wood of the stem [28]. During the last years, tapping pine trees to collect secretions of resin has resurgence [29,30], and among these, the activities related to this field, as pine cleaning activities during the summer period when fire risk is high [29]. Therefore, the revalorization of pine resin and gums derivatives are positive for good forest management practices. Concerning the plastic packaging industry, gum rosin is a natural and easily available material that has great potential in the development of blends with biopolymers. Gum rosin is a rigid and brittle solid that has a thermoplastic behavior [31]. Gum rosin is the non-volatile fraction of pine resin [32] and is composed mainly of abietic- and pimaric-type rosin acids that have characteristic hydrophenanthrene structures [33]. Rosin acids structure have conjugated double bonds and a carboxylic group which enable gum rosin the possibility to be chemically modified and be converted into a large number of derivatives such as salts, hydrogenated, esters, maleic anhydride adducts, and disproportionated rosins [33,34]. These modified rosins have different properties and are useful in several applications [33].

Gum rosin and its derivatives have been successfully used in the production of green plastics [34]. For instance, Arrieta et al. (2017) studied mixtures of triethylene glycol ester of gum rosin (TEGR) with linseed oil as natural additives in polyvinyl chloride (PVC). They have found that TEGR enhanced the tensile strength and the elastic modulus of the plastisol, and also, it provides a UV-blocking effect and contributes to increasing the thermal stability of PVC [29]. In 2019, Aldas et al. studied the effect of gum rosin and two derivatives (pentaerythritol rosin esters: LF and UT) on the performance of a commercial thermoplastic starch (Mater-Bi). They determined that gum rosin in 15 wt.% increases the elongation at break, toughness, and impact energy. Besides, LF in 10 wt.% increases the toughness, Young's modulus, and tensile strength of Mater-Bi. Meanwhile, UT in 15 wt.% increases the elongation at break of neat Mater-Bi and improves its processability performance by decreasing the processing temperature [19]. Then, the same authors deepen in a microscopic study of the mentioned materials and found the interactions of gum rosin and the two pentaerythritol rosin esters over the different phases of Mater-Bi [35]. In another area, Pavón et al. (2020) studied the effect of gum rosin on poly(ε-caprolactone) (PCL) processed by 3D-printing, it was determined that GR form homogeneous blends with PCL, plasticized the structure, and increase the hydrophobicity of the material [36]. Thus, gum rosin and its derivatives have acquired considerable interest as a sustainable additive, and the formulated materials which contain them can be used in diverse applications such as food packaging, mulch films, biodegradable and compostable films, and biocompatible materials.

Starch has proven to be an interesting food-grade polymeric system able to develop environmentally friendly packaging products that contribute to reducing petrochemical plastic consumption as well as waste generation, being widely blended with industrial wastes [14]. To extend starch applications in the food packaging field, it can be prepared in the thermoplastic form which requires water that acts as a destructuring agent and its further gelatinization. Then, the high water content joined with the heat applied during processing produces the starch granule swelling and the starch gelatinization through the disruption of the granule organization [37]. In previous work, TPS was prepared through melt extrusion from food-grade corn starch, glycerol, and water. Glycerol has been selected as a plasticizer not only because it is food-grade and contributes to the reduction in biopolymers intrinsic brittleness, but also it is a by-product of biodiesel production being important from an environmental point of view to revalorize industrial sub-products into useful plasticizer [10,37]. Then, TPS was blended, by melt extrusion, with five pine resin derivatives: gum rosin (GR), maleic anhydride-modified gum rosin (CM), pentaerythritol ester of gum rosin (LF), disproportionated gum rosin (RD), and glycerol ester of gum rosin (UG). Specimens of the materials were obtained from the injection molding process simulating the industrial conditions. The used pine resin derivatives showed their ability to stiffen TPS structure and thus improve the mechanical performance of the specimens, increase the thermal stability, and shift the TPS glass transition temperature to higher values, leading to materials able to be used for rigid packaging intended for hot food applications [18].

In the present work, to address the interest of these materials for the biodegradable food packaging industry, the injected molded TPS-resin materials were characterized focusing on this field of application. Thus, the thermo-mechanical behavior, surface wettability, and color parameters were evaluated. The changes in the crystallinity due to the aging process were evaluated by X-ray diffraction pattern after 24 months of aging. Finally, the materials were disintegrated under composting conditions test was conducted in all TPS-resin blends to show that the materials close the loop and fit well the concept of materials for the circular economy approach.

2. Materials and Methods

2.1. Materials

Native corn starch (food-grade) composed of 27% amylose was provided by Cargill (Barcelona, Spain).

The plasticizers used were distilled water and glycerol, added in 10 wt.% and 25 wt.% respectively. Glycerol (99% of purity), was purchased from Panreac (Barcelona, Spain).

Gum rosin and four gum rosin derivatives were used as additives and added in 10 wt.% in the thermoplastic starch formulation: Gum Rosin (GR, softening point of 76 °C and acid number 167), supplied by Sigma-Aldrich (Mostoles, Spain); Colmodif R-330 (CM, softening point of 123 °C and acid number 252); Lurefor 125 resin (LF, softening point of 125 °C and acid number 11.9); Residis 455 (RD, softening point of 74.6 °C and acid number 157) supplied by Luresa (Segovia, Spain) and Unik Gum G88 (UG, softening point of 87 °C and acid number 7) supplied by United resins (Figueira da Foz, Portugal). Figure 1 shows the corresponding chemical formula of the used materials.

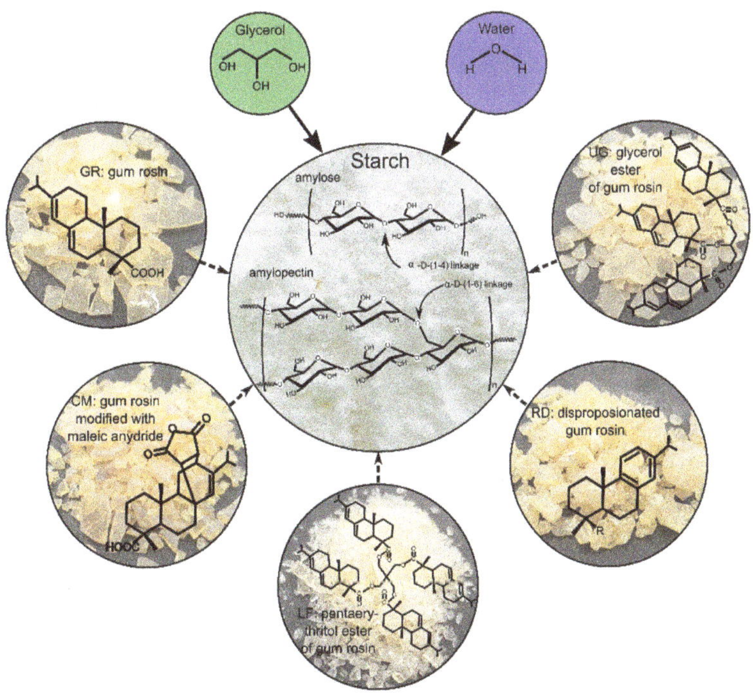

Figure 1. Chemical structure of gum rosin and gum rosin derivatives.

2.2. Methods

2.2.1. TPS-Resin Blends

TPS-resin blends were prepared by melt extrusion and the specimen samples were manufactured using injection molding, following the method already optimized in previous work [18]. A schematic representation of the thermoplastic starch (TPS) preparation starting from native corn starch, glycerol, and water, and further the development of TPS blended with gum rosin and its derivatives materials are shown in Figure 2. In brief, corn starch was manually mixed with water and glycerol, the resulting blend was hermetically stored in a sealed polyethylene bag for 24 h. To obtain TPS, the mixture was extruded in a co-rotating twin-screw extruder, at a temperature profile of 130, 110, 100, 90 °C (from die to hopper) and 20 rpm. The used extruder was from Dupra S.L (Castalla, Spain) equipped with a screw diameter of 25 mm and an L/D ratio of 24. After the extrusion process, each pine resin derivative was manually mixed with the corresponding amount of TPS, and the mixed materials were processed by a second extrusion. Neat TPS was also melt-extruded a second time to be used as a reference. The extruded materials were then pelletized. To obtain the specimens for further characterization the material was injected molded in an injection machine Sprinter-11, Erinca S.L. (Barcelona, Spain) (temperature profile of 130, 110, 100 ○C, from die to hopper). Before and after the processing and characterization, all materials were stored at 25 °C and 50 ± 5% of relative humidity (RH).

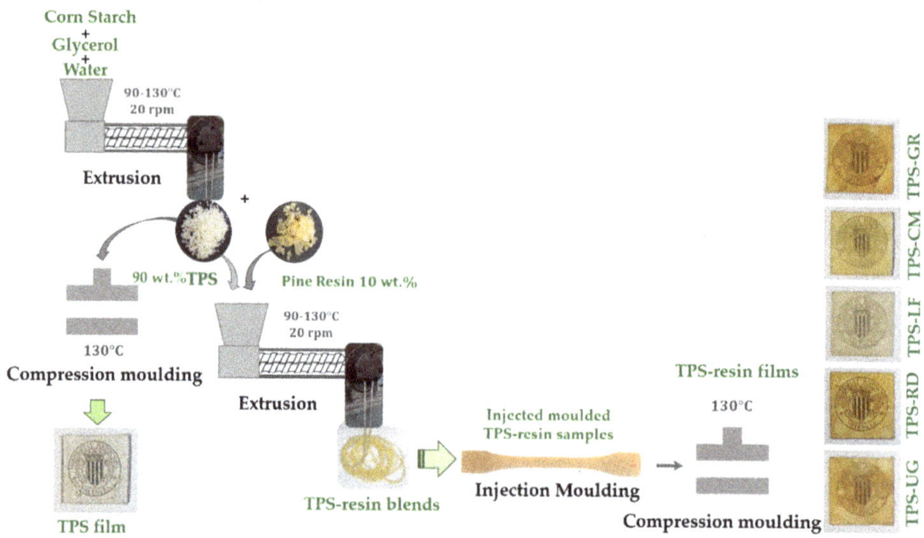

Figure 2. Schematic representation of TPS preparation and processing of TPS blends with gum rosin and gum rosin derivatives into injection molded and film samples.

2.2.2. Dynamic Mechanical Thermal Analysis

Dynamic Mechanical thermal analysis (DMTA) was conducted in DMA1 Mettler-Toledo (Schwerzenbach, Switzerland) using a single cantilever mode. For the analysis prismatic specimens with rectangular cross-section of $4.5 \pm 0.2 \times 1 \pm 0.2$ mm^2 and a length of 20 ± 2.0 of mm were used. The test was run with an oscillation frequency of 1 Hz at a constant heating rate of 2 °C/min from −100 to 117 °C, with a maximum deformation of 10 µm. The initial static force was 1N. As a result, the dynamic storage modulus (G') and loss factor (tan δ) curves as a function of temperature are reported. The glass transition temperatures (T_g) were taken as the maximum values of tan δ curves.

2.2.3. Water Uptake

Water uptake of the TPS-resin blends was determined using flexural test specimens (80 mm × 10 mm × 4 mm) and the parameters specified in ISO 62:2008 standard [38]. Before the test, the samples were dried at 40 °C in an air circulating oven model 2001245 Digiheat-TFT from J.P. Selecta S.A. (Barcelona, Spain). When dried, the samples were weighed (W_0) and then they were soaked in distilled water. Measurements of water uptake were taken at regular time intervals by removing the samples from the water tank, drying the excess water, and weighing each sample in an analytical balance AG 245 Mettler-Toledo (Barcelona, Spain) with a precision of 0.0001 g. After each measurement, the samples were returned to the water tank. The water absorption (c) was calculated by the difference between the sample weight after an immersion time t (W_t) and the initial dried sample weight (W_0) and according to Equation (1) [39].

$$c = \frac{(W_t - W_0)}{W_0} \times 100 \quad (1)$$

Three samples of each formulation were evaluated until the saturation weight (W_s) when no additional weight gain is observed with increasing time. The absorption curves with the mean values are reported.

2.2.4. Surface Characterization, Color, and Wettability

The color parameters of the CIE L*a*b* color space and the yellowness index (YI) were measured using a Colorflex-Diff2 458/08 colorimeter from HunterLab (Reston, VA, USA). Five different points were assessed at aleatory positions over the sample surface. Average values of five YI, L*, a*, and b* coordinates among the standard deviation were reported. Additionally, the total color differences (ΔE) were calculated using TPS as blank [40] and following Equation (2):

$$\Delta E = \sqrt{\Delta a^2 + \Delta b^2 + \Delta L^2} \quad (2)$$

The surface hydrophobicity of TPS-resin blends was studied through the water contact angle (WCA) of a sessile drop. Analyses were carried out in an optical goniometer EasyDrop-FM140 from Kruss Equipments (Hamburg, Germany) equipped with a camera. A water droplet (≈1.5 μL) was randomly deposited on the surface of the sample with a precision syringe. A capture of the droplet was taken with the camera and transferred to an image software (Drop Shape) for the measurements of the contact angle. Six contact angle measurements were done for each drop. The measurements were conducted at room temperature. The average WCA values are reported among the standard deviation, that in all the samples did not exceed ±3% [41].

The significant differences in the surface parameters were statistically assessed at a 95% confidence level according to Tukey's test using a one-way analysis of variance (ANOVA) employing OriginPro2015 software.

2.2.5. X-ray Diffraction (XDR)

XDR was used to analyze the influence of pine resin derivatives on starch retrogradation. Wide-angle X-ray Diffraction measurements were carried out using a Bruker D8 Advance X-Ray Diffractometer with a linear detector Lynxeye XE. The scattering angles (2θ) covered the ranges from 4° to 50° (θ is the Bragg angle) at a rate of 1°/min. The analyses were done using a 1 mm thick sample with a smooth surface. The XRD patterns of TPS-resin blends are reported in their initial state and after 24 months of storage.

2.2.6. Disintegration under Composting Conditions

The disintegration under composting conditions test was conducted following the parameters of the ISO-20200 standard for a thermophilic degradation period (90 days) [42]. The dry solid residue was prepared by combining 10% commercial compost (Mantillo, Spain), 30% rabbit food, 10% starch, 5% sugar, 1% urea, 4% corn oil, and 40% sawdust. Then, water was added to the mixture to adjust the final water content to 55%. The wet solid residue was placed in plastic containers.

TPS-resin squared films of side 25 cm with an average thickness of 2 mm were prepared for the disintegration study. The TPS-resin blends were compressed molded at 130 °C. Film samples were dried at 40 °C for 48 h before the test. Then, the samples were weighed and placed in wire mesh, which allowed the access of microorganisms and humidity, to facilitate their removal after treatment [43]. The samples were buried 5 cm deep in the wet solid residue in the plastic reactor and incubated under aerobic conditions (58 ± 2 °C) in an oven with circulating air. To guarantee aerobic conditions and relative humidity in the reactor, the compost was mixed gently, and water was added periodically [42,44].

Samples were taken out of the container at different disintegration days (1, 4, 7, 14, 21, 28, 49, 63, 77, and 90). The samples were washed with distilled water, dried in an oven at 40 °C for 48 h, and weighed. A visual evaluation was performed in all the samples when extracted from the composting medium; photographs are presented.

The disintegration degree at different days exposed to the compost medium was calculated by normalizing the sample weight to the initial weight. To determine the time at which 50% of the film is disintegrated (t_{50}) the disintegrability degree values were fitted using the Boltzmann equation (OriginPro 2015 software) [45,46] following Equation (3):

$$m = \frac{(m_i - m_\infty)}{1 + e^{\left(\frac{1-t_{50}}{d_t}\right)}} \quad (3)$$

where m_i is the initial mass value measured before the composting test and m_∞ is the final mass value measured after the final asymptotes of the disintegrability test. d_t is a parameter that describes the shape of the curve between the upper and lower asymptotes. The boundary and initial conditions m_i is 0% and m_∞ is 100%. t_{50} is known as the half-maximal degradation, and it is the time at which materials disintegrability reaches the average value between m_i and m_∞ [47].

3. Results

3.1. Dynamic Mechanical Thermal Analysis

Figure 3 shows the evolution of the loss factor (tan δ) and storage modulus (G′) with temperature for TPS and its blends with pine resin derivatives. TPS presents a characteristic loss factor curve (black line in Figure 3a) of thermoplastic starch, where a partially miscible system is detected. This is related to phase separation of the starch-glycerol system, resulting in glycerol-rich domains (β relaxation) and starch-rich domains (α relaxation) [48,49]. Two major transitions are seen in the tan δ curves, a narrow and high-intensity peak centered at −60 °C (T_β), corresponding to the glycerol rich phase, and a broad peak centered at 25 °C (T_α) corresponding to the starch-rich phase [49,50].

It is seen that the addition of pine resin derivatives to TPS did not change the β relaxation transition, as the glycerol content remains constant. However, a decrease in the broadening of the TPS α relaxation transition is observed. The reduction in the broadening suggests that the pine resin derivatives act as solvents to amylose and amylopectin and help to reduce the heterogeneity of the mixture [51]. The effect of this enhanced mixture is noted in the mechanical properties of the materials, reported in previous work [18]. It was determined that the tensile strength is reduced due to the addition of pine resin derivatives. However, Young's modulus does not change or have a significant increase (TPS-LF) with respect to TPS. The elongation at break increased significantly for TPS-GR and remained invariable with respect to neat TPS for the other formulations [18].

In the formulations that contain pine resin and derivatives, the last transition is detected between 50 °C to 100 °C. This transition is observed like a peak at 53 °C in TPS-GR, 58 °C in TPS-RD, 64.6 °C in TPS-UG, 71.6 °C in TPS-CM and 76 °C in TPS-LF in the tan δ curves (Figure 3). These temperatures are consistent with the glass transition temperatures (T_g) measured in previous work by differential scanning calorimetry (DSC) [18]. The values of this transition detected in DMA and those of T_g detected by DSC differ in some degrees because DMA works at higher frequencies than DSC [52]. Additionally, the differences in the temperature values where the peaks are located are due to the chemical structure of each pine resin derivative and it is linked to the degree of mobility that the pine resin additive allows the TPS chains to increase their chain mobility. GR is mainly composed of abietane-type acid while RD has pimarane-type acids in its structure. LF, UG, and CM, are chemically modified rosins with different degrees of modifications. LF is a pentaerythritol ester, UG is a glycerol ester and CM is a maleic anhydride-modified rosin [18]. The shifting of the T_g values to higher temperatures suggests a stiffening effect, and thus these materials result in interesting rigid packaging applications intended for hot food [18].

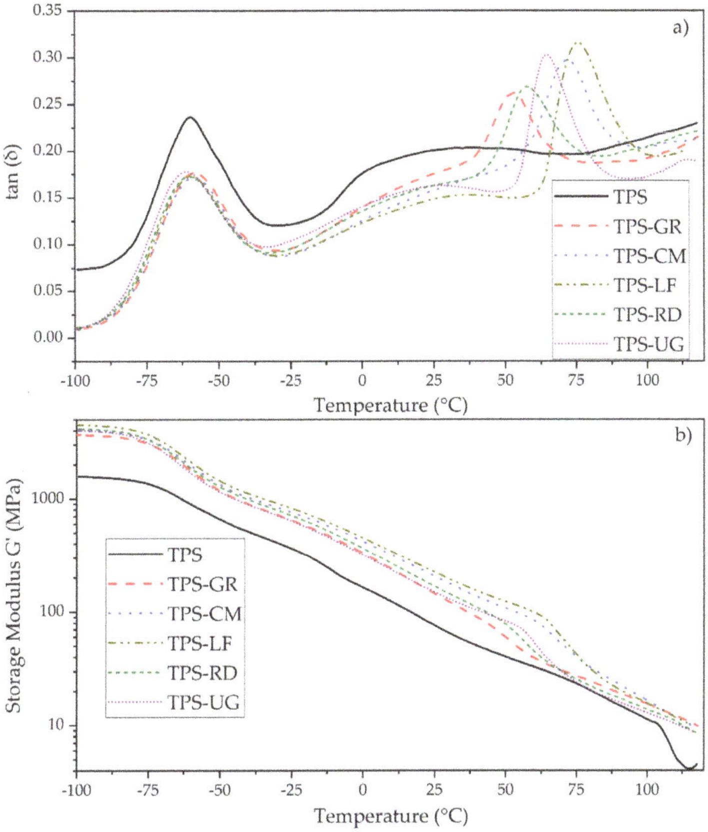

Figure 3. DMTA (**a**) Tan δ; (**b**) Storage modulus.

All the mention transitions are reflected with a corresponding decrease in the storage modulus (G′) curves (Figure 3b). At low temperatures (−100 to −70 °C), TPS storage modulus is 1500 MPa and the modulus of all TPS-pine resin derivative formulation is approximately 3900 MPa, a 260% higher than neat TPS. This indicates that in low temperatures, pine resin derivatives increase the elastic component of the matrix [19] by stiffening the structure. Then, the first fall in G′ is observed from −70 °C to −45 °C, which corresponds to glycerol glass transition temperature [52]. From −45 to 40 °C, it is seen that the storage modulus decreases proportionally with the increase of temperature. The behavior of the storage modulus is linked to the gain of mobility of molecular segments in the polymer when the temperature increase. At low temperatures, the molecules have low kinetic energy, therefore the mobility is reduced, and the storage modulus is presented as a plateau. When the temperature increase, so does the kinetic energy and the free volume in the molecular segments, which reduce the elastic component of the material (storage modulus) [53]. In this temperature range (−45 to 40 °C), pine resin derivatives continue to stiffen the TPS structure, increasing the elastic component of the TPS in a constant amount between 200 to 300%. This behavior shows a greater cohesion of the components and reduced heterogeneity, which could be attributed to a chemical interaction between pine resin derivatives and TPS functional groups [19,51]. These chemical interactions are seen in Fourier transform infrared spectroscopy (FTIR) carried out in previous work [18].

The second fall in G′ is observed between 45 °C and 75 °C, which shows a loss in the stiffness of TPS pine resin derivatives [54], as the materials reach their glass transition temperature. After this fall, the storage modulus of the TPS-pine resin derivatives formulations equals the storage modulus of neat TPS, which suggests that pine resin derivatives lose their rigidity due to the temperature.

3.2. Surface Characterization, Color, and Wettability

Surface color parameters for the CIEL*a*b* space are presented in Table 1. The formulations present significant differences ($p < 0.5$) in all the color parameters respect neat TPS due to the addition of the resin. However, in GR and RD the lightness increased by 55%, in CM and LF it increased 69% and UG produced an increase in the lightness of 80% in TPS.

Table 1. Water contact angle and color parameters for the CIEL*a*b* space of neat TPS and the formulations with 10 wt.% GR, CM, LF, RD, or UG.

	WCA	L*	a*	b*	ΔE	YI
TPS	53.0 ± 3.0 [a]	37.88 ± 0.91 [a]	−1.47 ± 0.06 [a]	5.93 ± 0.15 [a]	0.00 ± 0.00 [a]	21.04 ± 0.57 [a]
TPS-GR	80.9 ± 2.9 [b]	58.11 ± 1.05 [b]	−1.72 ± 0.08 [b]	17.55 ± 0.68 [b]	23.33 ± 1.19 [b]	44.6 ± 1.17 [b]
TPS-CM	70.5 ± 1.9 [c]	64.34 ± 0.70 [c]	−2.81 ± 0.16 [c]	23.68 ± 0.80 [c]	31.90 ± 1.01 [c]	53.47 ± 1.26 [c]
TPS-LF	86.4 ± 2.9 [d]	63.95 ± 0.47 [c]	−3.64 ± 0.10 [d]	12.78 ± 0.43 [d]	27.04 ± 0.54 [d]	29.35 ± 0.91 [d]
TPS-RD	77.9 ± 1.7 [b]	58.89 ± 0.78 [b]	−1.16 ± 0.18 [e]	20.85 ± 0.60 [e]	25.77 ± 0.90 [d]	51.89 ± 1.13 [d]
TPS-UG	86.0 ± 0.9 [d]	68.00 ± 0.94 [d]	−2.07 ± 0.09 [f]	17.44 ± 0.49 [b]	32.26 ± 0.97 [e]	39.46 ± 0.87 [e]

[a–f] Different letters within the same property show statistically significant differences between formulations ($p < 0.05$).

In all the materials, the a* coordinate (green to red), has negatives values between −1.16 and −1.72 showing the predominance of greenish tones over reddish ones. The b* coordinate (blue to yellow) has positive values between 5.93 and 23.68 which indicates that yellow shades were predominant over the blue ones [40]. All the resins increase the green shade in TPS from 17% to 148%, except for RD that reduces the green shade by 21%. And all the resins increase the yellow coloration of TPS by 100% or more of its initial value. Further, the yellowness index increased in the formulations due to the intrinsic yellowish coloration of the resins, being TPS-CM the formulation with the highest yellowness index and TPS-LF the formulation with the lowest value in this parameter.

The total color difference (ΔE) shows that the addition of any resin caused significant differences in the color of TPS. In all cases, these differences are higher than 2.0, which is the threshold of perceptible color differences for the human eye [10,36].

Materials intended for food packaging applications are required to protect foodstuff from humidity. Thus, to obtain information on the hydrophilic/hydrophobic nature of the TPS-resin-based materials, the surface wettability was determined by water contact angle measurements (Table 1). The neat TPS sample showed the lowest water contact angle value due to the number of hydroxyl groups from the native starch as well as glycerol, showing similar values to those reported for corn starch (with 25% amylose content) based formulations plasticized with glycerol [55]. Meanwhile, all TPS-resin blends showed higher WCA values than neat TPS. The surface wettability is dependent on the surface chemistry as well as topographical properties and it seems that the presence of either gum rosin or gum rosin derivatives increases the microstructural roughness (see SEM images in the Supporting Information Figure S1) of the materials. The highest WCA values were observed for TPS-LF and TPS-UG ($p > 0.05$) which were the resins with high molecular weight and high amounts of carbonyl groups. The carbonyl groups are positively interacting with starch hydroxyls groups [18] therefore, there are not hydrophilic groups available to interact with water at the surface of the material. The WCA value was followed by TPS-GR and TPS-RD ($p > 0.05$) which showed lower molecular weight than LF and UG and very similar molecular structures, and finally, the TPS-CM showed the lowest WCA among TPS-resins blend formulations, suggesting that there are some carbonyls and/or carboxylic groups of CM that are interacting with water at the surface. In fact, this formulation showed the less reduction of the -OH group band at 1648 cm^{-1} in the FTIR spectra (see Supporting Information Figure S2) which is related to the bound water, and thus suggest that the carbonyl and/or carboxylic groups of CM have more affinity with water than the rest of resins.

3.3. Water Uptake

The evolution of the water uptake of TPS and its formulations with resins are plotted as a function of time in Figure 4. It is observed that all the samples absorbed water. Briefly, the water uptake curves can be divided into two zones. In all the materials, regardless of the used resin, the water uptake is fast in the initial zone (t < 50 min). Whereas, in the second zone, the absorption rate decreases leading to a plateau that corresponds to the saturation [56]. It is observed, the addition of the pine resin derivatives had little effect on the water uptake capacity of TPS. However, the general trend suggests the equilibrium water uptake of TPS decreased with the addition of GR, LF, and RD, while stays in the same value with the addition of UG and CM. This result is related to the number and accessibility of polar groups [57]. In TPS-GR and TPS-RD there are fewer polar groups available to interact with water than in the other formulations, while in TPS-LF the accessibility to the groups is limited due to steric impediments. On the other hand, it is seen that the structure of TPS-CM and TPS-RD did not withstand the expansion and collapsed at 300 min of water uptake, which implies that the interaction inside their structure is weaker than the other formulations, which resisted until 480 min of water uptake before its failure. This could be related to the low toughness and low Young's modulus of these formulations reported in previous studies [18].

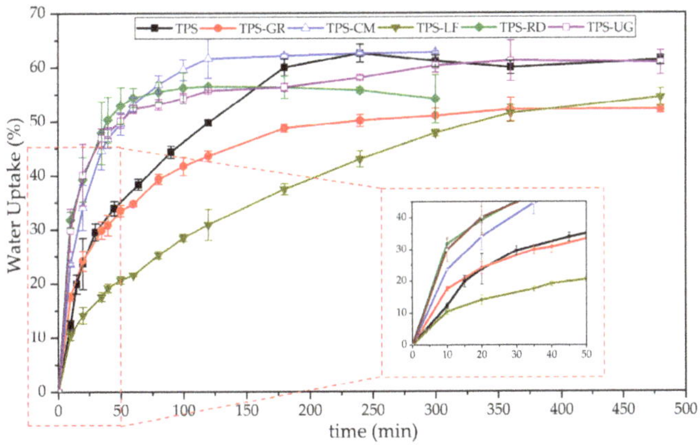

Figure 4. Water uptake of thermoplastic starch (TPS) and formulations with 10 wt.% of gum rosin (GR) and rosin derivatives (CM, LF, RD, UG) with an expanded area in the range from 0 to 480 min.

3.4. X-ray Diffraction Pattern (XDR)

A comparison between the X-ray scattering patterns of native corn starch, TPS, and TPS-pine resin derivatives, before and after being stored for 24 months is presented in Figure 5. Native starch (Figure 5a) show a predominant type A crystallinity, with peaks at 2θ of 15.2°, 17.2°, 18.1°, and 23°, typical of cereals [58]. The scattering pattern of TPS (Figure 5b) presents a broad amorphous baseline, typical of semi-crystalline polymers with a low degree of crystallinity [59]. Moreover, new peaks are observed at 2θ of 13°, 19° and 22°, which is indicative of V_H type crystallinity. The broad hump centered on 19° is characteristic of TPS [20]. The V_H type crystallinity is assigned to amylose complexed with glycerol formed during thermo-mechanical processing [58]. In the X-ray scattering patterns of TPS-pine resin derivatives before the storage (red curves in Figure 5) it is seen that the peaks corresponding to the A crystallinity (2θ of 15.2°, 17.2°, 18.1°, and 23°) have completely disappeared in TPS-CM, TPS-LF, TPS-RD, and TPS-UG, while residual native corn starch crystallinity is seen in TPS and TPS-GR at 2θ of 17°. This shows that CM, LF, RD, and UT favor the complete disruption of starch granules.

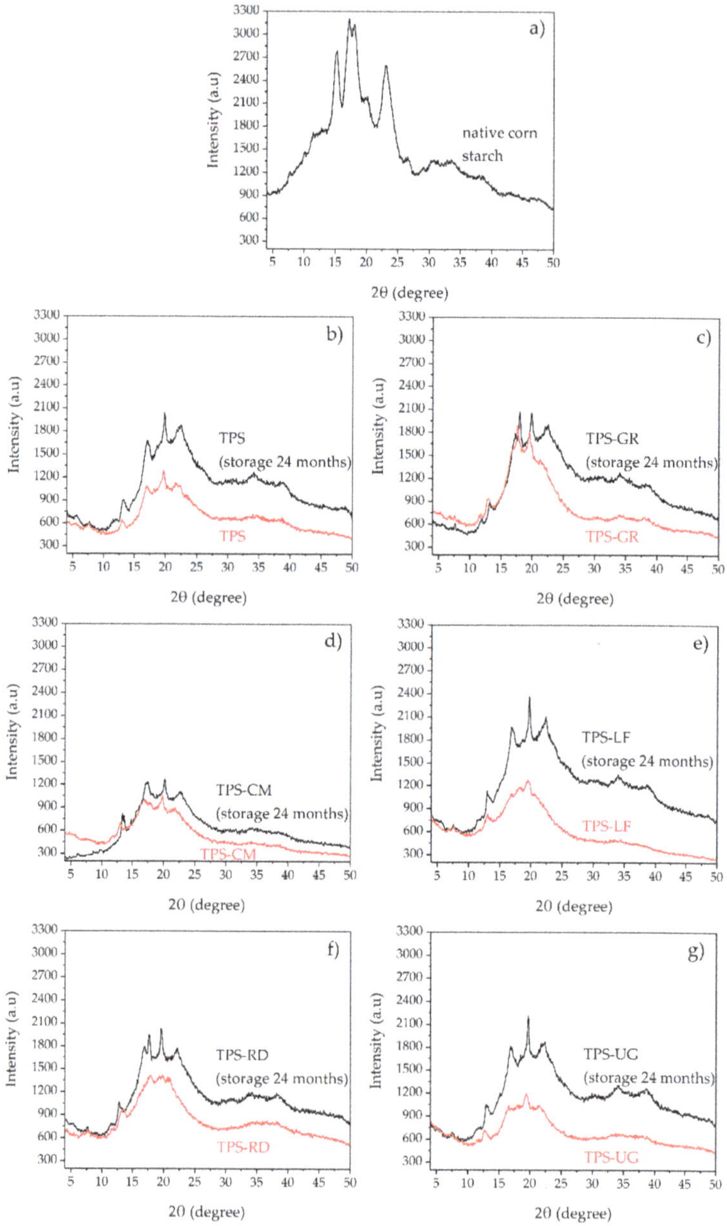

Figure 5. X-ray diffraction intensity scans for (**a**) native corn starch, (**b**) TPS, (**c**) TPS-GR, (**d**) TPS-CM, (**e**) TPS-LF, (**f**) TPS-RD, and (**g**) TPS-UG, in their initial state and after 24 months of storage.

After 24 months of storage (black curves in Figure 5), the formation of B-type crystallinity associated with retrogradation with peaks at 17°, 22°, and 30° is observed in all the formulations [59,60]. The scattering patterns area and intensity of peaks increased in the materials, which shows the crystallization of the structure due to aging. On one hand, LF and UG seem to have a negligible effect on the aging of TPS. On the other hand, the formulations with GR, CM, and RD, have a minimum change in the intensity of X-ray scattering patterns due to storage time. Therefore, it could be said that these resins restrain the aging of TPS due to retrogradation. This could be explained because the molecular structures of GR, CM, and RD are smaller than those of LF and UG (as seen in Figure 1), which allows them to hinder the molecular ordering of TPS. In fact, it has been reported that gums bind water and change the water distribution of the starch-based systems, thus weakening the recrystallization of starch molecular chains [61].

3.5. Disintegration under Composting Conditions

Since these materials are intended for sustainable packaging applications, the disintegrability under composting conditions mediated by thermophilic bacteria was assessed to get information regarding the possibility to dispose of these materials with organic wastes. Figure 6 shows the visual appearance of all formulations during the disintegration test, while Figure 7 shows the degree of disintegrability evaluated in terms of mass loss as a function of incubation time by using the Boltzmann equation by the corresponding half-maximal degradation time (see inset Table). The loss of transparency can be observed just after the first day of composting for all TPS-based formulations and these effects were greater after 4 days under composting conditions.

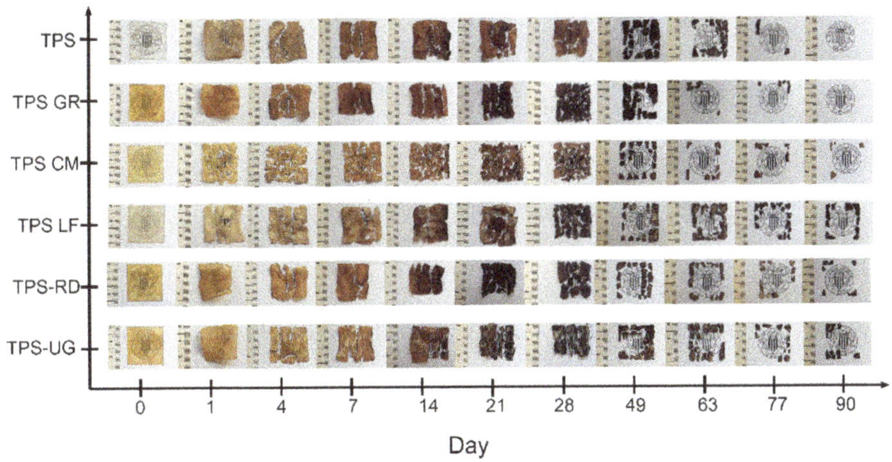

Figure 6. Visual appearance of thermoplastic starch (TPS) and formulations with 10 wt.% of gum rosin (GR) and rosin derivatives (CM, LF, RD, UG) during the disintegration test in controlled compost conditions in terms of the rosin derivatives and the elapsed time.

After one day under composting TPS became breakable and lost around 12% of the initial mass. Likewise, all TPS-resin formulations became smaller and breakable, losing between 14% and 19% of the initial mass. The higher mass observed after one day for neat TPS can be related to the fact that starch absorbed water and swelled [55]. The loss of transparency is linked to the changes in the refractive index because of water absorption and the beginning of the formation of low molecular weight compounds as a result of hydrolytic degradation [22,44]. Then, the smaller molecules become susceptible to the enzymatic degradation mediated by microorganisms. TPS changed its color becoming yellowish in 1 day and almost black after 21 days, in good accordance with the literature [22,62]. The presence of pine resin derivatives speeds up the color changes showing a more marked yellow color and reaching the black color between 21 and 49 days depending on the formulation. All formulations showed a similar kinetic pattern of disintegration (Figure 6). TPS-LF and TPS-UG formulation still showed some small pieces of materials after 90 days, while the rest of the formulations virtually disappeared in 77 days. These results are in accordance with the higher hydrophobicity showed by these formulations as it showed the higher WCA. Besides, TPS-LF was the last formulation in being completely disintegrated. The well chemical interaction through hydrogen bonding established in this formulation is also hindering the water diffusion through the bulk, as it was observed in water uptake results (Figure 4), delaying the overall hydrolysis process. Nevertheless, it should be underlined that all the developed TPS-resin formulations were disintegrated in less than 90 days under composting conditions, suggesting their perspective advantage for their industrial production for biodegradable food packaging applications.

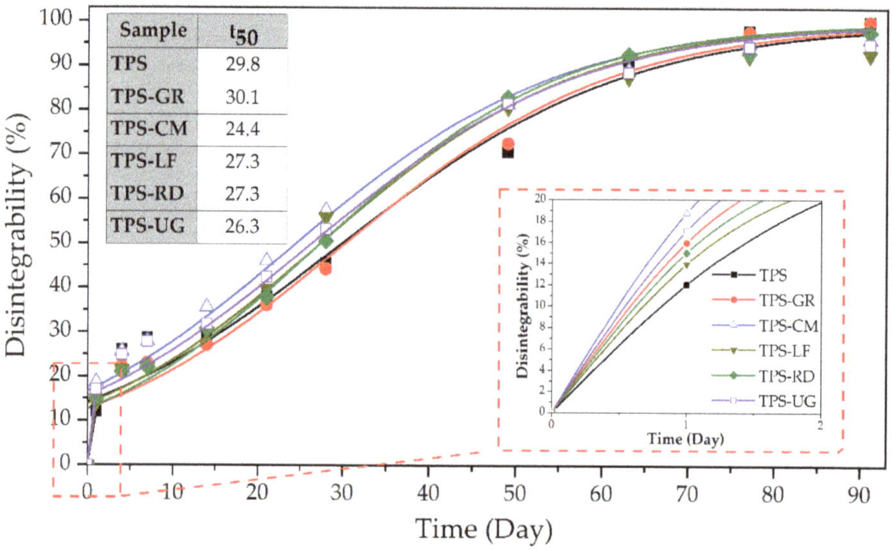

Figure 7. Disintegration degree of thermoplastic starch (TPS) and formulations with 10 wt.% of gum rosin (GR) and rosin derivatives (CM, LF, RD, UG) under controlled compost conditions as a function of time.

4. Conclusions

Thermoplastic starch (TPS) was blended with different pine resin derivatives and the obtained materials were characterized focusing on the field of food packaging. It was determined that pine resin derivatives stiffen the structure of thermoplastic starch from −100 °C to 40 °C, which suggests a good cohesion between pine resin derivatives and thermoplastic starch in this temperature range. Besides, the total color of the TPS was significatively influenced due to the addition of pine resin derivatives, which is a consequence of its intrinsic yellowish coloration. Regarding the hydrophobicity, it was determined that pine resin derivatives significatively increase the water contact angle (WCA) of TPS, providing a surface hydrophobicity improvement. WCA values were influenced by the hydrophilic chemical groups available in the structure of the respective pine resin derivative used in the blends. For instance, the blends with pentaerythritol resins (TPS-LF and TPS-UG) present the highest WCA values because the hydroxyl groups have an interaction with the hydroxyl groups in the starch structure, followed by TPS-GR and TPS-RD. Additionally, the water uptake analysis shows that TPS and TPS-pine resin derivatives blends present a high affinity with water, which causes a high water diffusion rate. However, it was seen that GR, LF, and RD decrease the equilibrium water uptake of

TPS because these resins present fewer polar groups available. X-ray diffraction analyses showed that CM, LF, RD, and UT favor the complete disruption of starch granules in the fabrication of TPS and that GR, CM, and RD restrain the aging of TPS. The disintegration under composting condition analyses shows that LF and UG delay the disintegration time which could be related to the availability of hydroxyl groups as mentioned in the WCA results. To end, it is worth mention that all the formulations were disintegrated in less than 90 days under composting conditions, which favors their use in biodegradable food packaging applications. Therefore, these materials are interesting for several sustainable rigid food packaging applications, since the increased T_g values allow to obtain rigid materials that do not suffer plastic deformation at room temperature, being fully biobased and biodegradable and, thus, aligned with the circular economy approach. Furthermore, the final properties of the material can be focused on a specific area depending on the used pine resin derivative. For applications that seek a film with higher transparency, low water uptake, and high mechanical resistance, the formulation TPS-LF is preferred. However, to restrain the aging due to retrogradation the formulations TPS-CM, TPS-RD, and TPS-GR are ideal. Finally, if it is desired to obtain a disintegration under composting conditions equal to pure TPS, GR can be used in the formulation.

Supplementary Materials: The following are available online at https://www.mdpi.com/article/10.3390/foods10061171/s1, Figure S1: Scanning electron microscopy (SEM) images of (a) TPS, (b) TPS-GR, (c) TPS-CM, (d) TPS-LF, (e) TPS-RD, and (f) TPS-UG. Figure S2: Fourier transform infrared spectroscopy (FTIR) spectra of: (a) TPS, (b) TPS-GR, (c) TPS-CM, (d) TPS-LF, (e) TPS-RD and (f) TPS-UG.

Author Contributions: Conceptualization, C.P. and J.L.-M.; data curation, M.A. and J.H.-F.; formal analysis, C.P. and M.A.; funding acquisition, J.L.-M. and M.P.A.; investigation, C.P., M.A., J.H.-F., and M.P.A.; methodology, C.P., M.A., and J.H.-F.; project administration, J.L.-M. and M.P.A.; resources, J.L.-M.; software, M.A.; supervision, J.L.-M. and M.P.A.; validation, C.P., J.L.-M., and M.P.A.; visualization, J.H.-F. and M.P.A.; writing–original draft, C.P.; writing–review and editing, M.A. and M.P.A. All authors have read and agreed to the published version of the manuscript.

Funding: This research was funded by MINECO, with the project: PROMADEPCOL: MAT2017-84909-C2-2-R.

Institutional Review Board Statement: Not applicable.

Informed Consent Statement: Not applicable.

Data Availability Statement: The data presented in this study are available on request from the corresponding author.

Acknowledgments: C.P. thanks Santiago Grisolía fellowship (GRISOLIAP/2019/113) from Generalitat Valenciana. M.A. thanks Secretaría de Educación Superior, Ciencia, Tecnología e Innovación (SENESCYT) and Escuela Politécnica Nacional (EPN).

Conflicts of Interest: The authors declare no conflict of interest.

References

1. Geyer, R.; Jambeck, J.R.; Law, K.L. Production, use, and fate of all plastics ever made. *Sci. Adv.* **2017**, *3*, 1–5. [CrossRef]
2. Hong, M.; Chen, E.Y.-X. Future Directions for Sustainable Polymers. *Trends Chem.* **2019**, *1*, 148–151. [CrossRef]
3. Schneiderman, D.K.; Hillmyer, M.A. 50th Anniversary Perspective: There Is a Great Future in Sustainable Polymers. *Macromolecules* **2017**, *50*, 3733–3749. [CrossRef]
4. Gkartzou, E.; Koumoulos, E.P.; Charitidis, C.A. Production and 3D printing processing of bio-based thermoplastic filament. *Manuf. Rev.* **2017**, *4*, 1–14. [CrossRef]
5. Hong, M.; Chen, E.Y.-X. Chemically recyclable polymers: A circular economy approach to sustainability. *Green Chem.* **2017**, *19*, 3692–3706. [CrossRef]
6. Plastics Europe Market Research Group (PEMRG). *Plastics-The Facts 2019 an Analysis of European Plastics Production, Demand and Waste Data*; PlasticsEurope: Brussels, Belgium, 2019.
7. Eichhorn, S.J.; Dufresne, A.; Aranguren, M.; Marcovich, N.E.; Capadona, J.R.; Rowan, S.J.; Weder, C.; Thielemans, W.; Roman, M.; Renneckar, S.; et al. Review: Current international research into cellulose nanofibres and nanocomposites. *J. Mater. Sci.* **2010**, *45*, 1–33. [CrossRef]

8. Wang, Z.; Ganewatta, M.S.; Tang, C. Sustainable polymers from biomass: Bridging chemistry with materials and processing. *Prog. Polym. Sci.* **2020**, *101*, 101197–101211. [CrossRef]
9. Pillai, C.; Paul, W.; Sharma, C.P. Chitin and chitosan polymers: Chemistry, solubility and fiber formation. *Prog. Polym. Sci.* **2009**, *34*, 641–678. [CrossRef]
10. Arrieta, M.P.; Peltzer, M.A.; López, J.; Garrigós, M.D.C.; Valente, A.J.; Jiménez, A. Functional properties of sodium and calcium caseinate antimicrobial active films containing carvacrol. *J. Food Eng.* **2014**, *121*, 94–101. [CrossRef]
11. Galbis, J.A.; De Gracia García-Martín, M.; Violante De Paz, M.; Galbis, E.; García-Martín, M.D.G.; De Paz, M.V.; Galbis, E. *Synthetic Polymers from Sugar-Based Monomers*; American Chemical Society: Washington, DC, USA, 2020; Volume 116, pp. 1600–1636.
12. Papageorgiou, G.Z. Thinking Green: Sustainable Polymers from Renewable Resources. *Polymers* **2018**, *10*, 952. [CrossRef] [PubMed]
13. Arrieta, M.P.; Samper, M.D.; Aldas, M.; López, J. On the Use of PLA-PHB Blends for Sustainable Food Packaging Applications. *Materials* **2017**, *10*, 1008. [CrossRef] [PubMed]
14. Singh, A.; Gu, Y.; Castellarin, S.D.; Kitts, D.D.; Pratap-Singh, A. Development and Characterization of the Edible Packaging Films Incorporated with Blueberry Pomace. *Foods* **2020**, *9*, 1599. [CrossRef]
15. Díaz-Galindo, E.P.; Nesic, A.; Bautista-Baños, S.; García, O.D.; Cabrera-Barjas, G. Corn-Starch-Based Materials Incorporated with Cinnamon Oil Emulsion: Physico-Chemical Characterization and Biological Activity. *Foods* **2020**, *9*, 475. [CrossRef] [PubMed]
16. Al-Hashimi, G.A.; Ammar, A.B.; Lakshmanan, G.; Cacciola, F.; Lakhssassi, N. Development of a Millet Starch Edible Film Containing Clove Essential Oil. *Foods* **2020**, *9*, 184. [CrossRef]
17. Leon-Bejarano, M.; Durmus, Y.; Ovando-Martínez, M.; Simsek, S. Physical, Barrier, Mechanical, and Biodegradability Properties of Modified Starch Films with Nut by-Products Extracts. *Foods* **2020**, *9*, 226. [CrossRef] [PubMed]
18. Aldas, M.; Pavon, C.; López-Martínez, J.; Arrieta, M.P. Pine Resin Derivatives as Sustainable Additives to Improve the Mechanical and Thermal Properties of Injected Moulded Thermoplastic Starch. *Appl. Sci.* **2020**, *10*, 2561. [CrossRef]
19. Aldas, M.; Ferri, J.M.; Lopez-Martinez, J.; Samper, M.D.; Arrieta, M.P. Effect of pine resin derivatives on the structural, thermal, and mechanical properties of Mater-Bi type bioplastic. *J. Appl. Polym. Sci.* **2020**, *137*, 48236–48250. [CrossRef]
20. Angellier, H.; Molina-Boisseau, S.; Dole, P.; Dufresne, A. Thermoplastic Starch-Waxy Maize Starch Nanocrystals Nanocomposites. *Biomacromolecules* **2006**, *7*, 531–539. [CrossRef]
21. Espigulé, E.; Puigvert, X.; Vilaseca, F.; Méndez, J.A.; Mutjé, P.; Gironès, J. Thermoplastic Starch-based Composites Reinforced with Rape Fibers: Water Uptake and Thermomechanical Properties. *BioResources* **2013**, *8*, 2620–2630. [CrossRef]
22. Sessini, V.; Arrieta, M.P.; Raquez, J.M.; Dubois, P.; Kenny, J.M.; Peponi, L. Thermal and composting degradation of EVA/Thermoplastic starch blends and their nanocomposites. *Polym. Degrad. Stab.* **2019**, *159*, 184–198. [CrossRef]
23. Sessini, V.; Arrieta, M.P.; Fernández-Torres, A.; Peponi, L. Humidity-activated shape memory effect on plasticized starch-based biomaterials. *Carbohydr. Polym.* **2018**, *179*, 93–99. [CrossRef] [PubMed]
24. Pawlak, F.; Aldas, M.; Parres, F.; López-Martínez, J.; Arrieta, M.P. Silane-Functionalized Sheep Wool Fibers from Dairy Industry Waste for the Development of Plasticized PLA Composites with Maleinized Linseed Oil for Injection-Molded Parts. *Polymers* **2020**, *12*, 2523. [CrossRef] [PubMed]
25. Pavon, C.; Aldas, M.; de la Rosa-Ramírez, H.; López-Martínez, J.; Arrieta, M.P. Improvement of PBAT Processability and Mechanical Performance by Blending with Pine Resin Derivatives for Injection Moulding Rigid Packaging with Enhanced Hydrophobicity. *Polymers* **2020**, *12*, 2891. [CrossRef] [PubMed]
26. de la Rosa-Ramirez, H.; Aldas, M.; Ferri, J.M.; López-Martínez, J.; Samper, M.D. Modification of poly (lactic acid) through the incorporation of gum rosin and gum rosin derivative: Mechanical performance and hydrophobicity. *J. Appl. Polym. Sci.* **2020**, *137*, 49346–49361. [CrossRef]
27. Moustafa, H.; El Kissi, N.; Abou-Kandil, A.I.; Abdel-Aziz, M.S.; Dufresne, A.; El Kissi, N.; Abou-Kandil, A.I.; Abdel-Aziz, M.S.; Dufresne, A. PLA/PBAT Bionanocomposites with Antimicrobial Natural Rosin for Green Packaging. *ACS Appl. Mater. Interfaces* **2017**, *9*, 20132–20141. [CrossRef]
28. Langenheim, J. Plant Resins. *Am. Sci.* **1990**, *78*, 16–24.
29. Arrieta, M.P.; Samper, M.; Jiménez-López, M.; Aldas, M.; López, J. Combined effect of linseed oil and gum rosin as natural additives for PVC. *Ind. Crop. Prod.* **2017**, *99*, 196–204. [CrossRef]
30. Rodríguez-García, A.; Martín, J.A.; López, R.; Sanz, A.; Gil, L. Effect of four tapping methods on anatomical traits and resin yield in Maritime pine (Pinus pinaster Ait.). *Ind. Crop. Prod.* **2016**, *86*, 143–154. [CrossRef]
31. Yadav, B.K.; Gidwani, B.; Vyas, A. Rosin: Recent advances and potential applications in novel drug delivery system. *J. Bioact. Compat. Polym.* **2015**, *31*, 111–126. [CrossRef]
32. Chang, R.; Rohindra, D.; Lata, R.; Kuboyama, K.; Ougizawa, T. Development of poly(ε-caprolactone)/pine resin blends: Study of thermal, mechanical, and antimicrobial properties. *Polym. Eng. Sci.* **2019**, *59*, 32–41. [CrossRef]
33. Yao, K.; Tang, C. Controlled Polymerization of Next-Generation Renewable Monomers and Beyond. *Macromolecules* **2013**, *46*, 1689–1712. [CrossRef]
34. Wilbon, P.A.; Chu, F.; Tang, C. Progress in Renewable Polymers from Natural Terpenes, Terpenoids, and Rosin. *Macromol. Rapid Commun.* **2013**, *34*, 8–37. [CrossRef]
35. Aldas, M.; Rayón, E.; López-Martínez, J.; Arrieta, M.P. A Deeper Microscopic Study of the Interaction between Gum Rosin Derivatives and a Mater-Bi Type Bioplastic. *Polymers* **2020**, *12*, 226. [CrossRef] [PubMed]

36. Pavon, C.; Aldas, M.; López-Martínez, J.; Ferrándiz, S. New Materials for 3D-Printing Based on Polycaprolactone with Gum Rosin and Beeswax as Additives. *Polymers* **2020**, *12*, 334. [CrossRef]
37. Sessini, V.; Arrieta, M.P.; Kenny, J.M.; Peponi, L. Processing of edible films based on nanoreinforced gelatinized starch. *Polym. Degrad. Stab.* **2016**, *132*, 157–168. [CrossRef]
38. International Standards Organization. *ISO 62:2008 Plastics—Determination of Water Absorption*; ISO (International Organization for Standardization): Geneva, Switzerland, 2008.
39. Liu, G.; Wu, G.; Chen, J.; Kong, Z. Synthesis, modification and properties of rosin-based non-isocyanate polyurethanes coatings. *Prog. Org. Coat.* **2016**, *101*, 461–467. [CrossRef]
40. Atodiresei, G.-V.; Sandu, G.; Tulbure, E.-A.; Vasilache, V.; Butnaru, R. Chromatic Characterization in CieLab System for Natural Dyed Materials, Prior Activation in Atmospheric Plasma Type DBD. *Rev. Chim.* **2013**, *64*, 165–169.
41. Fombuena, V.; Balart, J.; Boronat, T.; Sánchez-Nácher, L.; Garcia-Sanoguera, D. Improving mechanical performance of thermoplastic adhesion joints by atmospheric plasma. *Mater. Des.* **2013**, *47*, 49–56. [CrossRef]
42. International Standards Organization. *ISO 20200 Plastics—Determination of the Degree of Disintegration of Plastic Materials under Simulated Composting Conditions in a Laboratory-Scale Test*; ISO (International Organization for Standardization): Geneva, Switzerland, 2016.
43. Arrieta, M.; López, J.; Rayón, E.; Jiménez, A. Disintegrability under composting conditions of plasticized PLA–PHB blends. *Polym. Degrad. Stab.* **2014**, *108*, 307–318. [CrossRef]
44. Fortunati, E.; Puglia, D.; Santulli, C.; Sarasini, F.; Kenny, J.M. Biodegradation of Phormium Tenax/Poly (Lactic Acid) Composites. *J. Appl. Polym. Sci.* **2012**, *125*, 562–572. [CrossRef]
45. Arrieta, M.P.; Fortunati, E.; Dominici, F.; López, J.; Kenny, J.M. Bionanocomposite films based on plasticized PLA–PHB/cellulose nanocrystal blends. *Carbohydr. Polym.* **2015**, *121*, 265–275. [CrossRef]
46. Villegas, C.; Arrieta, M.P.; Rojas, A.; Torres, A.; Faba, S.; Toledo, M.J.; Gutierrez, M.A.; Zavalla, E.; Romero, J.; Galotto, M.J.; et al. PLA/Organoclay Bionanocomposites Impregnated with Thymol and Cinnamaldehyde by Supercritical Impregnation for Active and Sustainable Food Packaging. *Compos. Part B Eng.* **2019**, *176*, 107336–107348. [CrossRef]
47. Arrieta, M.P.; Parres, F.; López, J.; Jiménez, A. Development of a novel pyrolysis-gas chromatography/mass spectrometry method for the analysis of poly(lactic acid) thermal degradation products. *J. Anal. Appl. Pyrolysis* **2013**, *101*, 150–155. [CrossRef]
48. Da Róz, A.; Carvalho, A.; Gandini, A.; Curvelo, A. The effect of plasticizers on thermoplastic starch compositions obtained by melt processing. *Carbohydr. Polym.* **2006**, *63*, 417–424. [CrossRef]
49. Taguet, A.; Huneault, M.A.; Favis, B.D. Interface/morphology relationships in polymer blends with thermoplastic starch. *Polymer* **2009**, *50*, 5733–5743. [CrossRef]
50. Avérous, L.; Fringant, C.; Moro, L. Plasticized starch–cellulose interactions in polysaccharide composites. *Polymer* **2001**, *42*, 6565–6572. [CrossRef]
51. De Graaf, R.A.; Karman, A.P.; Janssen, L.P.B.M. Material Properties and Glass Transition Temperatures of Different Thermoplastic Starches after Extrusion Processing. *Starch-Staerke* **2003**, *55*, 80–86. [CrossRef]
52. Forssell, P.M.; Mikkilä, J.M.; Moates, G.K.; Parker, R. Phase and glass transition behaviour of concentrated barley starch-glycerol-water mixtures, a model for thermoplastic starch. *Carbohydr. Polym.* **1997**, *34*, 275–282. [CrossRef]
53. Patidar, D.; Agrawal, S.; Saxena, N. Storage modulus and glass transition behaviour of CdS/PMMA nanocomposites. *J. Exp. Nanosci.* **2011**, *6*, 441–449. [CrossRef]
54. Chartoff, R.P.; Sircar, A. Thermal Analysis of Polymers. *Encycl. Polym. Sci. Technol.* **2004**, 1–43. [CrossRef]
55. Chen, J.; Wang, X.; Long, Z.; Wang, S.; Zhang, J.; Wang, L. Preparation and Performance of Thermoplastic Starch and Microcrystalline Cellulose for Packaging Composites: Extrusion and Hot Pressing. *Int. J. Biol. Macromol.* **2020**, *165*, 2295–2302. [CrossRef] [PubMed]
56. Mathew, A.P.; Dufresne, A. Plasticized Waxy Maize Starch: Effect of Polyols and Relative Humidity on Material Properties. *Biomacromolecules* **2002**, *3*, 1101–1108. [CrossRef]
57. Edlund, U.; Albertsson, A.-C. Degradable Polymer Microspheres for Controlled Drug Delivery. *Polym. Phys.* **2007**, *157*, 67–112.
58. Ghanbari, A.; Tabarsa, T.; Ashori, A.; Shakeri, A.; Mashkour, M. Preparation and characterization of thermoplastic starch and cellulose nanofibers as green nanocomposites: Extrusion processing. *Int. J. Biol. Macromol.* **2018**, *112*, 442–447. [CrossRef]
59. Mendes, J.; Paschoalin, R.T.; Carmona, V.; Neto, A.R.S.; Marques, A.; Marconcini, J.; Mattoso, L.; Medeiros, E.; Oliveira, J. Biodegradable polymer blends based on corn starch and thermoplastic chitosan processed by extrusion. *Carbohydr. Polym.* **2016**, *137*, 452–458. [CrossRef]
60. Huneault, M.A.; Li, H. Preparation and properties of extruded thermoplastic starch/polymer blends. *J. Appl. Polym. Sci.* **2012**, *126*, 96–108. [CrossRef]
61. Zhao, Q.; Tian, H.; Chen, L.; Zeng, M.; Qin, F.; Wang, Z.; He, Z.; Chen, J. Interactions between soluble soybean polysaccharide and starch during the gelatinization and retrogradation: Effects of selected starch varieties. *Food Hydrocoll.* **2021**, *118*, 106765–106779. [CrossRef]
62. Ochoa-Yepes, O.; Medina-Jaramillo, C.; Guz, L.; Famá, L. Biodegradable and Edible Starch Composites with Fiber-Rich Lentil Flour to Use as Food Packaging. *Starch-Staerke* **2018**, *70*, 1700222–1700230. [CrossRef]

foods

MDPI

Article

Valorization of Liquor Waste Derived Spent Coffee Grains for the Development of Injection-Molded Polylactide Pieces of Interest as Disposable Food Packaging and Serving Materials

Enrique Terroba-Delicado [1], Stefano Fiori [2], Jaume Gomez-Caturla [1], Nestor Montanes [1], Lourdes Sanchez-Nacher [1] and Sergio Torres-Giner [3,*]

[1] Technological Institute of Materials (ITM), Universitat Politècnica de València (UPV), Plaza Ferrándiz y Carbonell 1, 03801 Alcoy, Spain; enterde@epsa.upv.es (E.T.-D.); jaugoca@epsa.upv.es (J.G.-C.); nesmonmu@upvnet.upv.es (N.M.); lsanchez@mcm.upv.es (L.S.-N.)

[2] R&D Department, Condensia Química S.A., Calle de la Cierva 8, 08184 Palau-soilità i Plegamans, Spain; sfiori@condensia.com

[3] Research Institute of Food Engineering for Development (IIAD), Universitat Politècnica de València (UPV), Camino de Vera s/n, 46022 Valencia, Spain

* Correspondence: storresginer@upv.es

Abstract: The present work puts the Circular Bioeconomy's concept into action, originally valorizing residues of spent coffee grains from the beverage liquor coffee industry to develop green composite pieces of polylactide (PLA). The as-received spent coffee grains were first milled to obtain the so-called spent coffee grounds (SCGs) that were, thereafter, incorporated at 20 wt.% into PLA by extrusion. Finally, the resultant green composite pellets were shaped into pieces by injection molding. Moreover, two oligomers of lactic acid (OLAs), namely OLA2 and OLA2$_{mal}$, the latter being functionalized with maleic anhydride (MAH), were added with SCGs during the extrusion process at 10 wt.%. The results show that, opposite to most claims published in the literature of green composites of PLA, the incorporation of the liquor waste derived SCGs increased the ductility of the pieces by approximately 280% mainly due to their high lipid content. Moreover, the simultaneous addition of OLA2 and OLA2$_{mal}$ further contributed to improve the tensile strength of the green composite pieces by nearly 36% and 60%, respectively. The higher performance of OLA2$_{mal}$ was ascribed to the chemical interaction achieved between the biopolyester and the lignocellulosic fillers by the MAH groups. The resultant green composite pieces are very promising as disposable food-serving utensils and tableware.

Keywords: PLA; coffee liquor; food packaging materials; waste valorization; Circular Bioeconomy

Citation: Terroba-Delicado, E.; Fiori, S.; Gomez-Caturla, J.; Montanes, N.; Sanchez-Nacher, L.; Torres-Giner, S. Valorization of Liquor Waste Derived Spent Coffee Grains for the Development of Injection-Molded Polylactide Pieces ofInterest as Disposable Food Packaging and Serving Materials. *Foods* **2022**, *11*, 1162. https://doi.org/10.3390/foods11081162

Academic Editor: Luís Manuel Lopes Rodrigues da Silva

Received: 28 March 2022
Accepted: 13 April 2022
Published: 16 April 2022

1. Introduction

In recent years, society has become more aware about the environmental issues produced by the extensive use of petrochemical polymers that are not compostable or recyclable. This has promoted the development of environmentally friendly polymer materials that can potentially replace conventional polymers [1]. Among these materials, green composites are particularly interesting, since they are formed by a biopolymer matrix and a natural filler [2–4]. Furthermore, green composite materials can strongly contribute to the development of the so-called "Circular Bioeconomy", since these are obtained from natural resources or, more sustainably, wastes and, after use, can fully disintegrate under different composting conditions [5]. In the area of biopolymers, polylactide (PLA) has made a large contribution, since this biopolyester is both fully bio-based, obtained from starch-rich materials, and compostable in controlled industrial facilities [6,7]. PLA is a linear biopolyester that shows high mechanical resistance and thermal stability, good transparency, and water-barrier performance. This thermoplastic shows a high application interest in food packaging [8–10]; however, in spite of this, the use of PLA-based articles in large-scale applications is still limited by its poor toughness and brittle behavior, low thermomechanical resistance, and relatively high cost [11]. In particular, its inherent high brittleness entails a considerable scientific challenge in the food packaging field [12].

In the novel Circular Bioeconomy context, lignocellulosic materials obtained from processing by-products and wastes of the food and agroforestry industries are also gaining more attention as cost-effective fillers due to their natural origin and favored biodegradability [13]. Thus, several agri-food waste derived fillers have been reported in recent years, for example, almond shell [14], peanut skin [15,16], argan nut shell [17,18], pineapple flour [19], coconut fibers [20], banana and plantain fibers [21], or pinecone [22]. In this regard, large quantities of by-products and wastes are generated during the 'fruit-to-cup' transformation of coffee beans, not only including leaf, flower, and cherry, but also coffee husks, coffee silver skin (CSS), and spent coffee grounds (SCGs) [23,24]. SCGs are the resultant dark-colored residual materials obtained during the treatment of coffee powder with hot water or steam for the instant (or hydrosoluble) coffee preparation. In this regard, it is considered that nearly 50% of the coffee production worldwide is processed for soluble coffee preparation, whereas only 0.33–0.45 kg of instant coffee is obtained from 1 kg of green coffee beans in the soluble coffee industry [25]. As a result, this process generates 550–670 g of SCGs per kg of coffee beans, which is equivalent to 6 million tons of SCGs per year. Polysaccharides are the most abundant components in SCGs (12.4 wt.% cellulose, 39.1 wt.% hemicellulose, and 23.8 wt.% lignin), followed by proteins (17.4 wt.%), lipids (2.3 wt.%), and ashes (1.3 wt.%, mainly potassium minerals), which varies depending on both the variety of coffee beans and extraction process [26]. However, the coffee liquor industry is not so popular due to the fact that it is limited to specific geographical areas. Liquor, also known as spirit, is an alcoholic beverage produced by the distillation of plants or grains, which have already experienced alcoholic fermentation. In this industry, coffee grains are naturally roasted and then macerated with ethanol to obtain the alcoholic beverage. Furthermore, unlike instant coffee, which uses pressurized hot water (~175 °C) to extract all of the soluble solids and volatile compounds [23], the coffee used in the liquor industry is not subjected to any drastic temperature or extraction processes. Therefore, the resultant SCGs can retain their natural compositions in proteins and lipids. The potential uses of SCGs include biodiesel production [27], elements to increase compost quality [28], water purification of pharmaceutic products [29], or as a biofuel on its own [30]. Nevertheless, SCGs can also be used as reinforcing fillers in the development of green composites [31–33]. Moreover, SCGs also contain polyphenols and free radicals that can offer antioxidant and anti-tumor activities [34]. However, similar to other lignocellulosic fillers or fibers, these tend to act as reinforcing elements due to their relatively high hardness. Thus, when they are incorporated into biopolymer matrices, hardness is increased, but the ductile properties are noticeably reduced [35].

Furthermore, one of the main drawbacks of lignocellulosic fillers when incorporated into polymer matrices is their poor filler-matrix interfacial adhesion, which increases the probability of particle aggregate formation. This low compatibility is based on the their poor chemical affinity, since PLA and other biopolyesters are highly hydrophobic, while lignocellulosic fillers present a more hydrophilic character [36]. As a result of this poor affinity, the thermal stability and mechanical strength and ductility of the resultant green composites of PLAs tend to decrease. In this regard, the use of plasticizers or compatibilizers can be very useful for improving the interfacial adhesion between the lignocellulosic fillers and polymer matrices [37]. In green composites, compatibilization can be achieved through a melt-grafting process of the natural fillers onto the polymer chains during extrusion using multi-functional additives, that is, with multiple reactive groups or sites [38]. Normally, these reactive compatibilizers are based on low-molecular-weight (M_W) polymers or oligomers containing multiple epoxy or maleic anhydride (MAH) groups [39]. During the so-called reactive extrusion (REX), some of the oxygen-based groups of the compatibilizer form new ester bonds with the terminal hydroxyl (—OH) or carbonyl (—COOH) groups of the biopolyester while, at the same time, other groups can react with the —OH groups present on the surface of the lignocellulosic filler, improving the compatibility by the formation of covalent bonds [14,40]. In PLA formulations, the oligomers of lactic acid (OLAs) can habitually favor compatibility through the plasticization of the biopolyester matrix, taking advantage of their similar chemical structure. In particular, 20 wt.% OLAs have been reported to reduce the glass transition temperature (T_g) of PLAs by 13 °C [41] as well as improve its mechanical ductile properties [42]. Moreover, it is possible to chemically modify OLAs by incorporating reactive functional groups into their structure. For example, Lowe et al. [43] reported the development of controlled M_W telechelic lactic acid oligomers with acrylate functionalities by transesterification. Additionally, the work of Grosse et al. [44] reported the addition of functional MAH groups to the structure of OLAs. Therefore, the resultant chemically modified OLAs are expected to serve as potential reactive compatibilizers for blends or composites based on PLA.

This work reports the preparation and characterization of injection-molded pieces of PLA/SCG green composites with the objective of developing a sustainable and cost-efficient material for food packaging and food contact disposables. To improve the compatibility between the PLA matrix and the SCG filler, and then the performance of the green composite, two novel OLAs were originally tested, namely OLA2 and $OLA2_{mal}$. Whereas OLA2 is designed to improve the impact strength of PLA-based formulations, $OLA2_{mal}$ is a chemically modified grade that presents several reactive MAH groups distributed along the oligomer chains.

2. Materials and Methods

2.1. Materials

PLA, with the commercial reference PURAPOL L130 grade and based on a 99% L-isomer, was provided by Corbion (Gorinchem, The Netherlands). It shows a density of 1.24 g·cm^{-3} and a melt flow rate (MFR) of 16 g/10 min (210 °C and 2.16 kg). This PLA grade is granted with the food contact status, whereas it is suitable for injection molding and is in compliance with the EN 13432 standard to be processed by industrial composting.

The spent coffee grains were supplied by Licores Sinc. S.A. (Alcoy, Spain), which is obtained as a by-product from the production of liquor coffee. In this process, coffee grains were first naturally roasted and then macerated for 10 days in alcohol to obtain the drink called coffee liquor. The received biomass obtained after the process of maceration was washed several times with clean water and, subsequently, vacuum dried in a dehumidifying stove (MCP Vacuum Casting System, Lubeck, Germany) for 1 week at 60 °C. Finally, to obtain the so-called SCGs, the grains were milled in a ZM 200 centrifugal mill from Retsch (Düsseldorf, Germany) at a speed of 12,000 rpm and sieved with a 250 μm mesh filter. Figure 1 presents the images of the as-received liquor waste-derived spent coffee grains (Figure 1a) and the resultant SCG particles (Figure 1b).

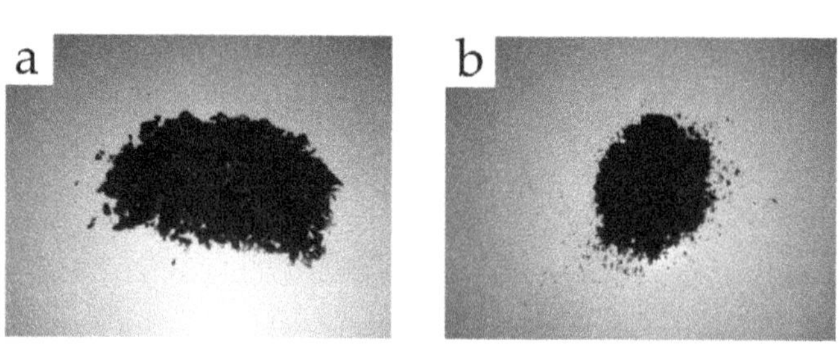

Figure 1. (**a**) As-received spent coffee grains from the production of liquor coffee, and (**b**) the resultant spent coffee grounds (SCGs) obtained after washing, drying, and milling.

Glyplast OLA2 and $OLA2_{mal}$ were provided by Condensia Química Inc. (Barcelona, Spain). Glyplast OLA2 is a liquid oligomer of lactic acid with an M_W comprised between

500–600 g·mol^{-1}. It has a viscosity of 90 mPa s at 40 °C and an ester content of >99%. Its density is 1.10 g·cm^{-3}, it has a maximum acid number of 2.5 mg KOH·g^{-1}, and a maximum water content of 0.1 wt.%. With regard to Glyplast OLA2$_{mal}$, it is an OLA2 derivative grafted with MAH. The MAH content ranges between 3 and 4 wt.%. Figure 2 depicts the chemical structure of both oligomers.

Glyplast OLA2
m = 6.9–8.3

Glyplast OLA2$_{mal}$
MAH content 3–4 wt.%

Figure 2. Chemical structures of the oligomers of lactic acid (OLAs): Gyplast OLA2 (**left**) and Gyplast OLA2$_{mal}$ (**right**).

2.2. *Preparation of the PLA/SCG Composites*

The PLA pellets and both OLAs, that is, OLA2 and OLA2$_{mal}$, were initially dried at 60 °C and 40 °C, respectively, for 48 h in a dehumidifying dryer MDEO to remove any residual moisture prior to processing. Then, the corresponding weight fraction (wt.%) of each component was mixed and pre-homogenized in a zipper bag. The corresponding formulations, shown in Table 1, were compounded in a twin-screw extruder from Construcciones Mecánicas Dupra, S.L. (Alicante, Spain). This extruder has a 25 mm diameter with a length-to-diameter ratio (L/D) of 24. The extrusion process was carried out at a rate of 22 rpm, using the following temperature profile (from hopper to die): 180–185–190–195 °C. The compounded materials were pelletized using an air-knife unit. In all cases, the residence time was approximately 1 min.

Table 1. Summary of the compositions according to the weight content (wt.%) of polylactide (PLA), spent coffee grounds (SCGs), and oligomers of lactic acid (OLAs).

Piece	PLA (wt.%)	SCG (wt.%)	OLA2 (wt.%)	OLA2$_{mal}$ (wt.%)
PLA	100	0	0	0
PLA + SCG	80	20	0	0
PLA + OLA2	90	0	10	0
PLA + OLA2$_{mal}$	90	0	0	10
PLA + OLA2 + SCG	70	20	10	0
PLA + OLA2$_{mal}$ + SCG	70	20	0	10

The resultant pellets after the extrusion process were shaped into standard samples in a Meteor 270/75 injector from Mateu & Solé (Barcelona, Spain). The temperature profiles in the injection molding unit were 185 °C (hopper), 190 °C, 195 °C, and 200 °C (injection nozzle). A clamping force of 75 tons was applied, while the cavity filling and cooling times were set to 1 and 10 s, respectively. Standard samples, with an average thickness of 4 mm, were obtained for characterization.

2.3. *Characterization of the PLA/SCG Composite Pieces*

2.3.1. Morphological Characterization

The morphology of the SCG particles and fractured samples obtained from the Charpy tests were studied by field emission scanning electron microscopy (FESEM) in a ZEISS ULTRA 55 microscope from Oxford Instruments (Abingdon, UK). Prior to observation, the samples were sputtered with a gold-palladium alloy in an EMITECH sputter coating SC7620 model from Quorum Technologies, Ltd. (Laughton, UK). The FESEM measurements were carried out using an acceleration voltage of 2 kV.

2.3.2. Mechanical Characterization

The tensile properties of PLA/SCG pieces sizing 150 mm × 10 mm × 4 mm were obtained using a universal testing machine ELIB 50 from S.A.E. Ibertest (Madrid, Spain), as recommended by ISO 527-1:2012. A 5-kN load cell was used and the cross-head speed was set to 5 mm·min^{-1}. The shore hardness was measured in a 676-D durometer from J. Bot Instruments (Barcelona, Spain), using the D-scale on rectangular samples with the dimensions of 80 mm × 10 mm × 4 mm and according to ISO 868:2003. The impact strength was also studied by means of injection-molded rectangular samples with the dimensions of 80 mm × 10 mm × 4 mm in a Charpy pendulum (1-J) from Metrotec S.A. (San Sebastián, Spain) on notched samples (0.25 mm radius V-notch), following the specifications of ISO 179-1:2010. All of the mechanical tests were performed at room temperature, and 6 samples of each material were tested and the corresponding values were averaged.

2.3.3. Thermal Analysis

The most relevant thermal transitions of the PLA/SCG green composites were obtained by differential scanning calorimetry (DSC) in a Mettler-Toledo 821 calorimeter (Schwerzenbach, Switzerland). The samples with an average weight of 6–7 mg were subjected to a thermal program divided into 3 stages: an initial heating from 25 °C to 180 °C, followed by a cooling to 0 °C, and a second heating to 300 °C. All the heating and cooling rates were set at 10 °C·min^{-1} and the tests were run in a nitrogen atmosphere with a flow-rate of 66 mL·min^{-1} using standard sealed aluminum crucibles with a capacity of 40 µL. The percentages of the crystallinity (X_c) and maximum crystallinity (X_{c_max}) were calculated during the second heating using the following equations:

$$X_c = \left[\frac{\Delta H_m - \Delta H_{cc}}{\Delta H_m^0 \times (1 - w)}\right] \times 100 \quad (1)$$

$$X_{c\_max} = \left[\frac{\Delta H_m}{\Delta H_m^0 \times (1 - w)}\right] \times 100 \quad (2)$$

where $\Delta H_m^0 = 93.7$ J/g is the theoretical enthalpy of a 100% crystalline PLA sample [45], the term $1 - w$ corresponds to the PLA weight fraction in the blend, and ΔH_m and ΔH_{cc} are, respectively, the melting and cold crystallization enthalpies.

The thermal degradation of the PLA/SCG pieces was assessed by thermogravimetric analysis (TGA) in a LINSEIS TGA 1000 (Selb, Germany). The samples with a weight ranging from 15 and 17 mg were placed in 70 µL alumina crucibles and subjected to a dynamic heating program from 40 °C to 700 °C at a heating rate of 10 °C·min^{-1} in an air atmosphere. The first derivative thermogravimetric (DTG) curves were also determined. All of the thermal tests were carried out in triplicate in order to obtain reliable results.

2.3.4. Thermomechanical Characterization

Dynamical mechanical thermal analysis (DMTA) was carried out in a DMA1 dynamic analyzer from Mettler-Toledo, working in single cantilever flexural conditions. Rectangular samples with the dimensions of 20 mm × 6 mm × 2.7 mm were subjected to a dynamic temperature sweep from 30 °C to 140 °C at a constant heating rate of 2 °C·min^{-1}. The selected frequency was 1 Hz and the maximum flexural deformation or cantilever deflection was set to 10 µm. All of the thermomechanical tests were performed in triplicate.

2.3.5. Color and Wetting Characterization

A Konica CM-3600d Colorflex-DIFF2 from the Hunter Associates Laboratory, Inc. (Reston, VA, USA) was used for the color measurement. The color coordinates (L*a*b*) were measured, with the illuminant D65 and observer 10°, according to the following criteria: L* = 0, darkness; L* = 100, lightness; a* represents the green (a* < 0)-to-red coordinate (a* > 0); and b* represents the blue (b* < 0)-to-yellow coordinate (b* > 0). The tests were run in triplicate.

Contact angle measurements were performed with an EasyDrop Standard goniometer model FM140 (KRÜSS GmbH, Hamburg, Germany), which was equipped with a video capture kit and analysis software (Drop Shape Analysis SW21; DSA1). Double distilled water was used as the test liquid. The tests were replicated 10 times.

2.3.6. Water Absorption Test

The water absorption capacity of the PLA/SCG pieces was evaluated by the water uptake method [37]. Injection-molded samples sizing 80 mm × 10 mm × 4 mm were first weighted in a balance and then placed inside a beaker filled with distilled water. The samples were wrapped with a metal grid. Subsequently, the samples were removed from the beaker and weighed at different time intervals for up to 14 weeks. For every measurement, the superficial moisture of the samples was removed using tissue paper. Three samples of each formulation were tested.

2.3.7. Biodisintegration Test

The biodegradability of the samples was evaluated using a disintegration test in controlled compost conditions, according to the guidelines of the ISO 20200 standard at the conditions of 58 °C and a relative humidity (RH) of 55%. Injection-molded samples with the dimensions of 30 mm × 30 mm × 4 mm were placed in a carrier bag and buried in soil with the following controlled composition: sawdust (40 wt.%), rabbit-feed (30 wt.%), ripe compost (10 wt.%), corn starch (10 wt.%), saccharose (5 wt.%), corn seed oil (4 wt.%), and urea (1 wt.%). To monitor the biodisintegration process, the samples were periodically unburied, washed with distilled water, dried, and weighed in an analytical balance. Pictures of the samples were also taken at each weight measurement, with the objective to visually evaluate the biodisintegration process. The weight loss, as a result of the disintegration in the controlled compost soil, was calculated by means of the following expression:

$$\text{Weight loss } (\%) = \left(\frac{W_0 - W_t}{W_0}\right) \times 100 \tag{3}$$

where W_t is the weight of the sample after a certain period of time in the controlled compost soil and W_0 is the initial dry weight of the sample. All of the tests were triplicated in order to ensure the reliability of the results.

2.4. Statistical Analysis

The results were subjected to analysis of variance (ANOVA) using Statgraphics Centurion XVII-64 software (Manugistics Corp., Rockville, MD, USA). To this end, significant differences were assumed with a significance level greater than 95% ($p < 0.05$).

3. Results and Discussion

3.1. Morphology of the SCG Particles

Particle interlocking is an essential mechanism in polymer composites. Figure 3 shows the morphology of the SCG particles observed by FESEM and their respective length and diameter histograms. In Figure 3a, one can observe that the lignocellulosic particles display a round-like morphology with a rough surface, which can be ascribed to the effect of the milling process on the particles with high hardness. These particles also displayed high porosity on their surface, which can be beneficial towards interactions with the biopolymer matrix by acting as anchoring points. In this regard, Mendes et al. [46] observed a very similar structure for the SCG powder when studying the composites of high-density polyethylene (HDPE). It was also reported that SCG particles tend to form small aggregates due to their hydrophilic characteristic. In this regard, Garcia-Garcia et al. [33] expressed that the individual particle size is in the 15–20 µm range, whereas these form aggregates sized 60–80 µm. Figure 3b,c present the collation of the particle histograms in terms of the length and diameter, respectively. The particles presented an average length of approximately 42.5 µm, whereas the average diameter was ~40 µm. In both cases, the dimensions were defined by a classical monomodal distribution. Furthermore, this particle morphology is based on a relatively low size, which can offer a positive effect on the overall properties. For instance, Crespo et al. [47] observed that large lignocellulosic particles, with diameter sizes exceeding 150 µm, led to an impairment and certain heterogeneity in the composites of vinyl plastisols.

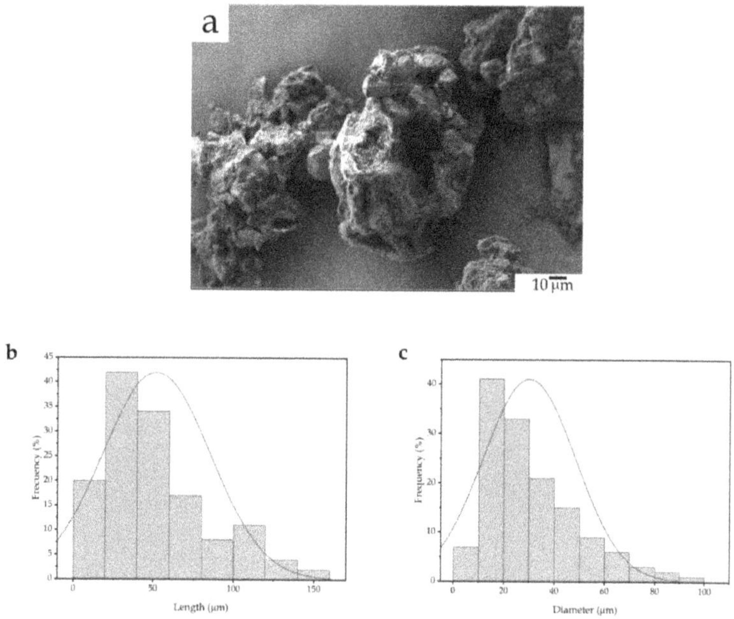

Figure 3. (**a**) Field emission scanning electron microscope (FESEM) image of the spent coffee ground (SCG) particles. Image taken with a magnification of 500× and a scale marker of 10 µm; (**b**) Histogram of the lengths of the SCG particles; and (**c**) histogram of the diameters of the SCG particles.

3.2. *Optical Properies of the PLA/SCG Composites*

The visual appearance is essential in terms of the impression that an end-product makes on the consumer. In this sense, Figure 4 shows the visual appearance of the pieces of PLA and its combinations with SCGs and the two-tested OLAs obtained by injection molding. It can be observed that the neat PLA piece and the pieces of PLA processed with both OLAs showed a relatively high contact transparency. The loss of transparency in the PLA sample is due to the semicrystalline nature of the biopolyester, in which the amorphous and crystalline regions have different light refraction indexes [48]. While the addition of OLA2 and $OLA2_{mal}$ kept the original appearance of the PLA piece, the samples containing SCGs developed a dark brown-to-black tonality. The latter optical effect is due to the intrinsic characteristic color of these SCG particles, as it can be observed in Figure 1, making the samples also completely opaque.

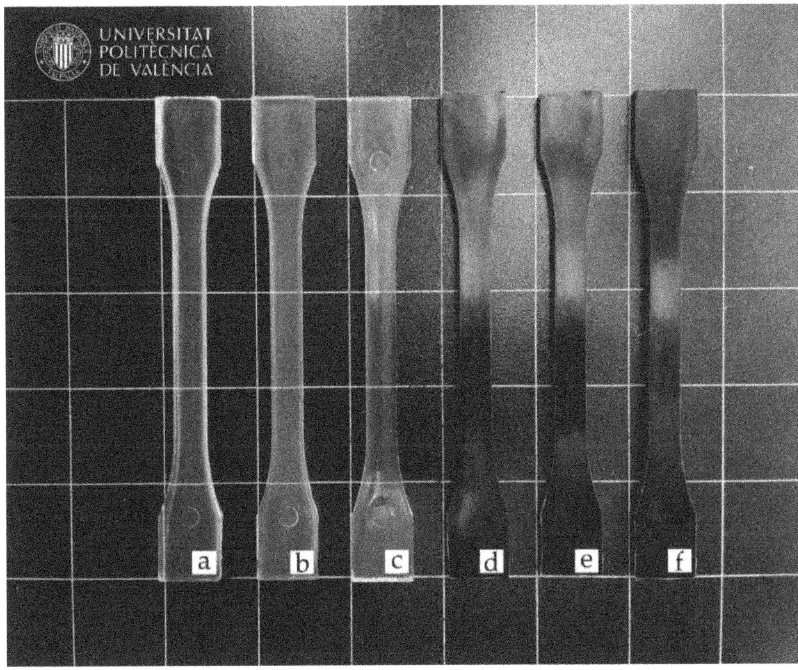

Figure 4. The visual appearance of the injection-molded pieces of the polylactide (PLA)/spent coffee grounds (SCGs) containing the oligomers of lactic acid (OLAs): (**a**) PLA; (**b**) PLA + OLA2; (**c**) PLA + $OLA2_{mal}$; (**d**) PLA + SCG; (**e**) PLA + OLA2 + SCG; and (**f**) PLA + $OLA2_{mal}$ + SCG.

Table 2 gathers the L*a*b* color coordinates obtained for these injection-molded pieces. Luminance (L*) is indicative of the brightness or lightness of the color of a sample. In this sense, the injection-molded PLA piece and the PLA/OLA2 and PLA/$OLA2_{mal}$ pieces exhibited very similar L* values, in the range of 45–48. This was expected due to the very similar nature and miscibility between the biopolymer and its oligomers. In the case of the green composite pieces, which were filled with SCGs, these presented significantly lower ($p < 0.05$) L* values than PLA, of approximately 26, not being significantly different ($p > 0.05$) among the two samples. Furthermore, the color coordinate a*, which is representative of a green (negative) or red color (positive), showed values in the range of −0.23 to 0.13 for the pieces of PLA, PLA/OLA2, and PLA/$OLA2_{mal}$. In the case of b*, which indicates a blue (negative) or yellow (positive) color, all of the unfilled pieces presented similar positive values between 1 and 2. Therefore, these pieces tend to show a slight yellow, pale color, showing no significant differences ($p > 0.05$) among them. On the contrary, all the SCG-containing pieces showed very positive a* values, in the range of 0.7–1.1, whereas the b* values were in the 4.9–5.6 range. The combination of the red and yellow colors thus quantified the above-qualified dark brown aspect of the green composite pieces. A similar color change was reported by Suaduang et al. [31] for the PLA/SCG composites containing up to 10 wt.% SCGs, showing a* and b* values from −0.9 to 7.5 and from −4.4 to 16.5, respectively.

Table 2. Luminance and color coordinates (L*a*b*) of the injection-molded pieces of polylactide (PLA)/spent coffee grounds (SCGs) containing the oligomers of lactic acid (OLAs).

Piece	L*	a*	b*
PLA	48.0 ± 0.5 [a]	−0.23 ± 0.02 [a]	1.89 ± 0.08 [a]
PLA + SCG	26.6 ± 0.1 [b]	1.10 ± 0.13 [b]	5.42 ± 0.19 [b]
PLA + OLA2	45.6 ± 0.1 [c]	0.13 ± 0.02 [c]	1.25 ± 0.08 [c]
PLA + $OLA2_{mal}$	47.4 ± 0.2 [c]	−0.14 ± 0.02 [d]	1.68 ± 0.09 [d]
PLA + OLA2 + SCG	26.3 ± 0.3 [b]	0.74 ± 0.02 [e]	4.91 ± 0.06 [e]
PLA + $OLA2_{mal}$ + SCG	25.8 ± 0.2 [b]	0.75 ± 0.04 [e]	5.56 ± 0.11 [f]

[a–f] Different letters in the same column indicate a significant difference among the samples ($p < 0.05$).

3.3. Mechanical Properties of the PLA/SCG Composites

Table 3 gathers the results obtained in the mechanical characterization of the PLA/SCG pieces. These results are of great interest to evaluate the effect of both SCGs and OLA2 and $OLA2_{mal}$ on PLA in terms of the mechanical resistance and ductile properties, which can be relevant for food packaging applications. It can be observed that the neat PLA piece showed a Young modulus (E) of 2913 MPa, a maximum tensile strength (σ_{max}) of 52.4 MPa, and an elongation at break (ε_b) of 10.4 %. These values agree with those previously reported by, for instance, Agüero et al. [49]. These are indicative of a rigid material with low ductility, particularly when compared to other biopolymer pieces, such as those of bio-based high-density polyethylene (bio-HDPE) [50] or polyethylene terephthalate (bio-PET) [51], showing values of ε_b of nearly 500%. The incorporation of the SCG particles into PLA decreased all of the mechanical resistance properties. Then, the E and σ_{max} values were reduced to 2367 MPa and 13.9 MPa, respectively, which indicates that a poor dispersion of the lignocellulosic fillers in the biopolyester matrix was attained [32]. However, contrary to most of the claims published in the literature of PLA composites based on lignocellulosic fillers derived from soluble coffee wastes [52], the ductile properties were improved. In particular, the ε_b value significantly increased to 39.6%, which approximately corresponds to a 280% increase in relation to the neat PLA piece. This ductility improvement suggests that the PLA matrix was plasticized by the SCG particles, since this waste-derived biomass can contain traces of water and ethanol and, more importantly, large amounts of lipids [53]. Among the lipids, the main plasticizing molecules correspond to organic compounds containing oxygen-based groups, such as fatty acids [54]. In any case, since both water and ethanol were removed by drying prior to processing the samples by extrusion to avoid the hydrolysis of PLA, plasticization can then be mainly ascribed to the presence of fatty acids. In this regard, it has been reported that the coffee oil content in SCGs is 10.98 wt.%, from which approximately 46% corresponds to linoleic acid [55]. In this sense, the work of Battegazzore et al. [56] reported the presence of several plasticizing compounds in natural fillers, such as hazelnut skin or cocoa by-products, which were also used to successfully increase the ductility of PLA.

From the above, it can be considered that coffee wastes derived from the liquor industry seem to possess a higher lipid content than those obtained from traditional soluble coffee. This key effect is related to differences in their processing technologies that, in the case of coffee liquor, does not imply the use of high temperature or extraction steps, apart from the natural roasting of the coffee grains. This result agrees with the findings reported by Suaduang et al. [31], who showed that contents of 7.5 and 10 wt.% of SCGs derived from the coffee bean roasting process, not yet being used for soluble or liquor coffee, resulted in a slight increase in the elongation at break of PLA films. In this former study, ε_b increased from 4.18% and 5.04% in the transversal (TD) and machine direction (MD), to 4.24% and 5.31% and to 5.33% and 6.63% for 7.5 and 10 wt.% loadings of SCGs, respectively. The higher improvement observed here can be then related to the higher SCG content used in the green composite, but also, more importantly, to the potential differences in the high fatty-acid content and remaining solvent traces due to the particular production process of coffee liquor and mild drying conditions. Indeed, as described above, other more conventional polymer composites based on SCGs derived from the soluble coffee industry showed that increasing the filler content reduced the ductility of the polymer. For instance, the work of Mendes et al. [46] reported a decrease in ε_b of HDPE of approximately 50%, in relation to neat HDPE, for a 30 wt.% SCG content.

Table 3. Mechanical properties of the injection-molded pieces of polylactide (PLA)/spent coffee grounds (SCGs) containing the oligomers of lactic acid (OLAs) in terms of tensile modulus (E), maximum tensile strength (σ_{max}), elongation at break (ε_b), Shore D hardness, and impact strength.

Piece	E (MPa)	σ_{max} (MPa)	ε_b (%)	Shore D Hardness	Impact Strength ($kJ \cdot m^{-2}$)
PLA	2913 ± 84 [a]	52.4 ± 3.1 [a]	10.4 ± 0.8 [a]	82.0 ± 0.5 [a]	27.7 ± 2.3 [a]
PLA + SCG	2367 ± 61 [b]	13.9 ± 0.7 [b]	39.6 ± 2.0 [b]	80.6 ± 0.9 [a]	18.6 ± 1.9 [b]
PLA + OLA2	2835 ± 49 [a]	35.3 ± 3.9 [c]	6.1 ± 0.4 [c]	76.8 ± 1.3 [b]	29.3 ± 3.6 [c]
PLA + $OLA2_{mal}$	3002 ± 172 [a]	24.0 ± 3.7 [d]	4.3 ± 0.6 [d]	82.8 ± 0.8 [a]	31.2 ± 2.1 [d]
PLA + OLA2 + SCG	2042 ± 43 [c]	18.9 ± 1.3 [e]	33.4 ± 3.0 [e]	81.4 ± 1.1 [a]	20.6 ± 2.2 [e]
PLA + $OLA2_{mal}$ + SCG	2291 ± 96 [d]	22.2 ± 0.9 [f]	20.7 ± 2.3 [f]	81.8 ± 1.1 [a]	18.9 ± 2.4 [b]

[a–f] Different letters in the same column indicate a significant difference among the samples ($p < 0.05$).

In relation to the two tested OLAs, the addition of OLA2 and $OLA2_{mal}$ clearly decreased the σ_{max} value of PLA to 35.3 MPa and 24.0 MPa, respectively, which is in agreement with previous studies [41,57], whereas the E values remained nearly unaltered, being not significantly different ($p > 0.05$). However, the ε_b value also slightly decreased to 6.1% and 4.3% for PLA+OLA2 and PLA+$OLA2_{mal}$, respectively. This result could be due to the fact that both OLAs reduce the cohesion of the macromolecular PLA chains. In this regard, Burgos et al. [57] reported that the E values of PLA decreased with the content of OLAs from 2500 MPa, for neat PLA, to 250 MPa for PLA containing 25 wt.% OLA content. Intermediate values were obtained when OLA2 and $OLA2_{mal}$ were added to the green composite. In particular, the OLA2-containing piece showed the lowest resistant properties, as the E value significantly decreased to 2042 MPa, while σ_{max} slightly, but significantly ($p < 0.05$), increased to 18.9 MPa in relation to the green composite sample. However, it also showed ductile properties, with an ε_b value of 33.4%. For the green composite piece with $OLA2_{mal}$, one can observe that it presented better mechanical resistant properties, showing values of E and σ_{max} of 2291 MPa and 20.7 MPa, respectively, but ε_b still resulted in a value of 20.7%. This difference observed in the mechanical response of the green composite pieces with the two OLAs tested here can be ascribed to the MAH content of $OLA2_{mal}$, which makes it more reactive and can lead to better chain-to-chain PLA interactions [58]. Moreover, the presence of the multiple MAH groups can provide certain grafting for the cellulose materials onto the backbone of PLA and, thus, acting as an interfacial compatibility agent with the SCG particles [32]. These results were evidenced by the σ_{max} increases of approximately 36% and 60% observed for the PLA+OLA2+SCG and PLA+$OLA2_{mal}$+SCG pieces, respectively, in comparison to the PLA+SCG piece. All in all, the combination of the three materials, namely SCGs, OLA2, and OLA2mal, yielded PLA pieces with varying mechanical performances in terms of rigidity and ductility. Therefore, the simultaneous addition of OLA2 and, more notably, $OLA2_{mal}$ increased the mechanical strength of the green composite PLA+SCG and also provided intermediate ductile properties.

Regarding Shore D hardness, it can be observed that most of the PLA and its green composite pieces developed in the present study showed values in the 80–82 range as being, in most cases, slightly lower than neat PLA. This fact agrees with the results obtained by Lascano et al. [41] and it supports the fact that SCGs, despite being hard lignocellulosic fillers [37], yield an overall softening effect on the PLA material due to their high lipid content. The only significant ($p < 0.05$) change was observed for the green composite containing PLA+OLA2, for which the hardness decreased to 76.8, due to the plasticizing effect exerted by this type of OLA that seems to ascribe more mobility to the PLA chains. In relation to the impact strength, the neat PLA piece showed a value of 27.7 kJ/m^2, which is a relatively low value, and characteristic of a brittle material [41]. The addition of SCGs at 20 wt.% significantly reduced this value further ($p < 0.05$) to 18.6 kJ/m^2, which could be attributed to the presence of SCG aggregates that embrittled the PLA matrix by creating local tensions [59]. In this regard, da Silva et al. [60] also reported a toughness decrease in PLA of 6% for contents of SCGs above 15 wt.%. The impact strength of the PLA pieces and the green composite pieces successfully increased after the addition of both OLAs. In the case of the neat PLA, without SCGs, OLA2 increased its impact strength by approximately 6%, while OLA2$_{mal}$ increased it by 12.6%. As reported earlier [41], this slight, but significant ($p < 0.05$), energy absorption enhancement during impact can be ascribed to the high solubility of OLAs in the PLA matrix that enables the inhibition of microcrack formation and growth. However, it is also worth noting that the combination of SCGs with OLA2 and OLA2$_{mal}$ yielded a slight improvement to the overall impact strength of the injection-molded green composite pieces, from which it can be inferred that, for these formulations, the OLAs tested in the present study mainly act as compatibilizers, rather than impact modifiers.

3.4. Morphology of the PLA/SCG Composites

In order to better ascertain the effect of the SCG particles and the two tested OLAs on PLA, the morphology of the fracture surfaces of the injection-molded pieces obtained after the impact tests were observed by FESEM. The resultant micrographs are presented in Figure 5. As it can be observed in Figure 5a, the neat PLA piece displayed the characteristic morphology of a polymer with brittle behavior due to the formation of a smooth fracture surface with the presence of microcracks (see the white arrows). A similar morphology was previously reported for PLA by Quiles-Carrillo et al. [61], confirming the aforementioned mechanical properties in relation to its low impact strength and ductility. Conversely, in Figure 5b, the fracture surface of the piece changed to a rougher morphology, as a result of the increase in ductility and the presence of the SCG fillers. However, the existence of some holes (identified in the image by white circles) corroborates the lack of compatibility between the PLA matrix and SCGs. These voids correspond to the lignocellulosic particles that were detached after impact [14]. Although a good dispersion of the SCG particles along the PLA matrix can be inferred due to the manifold presence of well-distributed holes, the suggested poor filler-to-matrix adhesion contributed to a reduced mechanical resistance, as shown above, during the mechanical analysis.

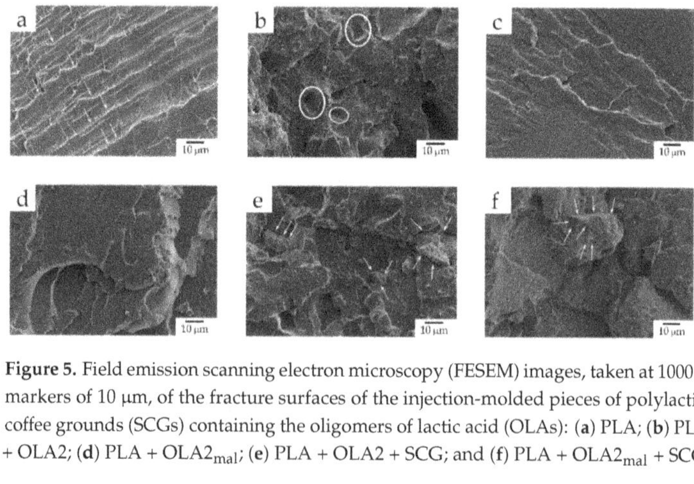

Figure 5. Field emission scanning electron microscopy (FESEM) images, taken at 1000× and with scale markers of 10 μm, of the fracture surfaces of the injection-molded pieces of polylactide (PLA)/spent coffee grounds (SCGs) containing the oligomers of lactic acid (OLAs): (**a**) PLA; (**b**) PLA+SCG; (**c**) PLA + OLA2; (**d**) PLA + OLA2$_{mal}$; (**e**) PLA + OLA2 + SCG; and (**f**) PLA + OLA2$_{mal}$ + SCG.

Figure 5c,d show the micrographs of the fracture surface for both the PLA+OLA2 and PLA+$OLA2_{mal}$ pieces, respectively, which yield very similar morphologies. In contrast with the fracture surface of the neat PLA piece, which presented some microcracks, one can observe the formation of several macrocracks along the biopolyester matrix. This feature, which was reported by Lascano et al. [41] when studying the effect of different OLA concentrations on PLA, can be appreciated, especially in the fracture surface of the PLA+$OLA2_{mal}$ piece. Therefore, OLA2 and, more noticeably, $OLA2_{mal}$, led to a rougher surface that corresponds to a fracture with a major energy absorption, which agrees with the mechanical properties reported above in terms of toughness. Interestingly, it is also worth mentioning that phase separation was not noticeable due to the great chemical affinity between PLA and OLAs. Finally, Figure 5e,f show the FESEM micrographs corresponding to the fracture surfaces of the PLA+OLA2+SCG and PLA+$OLA2_{mal}$+SCG pieces, respectively. It can be observed that, in both cases, the interfacial adhesion of the SCG particles with the surrounding PLA matrix was noticeably improved due to the decrease in the void density. Furthermore, the SCG particles seem to be more imbedded into the biopolyester, since the gap between the edge of the particles (see the white arrows) and the PLA matrix was very narrow. Thus, both OLAs can potentially increase the affinity of the SCG particles in PLA, especially $OLA2_{mal}$, due to the presence of the multiple MAH groups that can interact with both the backbone ester (R—COO—R) groups of PLA and —OH groups of SCGs [54]. This visual effect was more intense in the PLA+$OLA2_{mal}$+SCG sample, since the particle-matrix gap was almost imperceptible, which correlates better with the mechanical properties reported above.

3.5. Thermal Properties of the PLA/SCG Composites

The DSC thermograms corresponding to the cooling and second heating steps were analyzed to study the thermal properties of the green composites. Figure 6 shows the cooling (Figure 6a) and heating curves (Figure 6b), whereas Table 4 gathers the main thermal parameters obtained from these thermograms. One can observe that the neat PLA sample showed a T_g of 62.8 °C, which is noticeable from a change in the baseline of the heating thermogram. A very similar result was observed by Lascano et al. [41], who reported a PLA T_g value of 63 °C. In this case, however, cold crystallization was not observed during the second heating, due to the fact that complete crystallization was fully achieved during the slow cooling step carried out at 10 $°C \cdot min^{-1}$. During cooling, the PLA molecules crystallized from the melt, showing a low-intense crystallization temperature (T_c) at nearly 100 °C. The melting of the biopolymer finally occurred at a temperature of 173.3 °C, which is also close to the melting temperature (T_m) value previously obtained [41]. The addition of SCGs to PLA slightly reduced the value of T_g, from 62.8 °C to 61.1 °C, supporting the plasticizing effect exerted by the coffee oil on the biopolyester amorphous regions. The green composite sample also crystallized from the melt, showing a T_c value of 95.6 °C, whereas it cold crystallized further during the second heating at 99.3 °C, which corresponds to the cold crystallization temperature (T_{CC}). The higher intensity of the crystallization peak certainly points to the fact that the presence of SCGs increased the crystallization of the PLA molecules from the melt. Finally, the T_m value of the PLA sample was unaffected by the presence of SCGs, showing no significant difference ($p > 0.05$). In relation to the crystallinity, the neat PLA sample showed an X_C value of nearly 20%, which is very similar to that obtained by Rojas-Lema et al. [62], whereas the single addition of SCGs increased the crystallinity to approximately 36%. In the case of the maximum crystallization degree (X_{C_max}), which does not consider the crystal formed during heating and presents more accurate information about the effect of a given additive on the polymer crystallinity, this increase was higher, nearly 54%. This result indicates that the lignocellulosic fillers acted as heterogenous nuclei during the formation of the crystals, and considerably increased the number of crystals that were formed. This phenomenon was previously described by da Silva et al. [60], who also observed an increase in the PLA crystallinity by means of SCGs obtained as waste from the preparation of instant coffee. The authors reported an increase in the crystallinity of PLA from 9.10% to 12.7% for a 5 wt.% content of SCGs in the green composite.

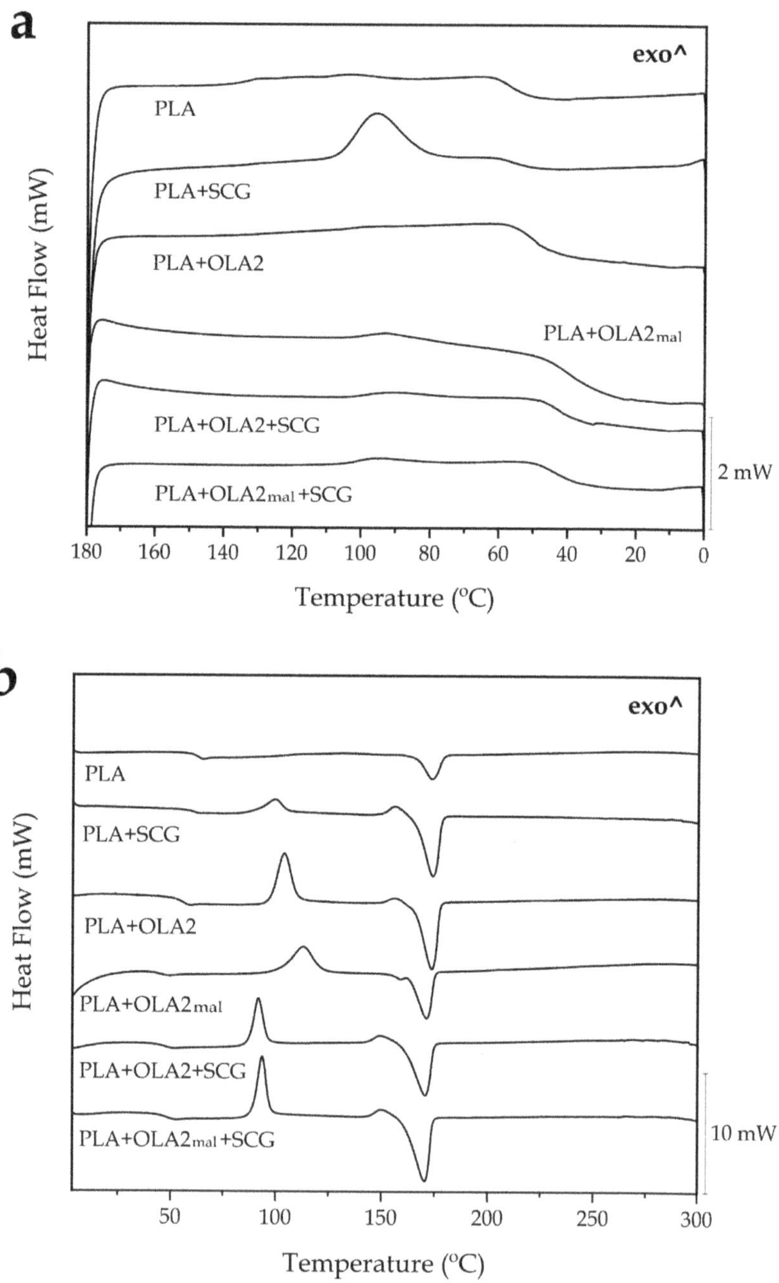

Figure 6. Differential scanning calorimetry (DSC) thermograms of the injection-molded pieces of polylactide (PLA)/spent coffee grounds (SCGs) containing the oligomers of lactic acid (OLAs) during cooling (**a**) and second heating (**b**).

Table 4. Thermal properties of the injection-molded pieces of polylactide (PLA)/spent coffee grounds (SCGs) containing the oligomers of lactic acid (OLAs) in terms of glass transition temperature (T_g), crystallization temperature (T_C), cold crystallization temperature (T_{CC}), melting temperature (T_m), crystallinity (X_C), and maximum crystallinity (X_{C_max}).

Piece	T_g (°C)	T_C (°C)	T_{CC} (°C)	T_m (°C)	X_C (%)	X_{C_max} (%)
PLA	62.8 ± 0.2 ^{a}	103.7± 0.1 ^{a}	-	173.3 ± 1.9 ^{a}	19.96 ± 0.3 ^{a}	19.96 ± 0.3 ^{a}
PLA + SCG	61.1 ± 0.3 ^{a}	95.6 ± 1.0 ^{b}	99.3 ± 0.9 ^{a}	173.6 ± 2.0 ^{a}	36.42 ± 0.7 ^{b}	53.90 ± 0.6 ^{b}
PLA + OLA2	55.4 ± 0.2 ^{b}	96.7 ± 0.5 ^{b}	103.7 ± 1.2 ^{a}	173.3 ± 1.9 ^{a}	7.59 ± 0.5 ^{c}	50.87 ± 0.5 ^{c}
PLA + $OLA2_{mal}$	45.3 ± 0.4 ^{c}	94.9 ± 0.2 ^{b}	112.9 ± 1.5 ^{b}	171.0 ± 2.4 ^{a}	3.91 ± 0.4 ^{d}	41.98 ± 0.3 ^{d}
PLA + OLA2 + SCG	47.3 ± 0.3 ^{c}	90.7 ± 0.3 ^{c}	91.9 ± 1.1 ^{c}	170.2 ± 2.5 ^{a}	21.04 ± 0.3 ^{e}	54.28 ± 0.1 ^{e}
PLA + $OLA2_{mal}$ + SCG	48.4 ± 0.2 ^{c}	100.7 ± 0.4 ^{d}	93.7 ± 1.0 ^{c}	170.1 ± 2.2 ^{a}	23.02 ± 0.2 ^{f}	61.44 ± 0.2 ^{f}

$^{a–f}$ The different letters in the same column indicate a significant difference among the samples ($p < 0.05$).

In regard to the effect of the OLAs, it can be observed that OLA2 and $OLA2_{mal}$ provoked a significant ($p < 0.05$) decrease in T_g, reducing it to 55.4 °C and 45.33 °C, respectively. This reduction has been attributed to the plasticizing effect exerted by OLAs [42,58]. Similar to the case of SCGs, for the samples containing OLA2 and $OLA2_{mal}$, cold crystallization also occurred, showing T_{CC} values of 103.7 °C and 112.9 °C, respectively. This suggests that the PLA crystallization was promoted, as a result of its high solubility and interaction with the PLA molecules [60]. In terms of melting, the slightly lower T_m value attained in the $OLA2_{mal}$-containing sample, which was reduced to 171 °C, confirms its higher interaction with the PLA matrix. However, although the crystallinity degree, that is, X_C, was not significantly ($p > 0.05$) affected by the addition of both OLAs, one can observe that the maximum crystallinity degree, that is, X_{C_max}, increased to approximately 51% and 42%, for OLA2 and $OLA2_{mal}$, respectively. In this regard, other previous studies reported that OLAs can also act as heterogeneous nucleating agents and facilitate the crystal formation in PLA, since they are still present in the form of thin solids with a melting point higher than the T_C value of PLA [63]. In regard to the combined additions of SCGs and the two tested OLAs, one can observe that T_g suffered a considerable decrease due to the plasticization of PLA by both the coffee oil and the presence of the oligomers. The T_{CC} values also decreased further to 91.9 °C and 93.7 °C for PLA+OLA2+SCG and PLA+$OLA2_{mal}$+SCG, respectively, which can be ascribed to the facilitated mobility of the amorphous phase. However, even though the PLA matrix was highly plasticized, the absence of crystallization from the melt may also suggest an impairment during the crystal formation due to the presence of these additives. In all cases, and similar to the PLA+SCG, PLA+OLA2, and PLA+$OLA2_{mal}$ samples, the green composites containing both additives showed significantly ($p < 0.05$) higher values of X_{C_max}, with values of approximately 54% and 61% for PLA+OLA2+SCG and PLA+$OLA2_{mal}$+SCG, respectively.

In terms of the thermogravimetric characterization, Figure 7 presents the variations of mass (Figure 7a) and the first derivatives (Figure 7b) with temperature to evaluate the degradation processes of each sample. Moreover, Table 5 gathers and presents the most relevant parameters related to the degradation processes. In this sense, it can be observed that neat PLA exhibited a single-step degradation process, showing a slightly superior thermal stability than its green composites with SCGs and blends with OLAs. In particular, PLA presented the values of $T_{5\%}$ and T_{deg} of 319.3 °C and 361.3 °C, showing a mass loss at T_{deg} of 57.2%. A very similar degradation profile was observed for PLA by Rojas-Lema et al. [62]. The addition of SCGs significantly ($p < 0.05$) reduced the thermal stability of PLA and generated a low-intense second degradation step at high temperatures. The first and main weight loss initiated at approximately 305 °C, which means a reduction of approximately 14 °C with respect to the neat PLA. This is related to the degradation of low-M_W components present in the lignocellulosic residue, such as hemicellulose [64]. The second step, which appeared at the 370–500 °C range, was less pronounced and it can be ascribed to the thermal degradation of lignin and cellulose present in SCGs [65]. The maximum degradation peak, represented by T_{deg}, also decreased by approximately 2 °C versus the neat PLA, although this difference was not significant ($p > 0.05$). A similar reduction in the thermal stability was observed by Mendes et al. [46] when studying the thermal properties of HDPE/SCG composites. The incorporation of OLA2 and $OLA2_{mal}$ into PLA also decreased the onset degradation, that is, $T_{5\%}$, to 287.7 °C and 289 °C, respectively. In the case of T_{deg}, a reduction to the values of 354.3 °C and 356.7 °C was, respectively, observed, meaning a significant ($p < 0.05$) loss of thermal stability. In this context, several studies reported the inherently lower thermal stability of OLAs in comparison to PLA [41,62]. Furthermore, $OLA2_{mal}$ yielded a slightly higher thermal stability than OLA2. This fact can be ascribed to the presence of MAH groups in this OLA grade that, similar to other properties, resulted in a higher interaction with the PLA matrix by forming a more stable macromolecule, in which the chain–scission process at a high temperature was delayed [58]. The combined systems of PLA+OLA2+SCG and PLA+$OLA2_{mal}$+SCG also showed the characteristic two-step degradation process of the green composite due to the presence of the lignocellulosic filler. Thermal degradation occurred more rapidly at the start of the process, showing $T_{5\%}$ values of 275 °C and 282 °C, respectively. However, the T_{deg} values were not significantly different ($p > 0.05$) than those of the neat PLA, that is, 361.3 °C, being slightly higher than those of the green composite without OLAs and the OLA-containing PLA samples. Finally, the residual mass, in all cases, was below 1 wt.%. This suggests a good synergetic effect between SCGs and OLAs, which was previously observed during the morphological analysis. In general terms, one can thus consider that the combined addition of SCGs derived from the liquor industry and OLAs can lead to materials with a relatively similar thermal stability than PLA, which are adequate for applications that deal with temperatures below 275 °C.

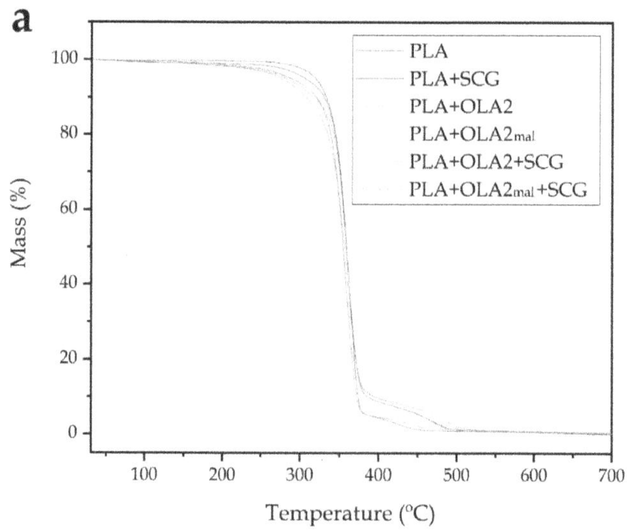

Figure 7. *Cont.*

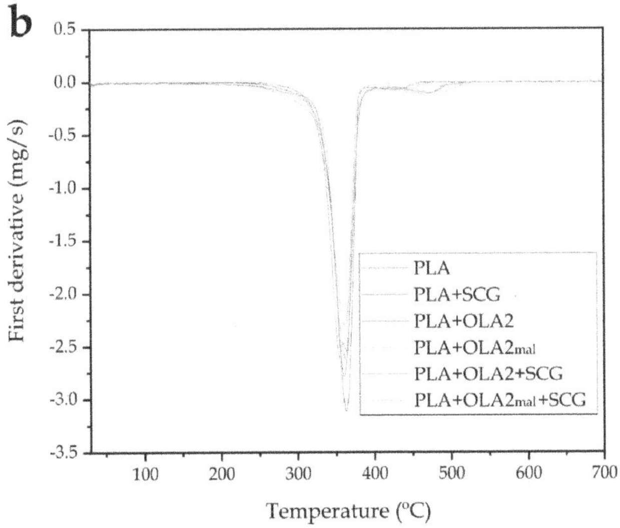

Figure 7. (**a**) Thermogravimetric analysis (TGA) curves and (**b**) first derivative (DTG) of the injection-molded pieces of polylactide (PLA)/spent coffee grounds (SCGs) containing the oligomers of lactic acid (OLAs).

Table 5. Main thermal degradation parameters of the injection-molded pieces of polylactide (PLA)/spent coffee grounds (SCGs) containing the oligomers of lactic acid (OLAs) in terms of the onset degradation temperature measured at a mass loss of 5 wt.% ($T_{5\%}$), degradation temperature (T_{deg}), mass loss at T_{deg}, and residual mass at 700 °C.

Piece	$T_{5\%}$ (°C)	T_{deg} (°C)	Mass Loss at T_{deg} (%)	Residual Weight (%)
PLA	319.3 ± 0.8 [a]	361.3 ± 2.4 [a]	57.2 ± 0.9 [a]	0.1 ± 0.1 [a]
PLA + SCG	305.3 ± 1.1 [a]	359.0 ± 1.3 [a]	51.4 ± 1.1 [b]	0.3 ± 0.1 [a]
PLA + OLA2	287.7 ± 1.4 [b]	354.3 ± 1.8 [b]	49.9 ± 1.0 [b]	0.4 ± 0.2 [a]
PLA + $OLA2_{mal}$	289.0 ± 0.5 [b]	356.7 ± 2.2 [b]	57.3 ± 0.8 [c]	0.6 ± 0.2 [b]
PLA + OLA2 + SCG	275.0 ± 1.2 [c]	361.3 ± 1.0 [a]	61.7 ± 1.0 [d]	0.2 ± 0.1 [a]
PLA + $OLA2_{mal}$ + SCG	282.0 ± 0.7 [c]	361.3 ± 1.5 [a]	57.3 ± 0.7 [e]	0.5 ± 0.1 [b]

[a–e] The different letters in the same column indicate a significant difference among the samples ($p < 0.05$).

3.6. Thermomechanical Properties of the PLA/SCG Composites

DMTA determines the mechanical properties as a function of temperature in dynamic conditions through the application of sinusoidal stress. Figure 8 shows the DMTA curves for all the PLA/SCG composites, while Table 6 gathers the most relevant thermomechanical parameters. In Figure 8a, one can observe the evolution of the storage modulus as a function of temperature. Regarding the neat PLA piece, the storage modulus suffers a pronounced drop in the temperature range between 50 °C and 70 °C, which is indicative of the α-relaxation process of the PLA chains when surpassing the glass transition region [41]. This can be observed in the difference between the storage modulus of 1263 MPa, at 35 °C, and the storage modulus value of 1.5 MPa, at 80 °C. Another important feature is the storage modulus change in the 90–100 °C range, which corresponds to the cold crystallization process [37]. This provoked a rearrangement of the PLA chains into a more ordered structure with higher thermomechanical resistance. The addition of SCGs to PLA significantly ($p < 0.05$) decreased the storage modulus of the biopolyester to 1150 MPa at 35 °C, which can be ascribed to the plasticization process by the remaining coffee oil in the filler. However, it yielded the cold crystallization process to occur at lower temperatures than in the case of the neat PLA (80–90 °C range). This confirms that the SCG particles also acted as nucleating agents during the crystallization process of PLA, as previously described during the DSC analysis, resulting in a storage modulus of 80 MPa at 100 °C. Similar behavior was observed by Quiles-Carrillo et al. [37] when PLA was reinforced with almond shell flour (ASF), a lignocellulosic filler. With the incorporation of OLA2 into the PLA matrix, the storage modulus presented a value of 1210 MPa at 35 °C, which is slightly, but still significantly ($p < 0.05$), lower than the value for PLA. Furthermore, in comparison with neat PLA, the storage modulus values of the OLA-containing PLA pieces were lower along the whole temperature range up to the glass transition region of the biopolymer, where they were exceeded. This result confirms that OLA2 reduces the stiffness of PLA due to its role as a plasticizing agent, as it was previously mentioned during the mechanical analysis. Similar behavior was observed in the $OLA2_{mal}$ sample, but with a more emphasized effect, reducing the storage modulus to 1130 MPa at 35 °C. It should also be noted that both OLAs reduced the temperature at which the cold crystallization phenomenon occurs. This is representative of the nucleating effect provided by the short-length OLA molecules, which were able to facilitate the packing of the PLA macromolecular structure [63]. The combination of OLA2 and $OLA2_{mal}$ with SCGs also led to a decrease in the storage modulus in all the temperature ranges, which can be mainly ascribed to the plasticizing effect of both the coffee oil and oligomers. The T_{CC} value was also reduced as a result of the combined nucleating effect of SCGs and OLA2 or $OLA2_{mal}$, which was previously observed during the DSC analysis. Moreover, during the thermomechanical analysis, it was observed that $OLA2_{mal}$ exerted a higher nucleating effect, as it decreased the crystallization process to the 70–80 °C temperature range.

Table 6. Thermomechanical properties of the injection-molded pieces of polylactide (PLA)/spent coffee grounds (SCGs) containing the oligomers of lactic acid (OLAs) in terms of the storage modulus at 35 °C, 80 °C, and 100 °C, and the dynamic damping factor (tan δ) peak.

Piece	Storage Modulus at 35 °C (MPa)	Storage Modulus at 80 °C (Mpa)	Storage Modulus at 100 °C (Mpa)	Tan δ Peak (°C)
PLA	1263 ± 30 [a]	1.5 ± 0.1 [a]	43 ± 5 [a]	73.1 ± 1.2 [a]
PLA + SCG	1150 ± 25 [b]	4.0 ± 0.2 [b]	80 ± 10 [b]	63.1 ± 1.4 [b]
PLA + OLA2	1210 ± 39 [c]	1.3 ± 0.2 [a]	53 ± 7 [c]	64.9 ± 2.5 [b]
PLA + $OLA2_{mal}$	1130 ± 31 [b]	58.0 ± 2.0 [c]	62 ± 4 [d]	54.4 ± 3.1 [c]
PLA + OLA2 + SCG	1037 ± 37 [d]	3.8 ± 0.3 [b]	80 ± 8 [b]	63.2 ± 1.7 [b]
PLA + $OLA2_{mal}$ + SCG	1115 ± 40 [b]	57.1 ± 2.5 [c]	51 ± 3 [c]	53.5 ± 1.6 [c]

[a–d] The different letters in the same column indicate a significant difference among the samples ($p < 0.05$).

Figure 8b presents the evolution of the dynamic damping factor (tan δ) with the temperature. This allows us to identify the α-relaxation of PLA, which is related to its T_g, observed as the maximum peak. The neat PLA shows a tan δ peak at 73.1 °C, which agrees with the previous values reported by the other authors [37,41]. The presence of SCGs in the PLA matrix decreased the tan δ peak to 63.1 °C due to the previously described plasticization process by the coffee oil, providing it with improved chain mobility and free volume. Moreover, it can be observed that OLA2 and $OLA2_{mal}$ also reduced the tan δ peaks to 64.9 and 54.4 °C, respectively, which successfully correlates with the T_g values attained in the DSC results shown above. Thus, the combined addition of OLA2 or $OLA2_{mal}$ with SCGs followed the same tendency as the single ones for OLA2 or $OLA2_{mal}$. In particular, the PLA+$OLA2_{mal}$+SCG sample presented a very similar tan δ peak value than the $OLA2_{mal}$ sample, that is, 53.5 °C, while the PLA+OLA2+SCG and PLA+OLA2 pieces presented values that exceeded 60 °C, specifically 63.2 °C in the case of the green composite. Interestingly, all of the green composite samples presented a maximum damping factor that was approximately 15% lower than in the unfilled PLA samples. This reduction is ascribed to the replacement in the injection-molded piece of the biopolymer with an amorphous region with a hard lignocellulosic filler, by which the material presents a reduced energy dissipation and toughness [36].

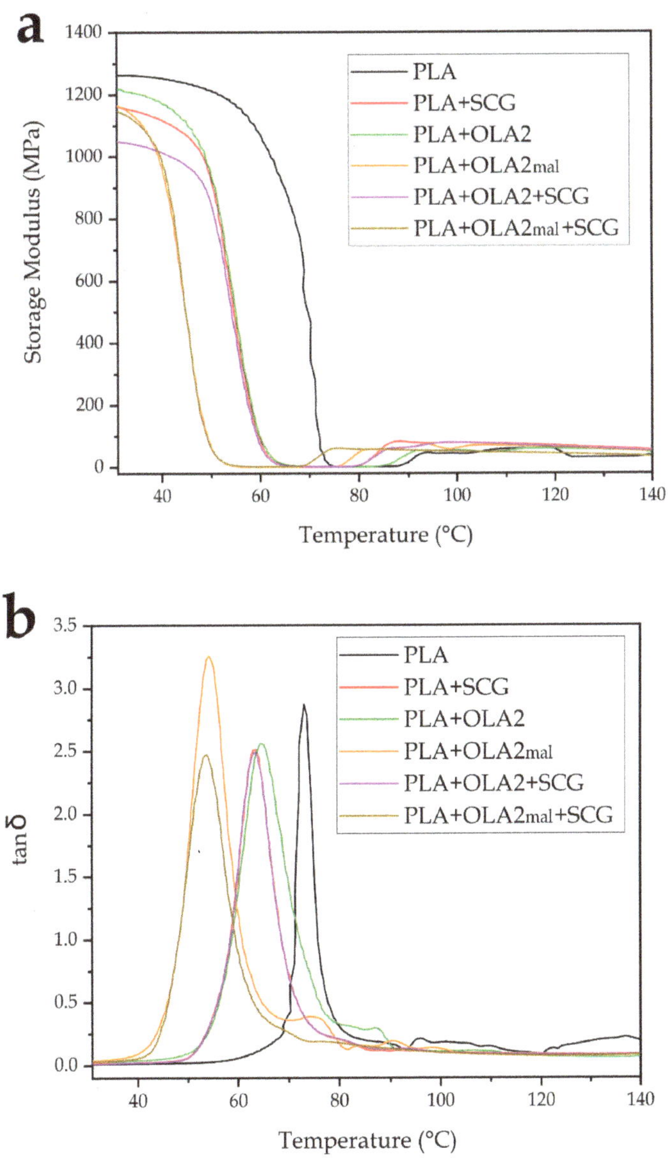

Figure 8. Plot evolution of (**a**) the storage modulus and (**b**) the dynamic damping factor (tan δ) of the injection-molded pieces of polylactide (PLA)/spent coffee grounds (SCGs) containing the oligomers of lactic acid (OLAs).

3.7. Water-Resistance Properties of the PLA/SCG Composites

Water contact angle measurements were conducted in order to obtain information about the hydrophilic behavior of the sample pieces and to evaluate the effect of the additives on the material surface. Figure 9 presents the water drops and their respective contact angles with the surface of the PLA pieces. It can be observed that all of the samples showed hydrophilicity since their contact angle was less than 90°. In particular, the neat PLA piece showed a contact angle of 87.1°, which is almost in the 90° limit and characteristic of the hydrophobic behavior of the biopolyester [66,67]. The addition of OLA2 and $OLA2_{mal}$ reduced the water contact angle to 83° and 81.8°, respectively, and thus made PLA more hydrophilic. This is due to the more hydrophilic nature of OLAs in comparison to PLA, as reported by Darie-Niţă et al. [66] when they studied different plasticizers for processing PLA films. In particular, the authors showed a reduction of about 12.5% in the contact angle in relation to the neat PLA for a 7 wt.% content of oligomers. In regard to the incorporation of the SCG particles, one can observe that the hydrophilicity of PLA also increased, reducing the value of the contact angle to 82.5°. This reduction can be related to the higher hydrophilicity of SCGs due to the high-content –OH groups in the lignocellulosic particles. This effect was previously observed by Laaziz et al. [67] when filling PLA with lignocellulosic almond particles. The combined additions that resulted in the PLA/OLA2/SCG and PLA/$OLA2_{mal}$/SCG pieces further decreased the contact angle to 79° and 80.4°, respectively. This is indicative of a synergistic effect between SCGs and OLAs, which both increase the polarity of PLA and its free volume. Moreover, it should be noted that the differences observed for the OLA2- and $OLA2_{mal}$-containing pieces can be related to their composition. In particular, it can be considered that $OLA2_{mal}$ achieved a lower contact angle value due to the presence of MAH, which increases the polarity of PLA to a higher extent than OLA2.

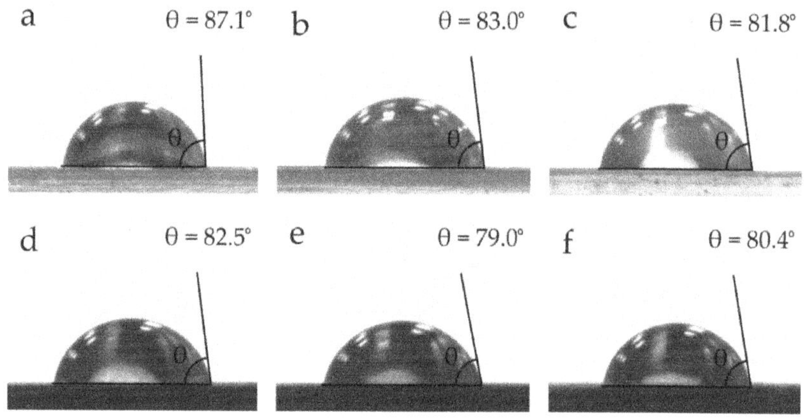

Figure 9. Water contact angle of the injection-molded pieces of polylactide (PLA)/spent coffee grounds (SCGs) containing the oligomers of lactic acid (OLAs): (**a**) PLA; (**b**) PLA + OLA2; (**c**) PLA + $OLA2_{mal}$; (**d**) PLA + SCG; (**e**) PLA + OLA2 + SCG; and (**f**) PLA + $OLA2_{mal}$ + SCG.

In addition to the contact angle measurements, the water uptake of the injection-molded pieces was evaluated through the water absorption test. Figure 10 presents the evolution of the water uptake of each sample after 14 weeks of immersion in distilled water. As expected, the neat PLA piece exhibited the lowest water absorption value during all of the immersion times, with a saturation value of approximately 0.67 wt.% at 14 weeks of immersion due to the hydrophobic nature of PLA. In this regard, Balart et al. [68] observed a very similar saturation content of water for PLA. The addition of SCGs to the PLA matrix significantly increased the water absorption capacity of the piece to a saturation value of 5.38 wt.%, which is more than 5 times the water absorption value of the neat PLA. This high increase is related to the hydrophilic nature of SCGs, as shown above during the water contact angle. In this sense, cellulose, hemicellulose, and lignin, all present in SCGs, contain multiple –OH groups that can readily interact with water molecules, forming hydrogen bonds and allowing water to enter the PLA's structure [68]. Similarly, Wu [32] reported high water absorption values for the PLA/SCG composites, reaching about a 12.5 wt.%

water absorption value for a 20 wt.% filler content after 60 days of immersion in water. On the other hand, it is worth mentioning that the presence of OLA2 and $OLA2_{mal}$ in the PLA matrix slightly increased the water uptake to values of 0.74 wt.% and 0.98 wt.%, respectively. This small increase may be related to the higher free volume attained in the PLA samples due to the plasticization caused by the oligomers, being slightly higher for $OLA2_{mal}$ due to its higher interaction with the biopolymer matrix as a result of the presence of the MAH groups, which are highly hydrophilic. Similar behavior was observed by Wu [32], who treated PLA with MAH and observed a water absorption value that was superior by approximately 4% to that of the neat PLA. Finally, the combined additions in the PLA+OLA2+SCG and PLA+$OLA2_{mal}$+SCG pieces yielded saturation values of 5.33 and 5.55 wt.%, respectively, which are very similar to the value obtained for the PLA+SCG piece. However, the water uptake rate was lower, particularly from week 2, which corroborates the above-reported morphological results in which it was indicated that the OLAs were mainly located at the filler-to-matrix interface and, then, could potentially reduce the water absorption on the lignocellulosic particle surfaces.

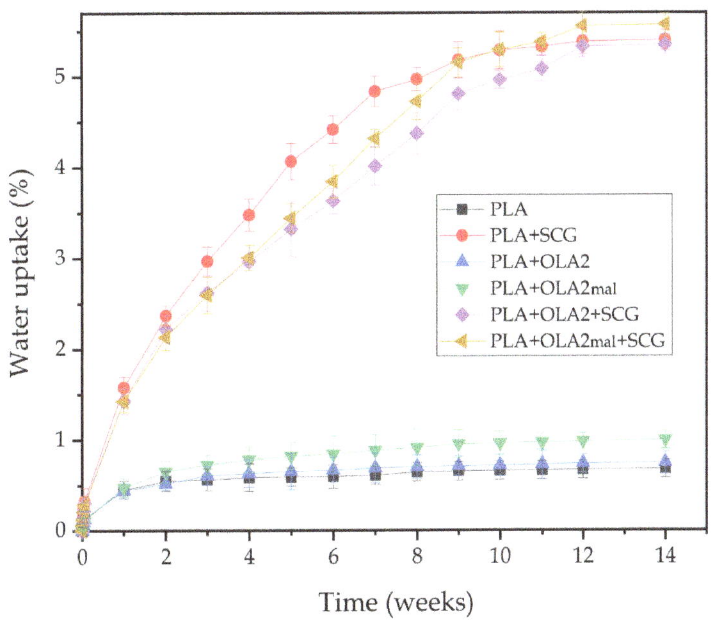

Figure 10. Water uptake of the injection-molded pieces of polylactide (PLA)/spent coffee grounds (SCGs) containing the oligomers of lactic acid (OLAs).

3.8. Disintegration in Controlled Compost Soil of the PLA/SCG Composites

Finally, the effect of both SCGs and OLAs on the biodegradation rate of PLA was ascertained. Figure 11 shows the percentage of weight loss as a function of the elapsed time during disintegration in the controlled compost soil of the injection-molded pieces. It can be observed that only the PLA, PLA+OLA2, and PLA+$OLA2_{mal}$ samples fully disintegrated during the test. This fact is related to the large thickness of the PLA pieces, that is, 4 mm. Nonetheless, all of the pieces achieved a high weight loss after 98 days in compost soil. The neat PLA presented a weight loss of 100% after 84 days due to a hydrolytic degradation process [68]. A similar disintegration profile was observed by Quiles-Carrillo et al. [69] when studying the compostability of PLA, achieving a mass decomposition of 100 wt.% after 56 days. These differences can be explained in terms of the compost composition and conditions. One can observe that the addition of 20 wt.% of SCGs highly decreased the biodegradation rate in comparison to the neat PLA, showing a weight loss of approximately

48% after 98 days. This delay in the disintegration rate can be due to the higher crystallinity that the PLA/SCG samples possess in relation to the neat PLA, as shown above during the DSC analysis. Since hydrolytic degradation occurs more rapidly in the amorphous regions thus, the higher the crystallinity, the lower the disintegration rate. In this regard, Balart et al. [68] also observed this effect for PLA/hazelnut shell flour (HSF) composites, where the lignocellulosic filler delayed the disintegration of PLA. The inclusion of OLA2 and $OLA2_{mal}$ into the PLA matrix increased the disintegration rate of the pieces during the first weeks, which could be ascribed to the fact that both oligomers plasticized the amorphous region of PLA and increased the free volume, favoring the hydrolysis of the biopolyester chains. Thus, after 49 days, PLA/OLA2 and PLA/$OLA2_{mal}$ presented a mass loss of approximately 60% and 65%, respectively, which was more rapid in comparison to the neat PLA that biodegraded by 55%. Finally, the combined additions of PLA with SCGs and OLA2 and $OLA2_{mal}$ showed disintegration rates lower than that of the neat PLA, presenting mass losses of nearly 65% and 60% at 98 days, respectively, and not fully degrading. This can be ascribed to the higher crystallinity of the green composite samples. However, their disintegration was superior to that observed for the PLA+SCG piece, because of the plasticization process.

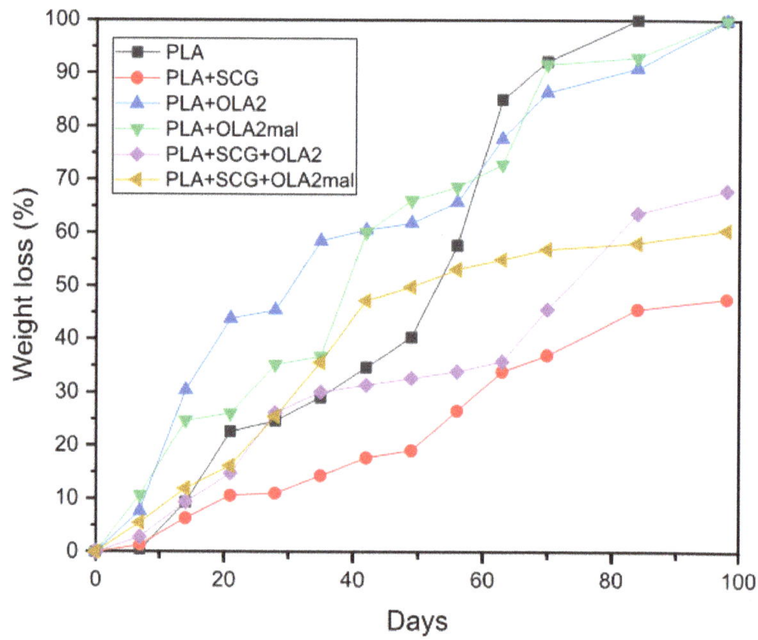

Figure 11. Evolution plot of the percentage of weight loss as a function of the elapsed time during disintegration in the controlled compost soil of the injection-molded pieces of polylactide (PLA)/spent coffee grounds (SCGs) containing the oligomers of lactic acid (OLAs).

Finally, Figure 12 shows the visual aspect of all the PLA pieces studied during the disintegration test, providing more information about their compostability profile. From the visual appearance of these samples, it can be confirmed that the SCG-containing pieces were not fully disintegrated after being buried for 98 days in compost soil. In regard to the neat PLA, it lost its transparent appearance only after 7 days of incubation in compost soil due to the hydrolysis process of the biopolyester [70]. After 14 days, the PLA pieces already started to develop a brown color, which evidences the beginning of disintegration. It then started to fragment and present microcracks, until its complete biodisintegration after 98 days. The PLA+SCG pieces did not present apparent disintegration signs until day 28, when it cracked and then started to fragment. It can also be observed that their disintegration rate was slower than that of PLA. Regarding the PLA+OLA2 and PLA+OLA2$_{mal}$ pieces, both started to show a dark brown color after 14 days, being quite fragmented at this time, which confirms the more rapid disintegration rate in comparison to the neat PLA. This confirms the faster disintegration profile observed in Figure 11, in which the two OLA-containing samples displayed a higher disintegration rate than neat PLA during the first 56 days, although the latter piece fully disintegrated earlier on. Concerning the simultaneous additions to produce the so-called PLA+SCG+OLA2 and PLA+SCG+OLA2$_{mal}$ pieces, these yielded similar disintegrated samples. However, the PLA+SCG+OLA2$_{mal}$ piece started to fragment earlier (day 14), in agreement with the disintegration profile shown above. These results suggest that, even though PLA is already a biodegradable polymer under compost soil conditions, the addition of OLAs can promote its disintegration. On the other hand, the incorporation of SCGs significantly delayed the biodisintegration of PLA, a fact that can be closely linked with the increase in the crystallinity that this food waste derived lignocellulosic filler provides.

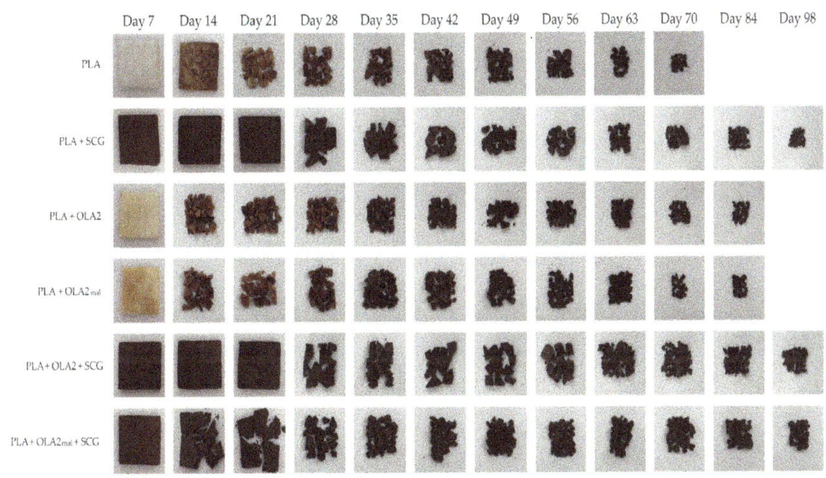

Figure 12. Visual aspect at selected disintegration times of the injection-molded pieces of polylactide (PLA)/spent coffee grounds (SCGs) containing the oligomers of lactic acid (OLAs).

4. Conclusions

The successful transition in the packaging industry of linear economies to circular and more sustainable bioeconomies certainly requires the development of high-performance materials derived from biomass and industrial wastes. The results presented in this work demonstrate, for the first time, the high potential of manufacturing PLA-based green composites through the valorization of wastes derived from the liquor coffee industry, referred to in the present study as SCGs. The use of these lignocellulosic fillers yielded PLA materials with an enhanced ductility and potentially lower cost due to the use of a residue with a low economical value. In terms of the mechanical performance, the addition of the liquor waste derived SCGs increased the ε_b value of the injection-molded PLA pieces from 10.4% to 39.6%, which represents an improvement of approximately 280%. This enhancement was related to the higher lipid content and remaining solvent traces present in the coffee wastes derived from the liquor preparation. In this industry, coffee grains are naturally roasted and then macerated with ethanol to obtain alcoholic beverages, so that they are not subjected to extraction processes with high temperatures. Furthermore, the simultaneous addition of OLAs, particularly the oligomer functionalized with multiple MAH groups, contributed to improve the impact strength and, more notably, the tensile strength of the green composite pieces. In particular, the OLA2$_{mal}$ sample yielded a σ_{max}

value of 20.7 MPa, respectively, corresponding to improvements of nearly 60% in relation to the PLA/SCG sample. The results attained in the present study were ascribed to the nucleation effect and interaction of the SCG fillers with the PLA matrix due to the presence of the multiple MAH groups in the oligomer, narrowing the filler-to-matrix gaps. In all cases, a relatively high thermal stability was attained, being adequate for applications that deal with temperatures below 275 °C, whereas the visual appearance of the pieces showed a typical dark-brown color that can resemble natural materials, such as wood. Finally, the resultant green composites showed slightly lower disintegration rates than PLA, but still biodegraded in compost conditions.

Therefore, it can be concluded that the use of coffee waste derived from the liquor industry is very promising for developing cost-effective green composites and it opens a new route towards the development of ductile PLA-based materials. These novel green composites are characterized by showing moderate tensile strength and ductility, thermal stability, being renewable and inexpensive due to being based on biomass derived from food wastes, and also biodegradable in industrial composting facilities. All of these features can open the door to their use in disposable food-serving utensils and tableware, such as plates, trays, spoons, forks, knives, cups, or straws. Moreover, a bright vision for the future might be anticipated for their applications in sustainable food packaging and related uses, for instance, single-use coffee cups and capsules that can be organically recyclable to reduce environmental pollution. Future works will focus on further optimizing the amount of SCGs, the development of other compatibilizers that lead to a higher performance, and also on the evaluation of the active properties of the resultant composites, such as their antioxidant capacity for food preservation. Moreover, the organoleptic properties and stability of the green composites over time will also be explored.

Author Contributions: The methodology, software, investigation, and writing—original draft were performed by E.T.-D., J.G.-C. and N.M.; S.F. contributed to the conceptualization and formal analysis; conceptualization and writing—review and editing were performed by S.T.-G.; supervision and funding and acquisition by L.S.-N. All authors have read and agreed to the published version of the manuscript.

Funding: The authors wish to thank the Spanish Ministry of Science and Innovation (MICI), grant number MEN PID2020-116496RB-C22, for funding this research.

Institutional Review Board Statement: Not applicable.

Informed Consent Statement: Not applicable.

Data Availability Statement: Data is contained within the article and also available on request.

Acknowledgments: S. Torres-Giner acknowledges the Spanish Ministry of Science and Innovation (MICI) for the funding received during his Ramón y Cajal contract (RYC2019-027784-I) and J. Gomez-Caturla wishes to thank UPV for his FPI grant (SP20200080). Condensia Condensia Química Inc. (Barcelona, Spain) and Licores Sinc S.A. (Alcoy, Spain) are both acknowledged for generously providing the raw materials. The authors are also grateful to the microscopy services at UPV for their help in collecting and analyzing the FESEM images.

Conflicts of Interest: The authors declare no conflict of interest.

References

1. Quiles-Carrillo, L.; Montanes, N.; Boronat, T.; Balart, R.; Torres-Giner, S. Evaluation of the engineering performance of different bio-based aliphatic homopolyamide tubes prepared by profile extrusion. *Polym. Test.* **2017**, *61*, 421–429. [CrossRef]
2. Faruk, O.; Bledzki, A.K.; Fink, H.-P.; Sain, M. Biocomposites reinforced with natural fibers: 2000–2010. *Prog. Polym. Sci.* **2012**, *37*, 1552–1596. [CrossRef]
3. Puglia, D.; Biagiotti, J.; Kenny, J. A review on natural fibre-based composites—Part II: Application of natural reinforcements in composite materials for automotive industry. *J. Nat. Fibers* **2005**, *1*, 23–65. [CrossRef]
4. Zinge, C.; Kandasubramanian, B. Nanocellulose based biodegradable polymers. *Eur. Polym. J.* **2020**, *133*, 109758. [CrossRef]
5. Arrieta, M.P.; López, J.; Rayón, E.; Jiménez, A. Disintegrability under composting conditions of plasticized PLA–PHB blends. *Polym. Degrad. Stab.* **2014**, *108*, 307–318. [CrossRef]

6. Dijkstra, P.J.; Du, H.; Feijen, J. Single site catalysts for stereoselective ring-opening polymerization of lactides. *Polym. Chem.* **2011**, *2*, 520–527. [CrossRef]
7. Sessini, V.; Palenzuela, M.; Damián, J.; Mosquera, M.E. Bio-based polyether from limonene oxide catalytic ROP as green polymeric plasticizer for PLA. *Polymer* **2020**, *210*, 123003. [CrossRef]
8. Quiles-Carrillo, L.; Montanes, N.; Lagaron, J.M.; Balart, R.; Torres-Giner, S. Bioactive multilayer polylactide films with controlled release capacity of gallic acid accomplished by incorporating electrospun nanostructured coatings and interlayers. *Appl. Sci.* **2019**, *9*, 533. [CrossRef]
9. Radusin, T.; Torres-Giner, S.; Stupar, A.; Ristic, I.; Miletic, A.; Novakovic, A.; Lagaron, J.M. Preparation, characterization and antimicrobial properties of electrospun polylactide films containing *Allium ursinum* L. extract. *Food Packag. Shelf Life* **2019**, *21*, 100357. [CrossRef]
10. Scarfato, P.; Di Maio, L.; Incarnato, L. Recent advances and migration issues in biodegradable polymers from renewable sources for food packaging. *J. Appl. Polym. Sci.* **2015**, *132*, 42597. [CrossRef]
11. Valerio, O.; Pin, J.M.; Misra, M.; Mohanty, A.K. Synthesis of glycerol-based biopolyesters as toughness enhancers for polylactic acid bioplastic through reactive extrusion. *ACS Omega* **2016**, *1*, 1284–1295. [CrossRef] [PubMed]
12. Bhardwaj, R.; Mohanty, A.K. Modification of brittle polylactide by novel hyperbranched polymer-based nanostructures. Biomacromolecules. *Biomacromolecules* **2007**, *8*, 2476–2484. [CrossRef] [PubMed]
13. Liu, R.; Peng, Y.; Cao, J.; Chen, Y. Comparison on properties of lignocellulosic flour/polymer composites by using wood, cellulose, and lignin flours as fillers. *Compos. Sience Technol.* **2014**, *103*, 1–7. [CrossRef]
14. Quiles-Carrillo, L.; Montanes, N.; Sammon, C.; Balart, R.; Torres-Giner, S. Compatibilization of highly sustainable polylactide/almond shell flour composites by reactive extrusion with maleinized linseed oil. *Ind. Crops Prod.* **2018**, *111*, 878–888. [CrossRef]
15. Zaaba, N.F.; Ismail, H.; Jaafar, M. Effect of peanut shell powder content on the properties of recycled polypropylene (RPP)/peanut shell powder (PSP) composites. *BioResources* **2013**, *8*, 5826–5841. [CrossRef]
16. Garcia-Garcia, D.; Carbonell-Verdu, A.; Jordá-Vilaplana, A.; Balart, R.; Garcia-Sanoguera, D. Development and characterization of green composites from bio-based polyethylene and peanut shell. *J. Appl. Polym. Sci.* **2016**, *133*, 43940. [CrossRef]
17. Essabir, H.; Hilali, E.; Elgharad, A.; El Minor, H.; Imad, A.; Elamraoui, A.; Al Gaoudi, O. Mechanical and thermal properties of bio-composites based on polypropylene reinforced with Nut-shells of Argan particles. *Mater. Des.* **2013**, *49*, 442–448. [CrossRef]
18. Essabir, H.; El Achaby, M.; Bouhfid, R.; Qaiss, A. Morphological, structural, thermal and tensile properties of high density polyethylene composites reinforced with treated argan nut shell particles. *J. Bionic Eng.* **2015**, *12*, 129–141. [CrossRef]
19. Kim, K.-W.; Lee, B.-H.; Kim, H.-J.; Sriroth, K.; Dorgan, J.R. Thermal and mechanical properties of cassava and pineapple flours-filled PLA bio-composites. *J. Therm. Anal. Calorim.* **2012**, *108*, 1131–1139. [CrossRef]
20. Torres-Giner, S.; Hilliou, L.; Melendez-Rodríguez, B.; Figueroa-Lopez, K.J.; Madalena, D.; Cabedo, L.; Covas, J.A.; Vicente, A.A.; Lagaron, J.M. Melt processability, characterization, and antibacterial activity of compression-molded green composite sheets made of poly(3-hydroxybutyrate-*co*-3-hydroxyvalerate) reinforced with coconut fibers impregnated with oregano essential oil. *Food Packag. Shelf Life* **2018**, *17*, 39–49. [CrossRef]
21. Adeniyi, A.G.; Ighalo, J.O.; Onifade, D.V. Banana and plantain fiber-reinforced polymer composites. Journal of Polymer Engineering. *J. Polym. Eng.* **2019**, *39*, 597–611. [CrossRef]
22. Arrakhiz, F.; Benmoussa, K.; Bouhfid, R.; Qaiss, A. Pine cone fiber/clay hybrid composite: Mechanical and thermal properties. *Mater. Des.* **2013**, *50*, 376–381. [CrossRef]
23. Mussatto, S.I.; Machado, E.M.; Martins, S.; Teixeira, J. Production, composition, and application of coffee and its industrial residues. *Food Bioprocess Technol.* **2011**, *4*, 661–672. [CrossRef]
24. Klingel, T.; Kremer, J.I.; Gottstein, V.; Rajcic de Rezende, T.; Schwarz, S.; Lachenmeier, D.W. A review of coffee by-products including leaf, flower, cherry, husk, silver skin, and spent grounds as novel foods within the European Union. *Foods* **2020**, *9*, 665. [CrossRef]
25. Mussatto, S.I.; Ballesteros, L.F.; Martins, S.; Teixeira, J.A. Extraction of antioxidant phenolic compounds from spent coffee grounds. *Sep. Purif. Technol.* **2011**, *83*, 173–179. [CrossRef]
26. Sluiter, A.; Hames, B.; Ruiz, R.; Scarlata, C.; Sluiter, J.; Templeton, D.; Crocker, D.L.A.P. Determination of structural carbohydrates. *Lab. Anal. Proced.* **2012**, *1617*, 1–16.
27. Mueanmas, C.; Nikhom, R.; Petchkaew, A.; Iewkittayakorn, J.; Prasertsit, K. Extraction and esterification of waste coffee grounds oil as non-edible feedstock for biodiesel production. *Renew. Energy* **2019**, *133*, 1414–1425. [CrossRef]
28. Liu, K.; Price, G. Evaluation of three composting systems for the management of spent coffee grounds. *Bioresour. Technol.* **2011**, *102*, 7966–7974. [CrossRef]
29. Lessa, E.F.; Nunes, M.L.; Fajardo, A.R. Chitosan/waste coffee-grounds composite: An efficient and eco-friendly adsorbent for removal of pharmaceutical contaminants from water. *Carbohydr. Polym.* **2018**, *189*, 257–266. [CrossRef]
30. Kang, S.B.; Oh, H.Y.; Kim, J.J.; Choi, K.S. Characteristics of spent coffee ground as a fuel and combustion test in a small boiler (6.5 kW). *Renew. Energy* **2017**, *113*, 1208–1214. [CrossRef]
31. Suaduang, N.; Ross, S.; Ross, G.; Pratumshat, S.; Mahasaranon, S. Effect of spent coffee grounds filler on the physical and mechanical properties of poly(lactic acid) bio-composite films. *Mater. Today Proc.* **2019**, *17*, 2104–2110. [CrossRef]

32. Wu, C.-S. Renewable resource-based green composites of surface-treated spent coffee grounds and polylactide: Characterisation and biodegradability. *Polym. Degrad. Stab.* **2015**, *121*, 51–59. [CrossRef]
33. García-García, D.; Carbonell, A.; Samper, M.; García-Sanoguera, D.; Balart, R. Green composites based on polypropylene matrix and hydrophobized spend coffee ground (SCG) powder. *Compos. Part B* **2015**, *78*, 256–265. [CrossRef]
34. Ramalakshmi, K.; Rao, L.J.M.; Takano-Ishikawa, Y.; Goto, M. Bioactivities of low-grade green coffee and spent coffee in different in vitro model systems. *Food Chem.* **2009**, *115*, 79–85. [CrossRef]
35. Jorda-Reolid, M.; Gomez-Caturla, J.; Ivorra-Martinez, J.; Stefani, P.M.; Rojas-Lema, S.; Quiles-Carrillo, L. Upgrading Argan Shell Wastes in Wood Plastic Composites with Biobased Polyethylene Matrix and Different Compatibilizers. *Polymers* **2021**, *13*, 922. [CrossRef]
36. Torres-Giner, S.; Montanes, N.; Fenollar, O.; García-Sanoguera, D.; Balart, R. Development and optimization of renewable vinyl plastisol/wood flour composites exposed to ultraviolet radiation. *Mater. Des.* **2016**, *108*, 648–658. [CrossRef]
37. Quiles-Carrillo, L.; Montanes, N.; Garcia-Garcia, D.; Carbonell-Verdu, A.; Balart, R.; Torres-Giner, S. Effect of different compatibilizers on injection-molded green composite pieces based on polylactide filled with almond shell flour. *Compos. Part B Eng.* **2018**, *147*, 76–85. [CrossRef]
38. Torres-Giner, S.; Montanes, N.; Boronat, T.; Quiles-Carrillo, L.; Balart, R. Melt grafting of sepiolite nanoclay onto poly(3-hydroxybutyrate-*co*-4-hydroxybutyrate) by reactive extrusion with multi-functional epoxy-based styrene-acrylic oligomer. *Eur. Polym. J.* **2016**, *84*, 693–707. [CrossRef]
39. Jafari, A.; Hassanajili, S.; Azarpira, N.; Karimi, M.B.; Geramizadeh, B. Development of thermal-crosslinkable chitosan/maleic terminated polyethylene glycol hydrogels for full thickness wound healing: In vitro and in vivo evaluation. *Eur. Polym. J.* **2019**, *118*, 113–127. [CrossRef]
40. Balart, J.; Fombuena, V.; Fenollar, O.; Boronat, T.; Sánchez-Nacher, L. Processing and characterization of high environmental efficiency composites based on PLA and hazelnut shell flour (HSF) with biobased plasticizers derived from epoxidized linseed oil (ELO). *Compos. Part B: Eng.* **2016**, *86*, 168–177. [CrossRef]
41. Lascano, D.; Moraga, G.; Ivorra-Martinez, J.; Rojas-Lema, S.; Torres-Giner, S.; Balart, R.; Boronat, T.; Quiles-Carrillo, L. Development of injection-molded polylactide pieces with high toughness by the addition of lactic acid oligomer and characterization of their shape memory behavior. *Polymers* **2019**, *11*, 2099. [CrossRef] [PubMed]
42. Martin, O.; Avérous, L. Poly(lactic acid): Plasticization and properties of biodegradable multiphase systems. *Polymer* **2001**, *42*, 6209–6219. [CrossRef]
43. Lowe, C.; Pechy, P.; Kausch, H.H. Synthesis of acrylate functional telechelic poly(lactic acid) oligomer by transesterification. *J. Mater. Sci. Mater. Med. Vol.* **2000**, *11*, 505–510.
44. Grosse, C.; Thévenon, M.F.; Noël, M.; Gerardin, P. Optimising wood chemical modification with lactic acid oligomers by screening of processing conditions and chemical additives. In Proceedings of the 47th IRG Annual Meeting, Lisbon, Portugal, 15–19 May 2016.
45. Torres-Giner, S.; Gimeno-Alcañiz, J.V.; Ocio, M.J.; Lagaron, J.M. Optimization of electrospun polylactide-based ultrathin fibers for osteoconductive bone scaffolds. *J. Appl. Polym. Sci.* **2011**, *122*, 914–925. [CrossRef]
46. Mendes, J.F.; Martins, J.T.; Manrich, A.; Luchesi, B.R.; Dantas, A.P.S.; Vanderlei, R.M.; Claro, P.C.; Neto, A.R.D.S.; Mattoso, L.H.C.; Martins, M.A. Thermo-physical and mechanical characteristics of composites based on high-density polyethylene (HDPE) e spent coffee grounds (SCG). *J. Polym. Environ.* **2021**, *29*, 2888–2900. [CrossRef]
47. Crespo, J.; Balart, R.; Sanchez, L.; Lopez, J. Mechanical behaviour of vinyl plastisols with cellulosic fillers. Analysis of the interface between particles and matrices. *Int. J. Adhes. Adhes.* **2007**, *27*, 422–428. [CrossRef]
48. Vidović, E.; Faraguna, F.; Jukić, A. Influence of inorganic fillers on PLA crystallinity and thermal properties. *J. Therm. Anal. Calorim.* **2017**, *127*, 371–380. [CrossRef]
49. Agüero, A.; Morcillo, M.d.C.; Quiles-Carrillo, L.; Balart, R.; Boronat, T.; Lascano, D.; Torres-Giner, S.; Fenollar, O. Study of the influence of the reprocessing cycles on the final properties of polylactide pieces obtained by injection molding. *Polymers* **2019**, *11*, 1908. [CrossRef]
50. Torres-Giner, S.; Torres, A.; Ferrándiz, M.; Fombuena, V.; Balart, R. Antimicrobial activity of metal cation-exchanged zeolites and their evaluation on injection-molded pieces of bio-based high-density polyethylene. *J. Food Saf.* **2017**, *37*, e12348. [CrossRef]
51. Montava-Jorda, S.; Lascano, D.; Quiles-Carrillo, L.; Montanes, N.; Boronat, T.; Martinez-Sanz, A.V.; Ferrandiz-Bou, S.; Torres-Giner, S. Mechanical recycling of partially bio-based and recycled polyethylene terephthalate blends by reactive extrusion with poly(styrene-*co*-glycidyl methacrylate). *Polymers* **2020**, *12*, 174. [CrossRef]
52. Ortiz-Barajas, D.L.; Arévalo-Prada, J.A.; Fenollar, O.; Rueda-Ordóñez, Y.J.; Torres-Giner, S. Torrefaction of Coffee Husk Flour for the Development of Injection-Molded Green Composite Pieces of Polylactide with High Sustainability. *Appl. Sci.* **2020**, *10*, 6468. [CrossRef]
53. Suaduang, N.; Ross, S.; Ross, G.M.; Wangsoub, S.; Mahasaranon, S. The physical and mechanical properties of biocomposite films composed of poly(lactic acid) with spent coffee grounds. *Key Eng. Mater.* **2019**, *824*, 87–93. [CrossRef]
54. Faludi, G.; Dora, G.; Imre, B.; Renner, K.; Móczó, J.; Pukánszky, B. PLA/lignocellulosic fiber composites: Particle characteristics, interfacial adhesion, and failure mechanism. *J. Appl. Polym. Sci.* **2014**, *131*. [CrossRef]
55. Khan, N.; Brown, J.B. The composition of coffee oil and its component fatty acids. *J. Am. Oil Chem. Soc.* **1953**, *30*, 606–609. [CrossRef]

56. Battegazzore, D.; Bocchini, S.; Alongi, J.; Frache, A. Plasticizers, antioxidants and reinforcement fillers from hazelnut skin and cocoa by-products: Extraction and use in PLA and PP. *Polym. Degrad. Stab.* **2014**, *108*, 297–306. [CrossRef]
57. Burgos, N.; Martino, V.P.; Jiménez, A.J. Characterization and ageing study of poly(lactic acid) films plasticized with oligomeric lactic acid. *Polym. Degrad. Stab.* **2013**, *98*, 651–658. [CrossRef]
58. Burgos, N.; Tolaguera, D.; Fiori, S.; Jiménez, A. Synthesis and Characterization of lactic acid oligomers: Evaluation of performance as poly(lactic acid) plasticizers. *Polym. Degrad. Stab.* **2014**, *22*, 227–235. [CrossRef]
59. Torres-Giner, S.; Montanes, N.; Fombuena, V.; Boronat, T.; Sanchez-Nacher, L. Preparation and characterization of compression-molded green composite sheets made of poly(3-hydroxybutyrate) reinforced with long pita fibers. *Adv. Polym. Technol.* **2018**, *37*, 1305–1315. [CrossRef]
60. da Silva, A.P.; Pereira, M.d.P.; Passador, F.R.; Montagna, L.S. PLA/Coffee Grounds Composites: A Study of Photodegradation and Biodegradation in Soil. Macromolecular Symposia. *Macromol. Symp.* **2020**, *394*, 2000091. [CrossRef]
61. Quiles-Carrillo, L.; Duart, S.; Montanes, N.; Torres-Giner, S.; Balart, R. Enhancement of the mechanical and thermal properties of injection-molded polylactide parts by the addition of acrylated epoxidized soybean oil. *Mater. Des.* **2018**, *140*, 54–63. [CrossRef]
62. Rojas-Lema, S.; Quiles-Carrillo, L.; Garcia-Garcia, D.; Melendez-Rodriguez, B.; Balart, R.; Torres-Giner, S. Tailoring the properties of thermo-compressed polylactide films for food packaging applications by individual and combined additions of lactic acid oligomer and halloysite nanotubes. *Molecules* **2020**, *25*, 1976. [CrossRef] [PubMed]
63. Xing, Q.; Zhang, X.; Dong, X.; Liu, G.; Wang, D. Low-molecular weight aliphatic amides as nucleating agents for poly(L-lactic acid): Conformation variation induced crystallization enhancement. *Polymer* **2012**, *53*, 2306–2314. [CrossRef]
64. Morrell, J. Degradation of lignocellulosic materials and its prevention. *JOM* **2014**, *66*, 580–587. [CrossRef]
65. Kim, S.-K.; Lee, T. Degradation of lignocellulosic materials under sulfidogenic and methanogenic conditions. *Waste Manag.* **2009**, *29*, 224–227. [CrossRef] [PubMed]
66. Darie-Niţă, R.N.; Vasile, C.; Irimia, A.; Lipşa, R.; Râpă, M. Evaluation of some eco-friendly plasticizers for PLA films processing. *J. Appl. Polym. Sci.* **2016**, *133*, 43223. [CrossRef]
67. Laaziz, S.A.; Raji, M.; Hilali, E.; Essabir, H.; Rodrigue, D.; Bouhfid, R. Bio-composites based on polylactic acid and argan nut shell: Production and properties. *Int. J. Biol. Macromol.* **2017**, *104*, 30–42. [CrossRef]
68. Balart, J.; Montanes, N.; Fombuena, V.; Boronat, T.; Sánchez-Nacher, L. Disintegration in compost conditions and water uptake of green composites from poly(lactic acid) and hazelnut shell flour. *J. Polym. Environ.* **2018**, *26*, 701–715. [CrossRef]
69. Quiles-Carrillo, L.; Montanes, N.; Lagaron, J.M.; Balart, R.; Torres-Giner, S. On the use of acrylated epoxidized soybean oil as a reactive compatibilizer in injection-molded compostable pieces consisting of polylactide filled with orange peel flour. *Polym. Int.* **2018**, *67*, 1341–1351. [CrossRef]
70. Quiles-Carrillo, L.; Montanes, N.; Pineiro, F.; Jorda-Vilaplana, A.; Torres-Giner, S. Ductility and toughness improvement of injection-molded compostable pieces of polylactide by melt blending with poly(ε-caprolactone) and thermoplastic starch. *Materials* **2018**, *11*, 2138. [CrossRef]

foods

MDPI

Article

Pectin-Based Films with Cocoa Bean Shell Waste Extract and ZnO/Zn-NPs with Enhanced Oxygen Barrier, Ultraviolet Screen and Photocatalytic Properties

Ana Cristina Mellinas, Alfonso Jiménez and María Carmen Garrigós *

Department of Analytical Chemistry, Nutrition & Food Sciences, University of Alicante, San Vicente del Raspeig, ES-03690 Alicante, Spain; cristina.mellinas@ua.es (A.C.M.); alfjimenez@ua.es (A.J.)
* Correspondence: mc.garrigos@ua.es; Tel.: +34-965-903-529

Received: 28 September 2020; Accepted: 27 October 2020; Published: 29 October 2020

check for updates

Abstract: In this work, pectin-based active films with a cocoa bean shell extract, obtained after waste valorisation of residues coming from the chocolate production process, and zinc oxide/zinc nanoparticles (ZnO/Zn-NPs) at different concentrations, were obtained by casting. The effect of the active additive incorporation on the thermal, barrier, structural, morphological and optical properties was investigated. Moreover, the photocatalytic properties of the obtained films based on the decomposition of methylene blue (MB) in aqueous solution at room temperature were also studied. A significant increase in thermal and oxidative stability was obtained with the incorporation of 3 wt% of ZnO/Zn-NPs compared to the control film. The addition of 5 wt% cocoa bean shell extract to pectin significantly affected the oxygen barrier properties due to a plasticizing effect. In contrast, the addition of ZnO/Zn-NPs at 1 wt% to pectin caused a decrease in oxygen transmission rate per film thickness (OTR.e) values of approximately 50% compared to the control film, resulting in an enhanced protection against oxidation for food preservation. The optical properties were highly influenced by the incorporation of the natural extract but this effect was mitigated when nanoparticles were also incorporated into pectin-based films. The addition of the extract and nanoparticles resulted in a clear improvement (by 98%) in UV barrier properties, which could be important for packaged food sensitive to UV radiation. Finally, the photocatalytic activity of the developed films containing nanoparticles was demonstrated, showing photodegradation efficiency values of nearly 90% after 60 min at 3 wt% of ZnO/Zn-NPs loading. In conclusion, the obtained pectin-based bionanocomposites with cocoa bean shell waste extract and zinc oxide/zinc nanoparticles showed great potential to be used as active packaging for food preservation.

Keywords: pectin; cocoa bean shell waste; zinc oxide nanoparticles; photocatalytic activity

1. Introduction

The use of biopolymers in films for food packaging has emerged as an alternative to plastic commodities by their bio-based origin, biodegradable character and possibilities to host multifunctional active agents [1]. In particular, the formulation of edible films obtained from lipids, proteins and polysaccharides as environmentally friendly materials for food packaging has been reported in the last few years [2–4]. Among them, polysaccharides are widely used in this application due to their large availability in nature, easy processing, cost-effectiveness and excellent biocompatibility [5]. The development of edible films based on polysaccharides has been reported to show good mechanical properties as well as environmentally friendly disposal after use. Biofilms based on cellulose [6] and derivatives [7], starch [8], chitosan [9,10] and pectin [11,12] have been proposed for food packaging applications.

Pectin is considered one of the most complex macromolecules in nature and it is formed by a group of structural polysaccharides, mostly containing galacturonic acid units [13]. Pectin is present in the primary cell walls of many plants, contributing rigidity to their structure, and it is frequently combined with lignin, hemicelluloses or cellulose. Pectin is basically composed of α-(1,4)-linked d-galacturonic acid [14] and its properties are influenced by the methyl esterification degree, which depends on the plant origin and the processing conditions [15]. In general, pectin shows good biodegradation performance, biocompatibility and non-toxicity, making it a good biomaterial for applications such as pharmaceutics, food packaging and cosmetics [16–18]. Nevertheless, pectin-based films also show some shortcomings, such as their poor barrier properties to water and gases [19,20]. The incorporation of hydrophobic compounds, such as essential oils, to the polymer matrix [19] or the development of blends using other polymers, such as chitosan [2,21,22], polyethylene glycol [23] and natural proteins [24,25], have been proposed as adequate strategies to improve the barrier properties of neat pectin. Moreover, the reinforcement with nanomaterials to produce pectin-based nano-biocomposites with improved properties has been recently proposed [26–29].

The use of sustainable and functional food packaging materials with natural additives or extracts has been proposed to improve the packaging functionalities as well as food quality and safety [30–32]. These functional additives are essential in the reduction or even complete elimination of the main food spoilage causes, such as rancidity, colour loss/change, nutrient losses, dehydration, microbial proliferation and off-odour production [33]. In this context, cocoa bean shell (CBS) is the major by-product obtained from the cocoa bean during the processing of chocolate and other food products [34]. CBS extracts could be a rich source of bioactive compounds, including polysaccharides, proteins and phenolic compounds, with high potential antioxidant/antimicrobial properties [35].

The interesting functionalities of zinc oxide nanoparticles (ZnO-NPs) such as antibacterial, antifungal, UV filtering properties, high catalytic and photochemical activities and non-toxicity to the environment have been reported by some authors [36,37]. In addition, these nanoparticles have been recently approved by the European Food Safety Authority (EFSA) to be used as a transparent ultraviolet (UV) light absorber in polyolefins for food packaging applications [38]. The addition of ZnO-NPs to polymer matrices such as polyurethane and chitosan has been reported to decrease water vapor and oxygen transmission rate compared to the neat films [39] while some authors reported the use of ZnO-NPs in modified atmosphere packaging (MAP) [40]. Few reports have been found combining ZnO-NPs and pectin. The development of chitosan/pectin/ZnO porous films for wound healing [41] and pectin/alginate/ZnO-NPs films as active packaging for food preservation [42] has been recently reported. Moreover, pectin-based gels with ZnO NPs have been prepared for drug delivery applications [43] and as adsorbents for cationic dyes [44]. This work reports for the first time the development and characterisation of pectin-based active films with ZnO/Zn-NPs and a cocoa bean shell extract (CBSE) obtained from food by-products. The effect of these additives at different content levels was evaluated by studying the structural, morphological, barrier, thermal, optical and photocatalytic properties of the developed films.

2. Materials and Methods

2.1. Materials and Reagents

Cocoa bean shell residue was obtained as a by-product from a local chocolate producer. Pectin from citrus peel (galacturonic acid, ≥74%) and all other chemicals and reagents were purchased from Sigma-Aldrich (Madrid, Spain) and they were of analytical grade.

2.2. CBS Extract Preparation

Microwave-assisted extraction (MAE) was used to obtain CBS extracts from cocoa bean shell wastes by following a previously reported method [45]. The aim of this procedure was to obtain valuable fractions rich in active compounds (mainly polyphenols and polysaccharides) to be further used in the preparation of the pectin-based films. Prior to extraction, CBS was mixed with distilled water (0.04 g/mL) in a round-bottom flask and pH was adjusted to 2 with hydrochloric acid (1 M) and to 12 using sodium hydroxide (1 M) to obtain CBSE-2 and CBSE-12 extracts, respectively. Samples were subjected to extraction by using microwave equipment (Flexiwave, Milestone srl, Sorisole, Italy) at 100 °C for 5 min. The obtained mixture was then centrifuged for 20 min at 5300 rpm and 4 °C to recover the active compounds. The solid was separated and the different extracts were stored at −20 °C until further use.

2.3. Synthesis of ZnO/Zn-NPs

For the synthesis of ZnO/Zn-NPs, a previously optimised method was followed. Ten milliliters of a zinc chloride aqueous solution (0.37 M) were mixed with 50 mL of CBSE-12 solution (0.01 g/mL) and this mixture was kept under constant stirring (400 rpm). This solution was submitted to microwave heating (Flexiwave, Milestone srl, Sorisole, Italy) at 100 °C for 5 min and further centrifuged for 15 min at 5300 rpm and 4 °C. Finally, the solid was recuperated and calcined at 550 °C in a muffle furnace (JP Selecta, Barcelona, Spain) for 2 h and stored in darkness under vacuum conditions until use.

2.4. Pectin-Based Film Preparation

The casting technique was used for the development of pectin-based films. One gram of pectin was dissolved in distilled water (50 mL) under stirring at 70 °C. After complete dissolution, sorbitol (0.2 wt%) was added as plasticiser and pH was set to 4.5 with HCl 1 M. This mixture was stirred for 1 h at 400 rpm. Then, different concentrations of ZnO/Zn-NPs (1 and 3 wt%) were dispersed in 30 mL of distilled water and they were further sonicated for 30 min. A CBSE-2-based solution (0.05 wt%) was also prepared and added to the pectin solution, with or without ZnO/Zn-NPs, and sonicated for an additional 30 min. All solutions were finally casted into Petri dishes (15 cm diameter) and dried at 25 °C and 40% relative humidity in a climatic chamber (Dycometal, Barcelona, Spain) for 48 h. A control pectin film (PC) without the addition of any active additive was also prepared. The obtained formulations used in this work are presented in Table 1.

Table 1. Moisture, thickness, barrier properties and photocatalytic activity of pectin-based films.

Formulation	CBSE-2 (wt%)	ZnO/Zn-NPs (wt%)	Moisture (%)	Thickness * (mm)	OTR.e (cm^3 mm m^{-2} day)	Degradation Efficiency (%)
PC (control)	-	-	18 ± 3 ^{a}	0.087 ± 0.022 ^{a}	0.51 ± 0.07 ^{a}	4 ± 2 ^{a}
PC-1ZnO	-	1	13 ± 1 ^{a}	0.073 ± 0.031 ^{a}	0.32 ± 0.02 ^{b}	72 ± 7 ^{b}
PC-3ZnO	-	3	14 ± 3 ^{a}	0.070 ± 0.024 ^{a}	0.51 ± 0.09 ^{a}	87 ± 4 ^{c}
PC-5E	5	-	13 ± 1 ^{a}	0.078 ± 0.016 ^{a}	0.63 ± 0.01 ^{c}	39 ± 6 ^{d}
PC-5E-1ZnO	5	1	17 ± 1 ^{a}	0.071 ± 0.011 ^{a}	0.36 ± 0.02 ^{b}	84 ± 4 ^{bc}
PC-5E-3ZnO	5	3	16 ± 4 ^{a}	0.085 ± 0.015 ^{a}	0.35 ± 0.13 ^{ab}	88 ± 2 ^{c}

Mean ± SD, $n = 3$ (* $n = 10$ for thickness). Different superscripts within the same column indicate significant differences between formulations ($p < 0.05$). CBSE-2: cocoa bean shell waste extract obtained at pH = 2. ZnO/Zn-NPs: zinc oxide/zinc nanoparticles. OTR.e: oxygen transmission rate per film thickness.

2.5. Pectin-Based Film Characterisation

2.5.1. Moisture Content (MC)

The MC of pectin-based films was determined, in triplicate, according to the method proposed by Younis and Zhao [46]. Film samples were first conditioned in a desiccator containing $CaCl_2$ at 10%

relative humidity (RH) for 2 days, cut into 2 × 2 cm^2 pieces and weighed (W_i). Then, films were dried at 105 °C for 24 h and weighed again (W_f). The MC was calculated according to Equation (1):

$$MC\ (\%) = \left(\frac{W_f - W_i}{W_i}\right) \times 100 \quad (1)$$

Thermogravimetric analysis (TGA) tests were performed, in triplicate, with a TGA/SDTA 851 Mettler Toledo thermal analyser (Schwarzenbach, Switzerland). Approximately 6 mg of each sample were heated from 25 to 700 °C at 10 °C/min under a nitrogen atmosphere (flow rate 50 mL/min).

Differential scanning calorimetry (DSC) tests were carried out to determine glass transition temperature (T_g) in all pectin-based films by using a TA DSC Q-2000 instrument (New Castle, DE, USA) under a nitrogen atmosphere (flow rate 50 mL/min). Four-milligram samples were initially submitted to −90 °C in isothermal mode for 3 min. The temperature program followed consisted of a first heating from −90 to 150 °C, then cooling to −90 °C and a further second heating to 150 °C, with all these stages at a 10 °C/min heating/cooling rate. Three replicates of each sample were performed.

The oxidation onset temperature (OOT) was also determined, in triplicate, by DSC in order to evaluate the oxidative stability of pectin-based films. Samples were heated up at 10 °C/min under a pure oxygen atmosphere (50 mL/min) from 25 °C up to the observation of the exothermic oxidation peak. OOT was calculated as the temperature for the intersection between the DSC baseline and the slope of the exothermic peak in each case.

2.5.2. Oxygen Transmission Rate (OTR)

An oxygen permeation analyser (8500 model Systech, Metrotec S.A, Spain) was used for OTR tests. Pure oxygen (99.9%) was introduced into the upper half of the diffusion chamber while nitrogen was injected into the lower half, where an oxygen sensor was located. Films were cut into 14 cm diameter circles for each formulation and they were clamped in the diffusion chamber at 25 °C before testing. In order to calculate the oxygen transmission rate per film thickness (OTR.e) values, the thickness of films at 10 different positions was measured with a digital micrometer (±0.001 mm, Mitutoyo, Japan). Tests were performed in triplicate.

2.5.3. Optical Properties

Ultraviolet–visible (UV–Vis) spectra of pectin-based films were obtained, in triplicate, with a Biomate-3 Spectrophotometer (Thermospectronic, Mobile, AL, USA). Film pieces, 10 × 30 mm^2, were placed on the internal side of the spectrophotometer cells and transmittance values at 280 and 600 nm were used to evaluate their optical properties.

Colour modifications on pectin-based films caused by the addition of the active additives were studied by using a VIS Minolta CM-2600d portable reflection spectrophotometer. Colour values were expressed as L* (lightness), a* (red/green) and b* (yellow/blue) coordinates in the CIELab colour space. These parameters were determined at five random different locations around the films' surface and the average values were calculated. Since samples were transparent, these measurements were taken over a white background. Total colour difference (ΔE) was calculated according to Equation (2).

$$\Delta E = \sqrt{\left(L_i^* - L\right)^2 + \left(a_i^* - a\right)^2 + \left(b_i^* - b\right)^2} \quad (2)$$

where L, a and b correspond to the coordinates of the control sample (pure pectin film) and L_i, a_i and b_i are the coordinates of each different formulation.

2.5.4. Attenuated Total Reflectance–Fourier Transform Infrared Spectroscopy (ATR–FTIR)

FTIR spectra of all film samples were recorded with a Bruker Analitik IFS 66/S (Ettlingen, Germany) infrared spectrophotometer equipped with a KBr beam splitter, a deuterated triglycine sulphate (DTGS) detector and Bruker OPUS software (Version 3.1). The analysis was performed in triplicate in the attenuated total reflectance (ATR) mode by using a Golden Gate accessory with a diamond crystal. Absorption spectra were obtained in the 4000–500 cm^{-1} range using 64 scans and a resolution of 4 cm^{-1}.

2.5.5. X-ray Diffraction (XRD)

XRD patterns of pectin-based films were recorded on a Bruker (Billerica, MA, USA) D8-Advance diffractometer equipped with a Goebel mirror for non-planar samples, a high-temperature chamber (up to 900 °C), a Kristalloflex K 760-80F X-ray generator (power 3000 W, voltage 20–60 kV and current 5–80 mA) and an X-Ray tube with a copper anode. Data were recorded by using Cu Kα radiation (1.5406 Å).

2.5.6. Field Emission Scanning Electron Microscopy (FESEM)

The surface and the cross-section of all pectin-based films were studied by FESEM (Supra 25-Zeiss, Jena, Germany) in order to evaluate their homogeneity and the influence of the addition of the active additives on the polymer morphology. Samples were coated with a gold layer prior to analysis in order to increase their electrical conductivity.

2.5.7. Photocatalytic Activity

The photocatalytic activity of the pectin-based films was evaluated by following a method based on the decomposition of methylene blue (MB) in aqueous solution at room temperature [8]. In brief, 50 mg of films were suspended in a beaker containing 50 mL of a 15 mg/L MB aqueous solution. Prior to irradiation, these suspensions were sonicated in the dark for 30 min to achieve the maximum adsorption of the dye onto the film surface. Then, the suspension was horizontally irradiated with a UV lamp under constant orbital stirring. After 60 min of irradiation, 5 mL of the dye solution were taken and the MB concentration in the solution was determined by UV–Vis spectroscopy at 664 nm. Tests were performed in triplicate for each formulation. The degradation efficiency (%) of MB was calculated with Equation (3):

$$Degradation\ efficiency\ \% = \left(\frac{C_i - C_t}{C_i}\right) \times 100 \tag{3}$$

where C_i is the initial MB concentration (15 mg/L) and C_t is the dye concentration at time t (mg/L).

2.6. Statistical Analysis

Statistical analysis of results was performed with the Statgraphics Centurion XVI statistical software. An analysis of variance (ANOVA) was carried out. Differences between values were assessed based on confidence intervals by using Tukey's test at a $p \leq 0.05$ significance level.

3. Results

3.1. Moisture Content

The moisture content determined in pectin-based films was related to the amount of water adsorbed onto their surface [47] and the obtained results are shown in Table 1. Nanocomposite active films exhibited a slightly lower moisture content although no significant differences were obtained compared to the neat pectin film ($p > 0.05$). This slight decrease in moisture content could be due to the formation of hydrogen bonds between the additives and the pectin structure which could reduce the diffusion of water molecules through the material [48].

3.2. Thermal Properties

TGA and DSC tests were performed to study the influence of the addition of ZnO/Zn-NPs and CBSE-2 on the thermal properties of the obtained films. The most relevant results are shown in Table 2. The thermograms obtained for all formulations showed three different regions, as can be observed in the differential thermogravimetric (DTG) curves (Figure 1). The first thermal step, in the 50–150 °C range, was attributed to the desorption of water linked to hydrophilic groups in the pectin structure [19,23,49]. The second degradation step was observed between 200 and 400 °C, corresponding to the main thermal degradation of pectin. For the neat polymer, the maximum degradation peak was observed at 231 ± 1 °C (Table 2), as expected from the decomposition of galacturonic acid chains followed by the decarboxylation of the ring and acid side groups, with the formation of several low molecular weight gases and solid char [49,50]. Finally, slight weight loss at temperatures higher than 400 °C was observed in all formulations due to the degradation of the organic matter and C–C bonds remaining after the pyrolysis process [51].

Table 2. Thermal properties of pectin-based films.

Formulation	T_{max} (°C)	T_g (°C)	OOT (°C)
PC	231 ± 1 ^{a}	52 ± 1 ^{a}	202 ± 1 ^{a}
PC-1ZnO	229 ± 1 ^{a}	51 ± 2 ^{a}	202 ± 1 ^{a}
PC-3ZnO	239 ± 2 ^{b}	51 ± 1 ^{a}	214 ± 2 ^{b}
PC-5E	229 ± 1 ^{a}	53 ± 2 ^{a}	206 ± 1 ^{c}
PC-5E-1ZnO	230 ± 1 ^{a}	53 ± 1 ^{a}	205 ± 1 ^{c}
PC-5E-3ZnO	229 ± 1 ^{a}	52 ± 1 ^{a}	207 ± 2 ^{c}

Mean ± SD, n = 3. Different superscripts within the same column indicate significant differences between formulations ($p < 0.05$). OOT. Oxygen onset temperature. T_{max}. Temperature of maximum degradation rate. T_g. Glass transition temperature.

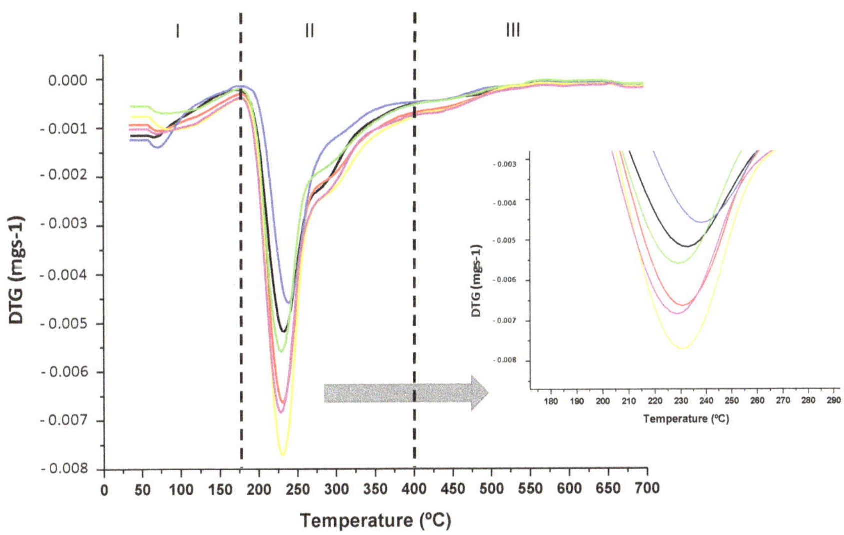

Figure 1. DTG curves of pectin-based films: PC (------), PC-1ZnO (------), PC-3ZnO (------), PC-5E (------), PC-5E-1ZnO (------) and PC-5E-3ZnO (------).

All composite films showed a similar behaviour in their thermal degradation profile, suggesting that the addition of the active compounds did not significantly influence the pectin thermal degradation under a nitrogen atmosphere. However, in the case of the PC-3ZnO film, a significant increase ($p < 0.05$) in the maximum degradation temperature (239 ± 2 °C, Table 2) was observed (see zoomed-in area in

Figure 1) which could be attributed to the strong interaction between the ZnO/Zn-NPs and the pectin structure via hydrogen bonding, as has been recently reported in other composites with zinc oxide nanoparticles and biopolymer matrices [52], resulting in a significant increase in the thermal stability in this formulation [53]. In addition, it should be noted that, due to the high thermal conductivity of ZnO/Zn-NPs, they could act as heat barriers, facilitating dissipation in the inner part of the composite matrix and likewise resulting in an improvement in the thermal stability of the overall nanocomposite, as has been recently reported [54].

Glass transition temperatures (T_g) were also obtained for all films and results are shown in Table 2. T_g depends on the polymer structural arrangement and gives an indication of the torsion oscillation of the carbon backbone in the polymers' structure [55]. No significant differences ($p > 0.05$) between the different active films and neat pectin were observed, suggesting that the addition of ZnO/Zn-NPs and CBSE-2 did not result in significant changes in the intrinsic pectin structure. A similar behaviour was observed by other authors in different polymer matrices after the incorporation of natural extracts and nanoparticles [11,56,57].

OOT measurements represent a very useful tool to evaluate the effect of the addition of different additives in the resistance of a polymer material to oxidative degradation processes [58]. The OOT values obtained for all pectin-based films are also shown in Table 2. An increase of around 10 °C was obtained with the incorporation of 3 wt% of ZnO/Zn-NPs compared to the control film, suggesting that the addition of these nanoparticles allows the thermo-oxidative stability of the polymer matrix to be increased. This behaviour is in agreement with the significant ($p < 0.05$) increase in thermal resistance observed for this formulation by TGA. Lower significant differences ($p < 0.05$) in OOT values (around 4 °C) were also observed between neat poly(ε-caprolactone) (PCL) and PC-5E, PC-5E-1ZnO and PC-5E-3ZnO films. These results can be explained by the presence of some antioxidant compounds in CBSE-2, mainly theobromine, caffeic acid, epicatechin and protocatechuic acid, showing that this extract has a high antioxidant capacity according to previous work [35,45], and reported free radical scavenging activity of zinc oxide nanoparticles [59]. Some authors have already reported the antioxidant effect of ZnO/Zn-NPs and they have justified this activity against oxygen by the large number of active sites present in these nanoparticles, which are able to fix oxygen molecules to their surface, with the consequent reduction of the oxidation reactions [60]. This effect could be really important for the potential application of these films for food packaging when the packaged food could suffer oxidative degradation. It has also been reported that the combination of Zn-based nanoparticles and some other compounds with well-known antioxidant activity could result in synergistic effects with an increase in the resistance to oxygen-induced damage in packaged food [61]. The positive effect of the addition of antioxidant compounds on the increase in OOT values has been previously reported in other polymer matrices, such as poly(lactic acid) (PLA) [62], poly(hydroxybutyrate) (PHB) [63] and polypropylene (PP) [64]. All these authors attributed this increase in OOT values to the prevention of the formation of free radicals caused by the chemical reactions with antioxidant compounds, delaying in this way the overall material oxidation.

3.3. *Oxygen Transmission Rate (OTR)*

The determination of barrier properties in films intended to be used in food packaging is essential to extend the food shelf-life by preventing humidity or diffusion of oxygen, ethylene, aroma or undesired flavours [22] which could produce adverse reactions and modify the organoleptic properties and/or quality of the food product [65]. Table 1 shows the results obtained for OTR.e in all pectin-based films. Significant differences ($p < 0.05$) were found between PC and PC-5E films, which were related to a plasticising effect due to the incorporation of active compounds in CBSE-2 into the pectin matrix, facilitating the diffusional process of oxygen molecules through the polymer matrix, as already reported for similar biopolymer matrices [66]. The addition of the active extract could contribute to an increase in the molecular mobility and the formation of a less packed network structure, separating pectin chains by creating voids and channels for oxygen molecules to pass through the polymer matrix, resulting in a higher oxygen transmission in the PC-5E formulation, as already reported by other authors [67,68].

In contrast, the addition of ZnO/Zn-NPs at 1 wt% to pectin significantly improved ($p < 0.05$) the oxygen barrier in all cases, causing a decrease in OTR.e values of approximately 50% compared to the control film (Table 1). This effect was associated with the formation of hydrogen bonds between ZnO/Zn-NPs and the pectin matrix, resulting in a decrease in the diffusion of oxygen molecules by producing a tortuous pathway through the films' structure [69,70]. Similar results were obtained for pectin-based films after the incorporation of montmorillonite [71] and carbon nanotubes [72]. However, when the concentration of ZnO/Zn-NPs was increased up to 3 wt%, the expected improvement in oxygen barrier properties was not observed, with similar OTR.e values not significantly differing ($p > 0.05$) from those obtained for the control film (Table 1). Although it is supposed that the formation of hydrogen bonds would be more favoured by the addition of larger ZnO/Zn-NPs amounts, high concentrations of these nanoparticles could result in some agglomerations, generating holes inside the nanocomposite structure that would allow an easier diffusion of oxygen [57,73]. Finally, a decrease in OTR.e value was observed for the PC-5E-3ZnO film, which might be due to the improvement in the dispersion of ZnO/Zn-NPs in the polymer matrix after the addition of CBSE-2, as already reported for other metallic nanoparticles [74], although no significant differences ($p > 0.05$) in OTR.e values were observed with the control film.

3.4. Optical Properties

The visual appearance of films for food packaging applications is a critical issue in most cases and the addition of compounds that could be structurally bound to the polymer matrix could result in changes in the optical properties of the resulting films. In this study, all pectin-based films were visually transparent (Figure 2), smooth surfaced, homogeneous and flexible. The CIELab colour parameters (L*, a*, b*) and total colour differences (ΔE) obtained for pectin-based films are shown in Table 3. As can be observed, the b* parameter, which is related to blue–yellow colour changes, was not significantly affected ($p > 0.05$) by the incorporation of the active compounds. Regarding L* and a* parameters, no significant differences ($p > 0.05$) between PC, PC-1ZnO and PC-3ZnO films were found. However, significant differences ($p < 0.05$) were observed in these parameters with the incorporation of CBSE-2 in pectin-based films. A significant increase ($p < 0.05$) in a* values due to the intrinsic reddish colour of the cocoa extract was obtained. Moreover, the addition of ZnO/Zn-NPs to active films containing CBSE-2 produced a general significant decrease ($p < 0.05$) in a* values, which was more apparent at high concentrations of the nanoparticles, which might be due to the whitish coloration of ZnO/Zn-NPs [75]. Lightness was also affected by the incorporation of CBSE-2 in pectin-based films, with a significant decrease ($p < 0.05$) in L* values. The addition of ZnO/Zn-NPs and CBSE-2 (PC-5E-1ZnO and PC-5E-3ZnO formulations) resulted in some lower but not significant ($p > 0.05$) L* values compared to the control film. A similar behaviour was recently reported for pectin-based nanocomposite films with the addition of active compounds, contributing to noticeable colour changes in the developed films [76].

Table 3. Colour parameters and transparency of pectin-based films.

Formulation	L*	a*	b*	ΔE	T_{280} (%)	T_{660} (%)
PC	85.4 ± 2.6^{a}	10.1 ± 3.1^{a}	3.5 ± 1.9^{a}		25 ± 2^{a}	87 ± 3^{a}
PC-1ZnO	86.0 ± 2.6^{a}	8.2 ± 0.6^{a}	3.6 ± 2.0^{a}	2.6 ± 0.6^{a}	2 ± 1^{b}	77 ± 4^{b}
PC-3ZnO	85.4 ± 2.4^{a}	9.8 ± 0.5^{a}	3.7 ± 1.8^{a}	2.8 ± 0.1^{a}	2 ± 1^{b}	77 ± 3^{b}
PC-5E	80.6 ± 2.6^{b}	20.0 ± 2.2^{b}	3.0 ± 2.3^{a}	10.3 ± 2.7^{b}	3 ± 1^{b}	74 ± 3^{b}
PC-5E-1ZnO	82.4 ± 2.4^{ab}	16.9 ± 0.7^{c}	3.4 ± 1.8^{a}	7.1 ± 0.9^{ab}	1 ± 1^{b}	75 ± 3^{b}
PC-5E-3ZnO	83.9 ± 2.4^{ab}	14.0 ± 0.4^{d}	3.6 ± 1.8^{a}	4.3 ± 0.6^{ab}	1 ± 1^{b}	77 ± 3^{b}

Mean ± SD, $n = 5$. Different superscripts within the same column indicate significant differences between formulations ($p < 0.05$).

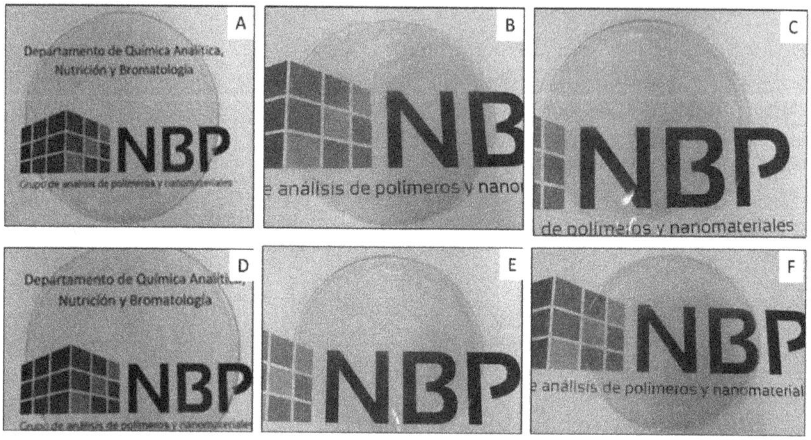

Figure 2. Visual appearance of pectin-based films: (**A**) PC, (**B**) PC-1ZnO, (**C**) PC-3ZnO, (**D**) PC-5E, (**E**) PC-5E-1ZnO and (**F**) PC-5E-3ZnO.

The developed pectin-based films were analysed by using UV–Vis spectrophotometry at 660 nm and 280 nm in order to evaluate transparency and UV blocking ability, respectively. High transparency values were obtained (Table 3) for all pectin-based films with transmittance values higher than 75%, considering the obtained films as optically clear [68], although a significant decrease ($p < 0.05$) was observed with ZnO/Zn-NPs with the addition of CBSE-2. These differences in transparency found between the control and active films may be due to the formation of more compact films by the interaction of the pectin matrix with the active compounds present in CBSE-2, mainly polyphenols [45], and the presence of ZnO/Zn-NPs contributing to the reduction of light intensity passing through the films [77]. The incorporation of ZnO/Zn-NPs also increases the amount of inorganic material not homogeneously distributed in the polymer matrix, resulting in a reduction in transparency [70].

Transmittance values obtained at 280 nm for control films (25 ± 2%, Table 3) indicated that neat pectin was a fairly good UV barrier by absorbing around 75% of the incident UV radiation [56]. A significant improvement ($p < 0.05$) in UV barrier properties was obtained for the active films (Table 3) with the addition of CBSE-2 and ZnO/Zn-NPs, with about 98% of the UV radiation screened by these films. This behaviour was related to the presence of phenolic compounds in cocoa extracts [78] and the studied nanoparticles [57,79]. In conclusion, the addition of CBSE-2 and ZnO/Zn-NPs to pectin-based films has shown great potential for UV light prevention to avoid photocatalytic reactions in packaged food.

3.5. ATR-FTIR Analysis

The FTIR spectra of neat pectin, CBSE-2 and active pectin-based films exhibited distinctive peaks in the 4000–500 cm^{-1} range. The PC film (Figure 3) showed a broad peak centred at 3281 cm^{-1} which was attributed to O–H stretching vibrations [11], while bands centred at 2894 and 2849 cm^{-1} were related to C–H stretching vibrations of methylene groups and methyl group of pectin polymer chains, respectively [80]. The intense bands observed at 1738 and 1603 cm^{-1} were associated with the ester stretching vibrations of the -$COCH_3$ group and the asymmetric stretching vibrations of the carboxylate anion –COO^-, giving an indication of the high methoxylation of the pectin molecules [19,81]. Some other bands were also obtained in the 1360–800 cm^{-1} range, which were related to the stretching vibrations of the C–O–C and C–C bonds of the carbohydrate ring [82].

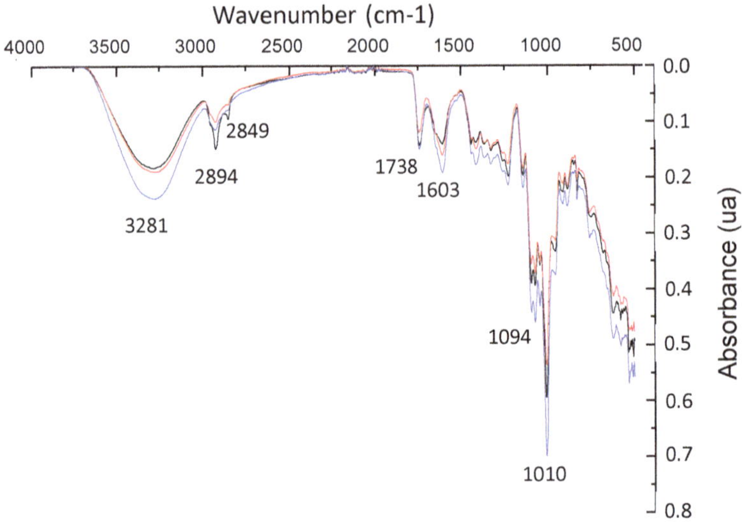

Figure 3. FTIR spectra of PC (------), PC-1ZnO (------) and PC-3ZnO (------) films.

Figure 3 also shows the FTIR spectra obtained for PC-1ZnO and PC-3ZnO films. The a of ZnO/Zn-NPs did not produce significant changes in the structure of the polymer matr decrease in the intensity of the 2894–2849 cm^{-1} bands at both nanoparticle concentrations was o which was related to the formation of hydrogen bonds between ZnO/Zn-NPs and the pecti A significant shift (3256 cm^{-1}) and increase in intensity of the band related to O-H stretching v was also shown at 3 wt% of ZnO/Zn-NPs in PC-3ZnO films [43].

The FTIR spectra of PC and PC-5E films and CBSE-2 are compared in Figure 4. As can no important changes in the spectrum of neat pectin were observed with the addition of CBSE-2. However, a significant shift of the band at 3275 cm^{-1} was found in the PC-5E film c to the CBSE-2 spectrum, which was associated with the interaction between the hydroxyl g the polyphenolic compounds and polysaccharides present in the cocoa extract [83] with t matrix. The typical absorption peaks of polysaccharides at 1014, 1070 and 1119 cm^{-1} were in the CBSE-2 spectrum and they were assigned to the –C–OH, –C–C– and –C–O– vibratio respectively [84]. Finally, two bands centred at 1628 and 1642 cm^{-1} with similar intensity wer in CBSE-2 spectra, confirming the presence of some pectin in CBSE-2, as these bands are us to determine the esterification degree of pectin [83,85,86], suggesting a value around 50%.

Finally, Figure 5 shows the FTIR spectra obtained for PC, PC-5E, PC-5E-1ZnO and PC films. No noticeable changes were obtained, suggesting that the addition of CBSE-2 and Zn did not produce significant changes in the pectin structure. A similar behaviour was re other authors with the incorporation of ZnO-NPs [87] or natural extracts derived from plan polymer matrices.

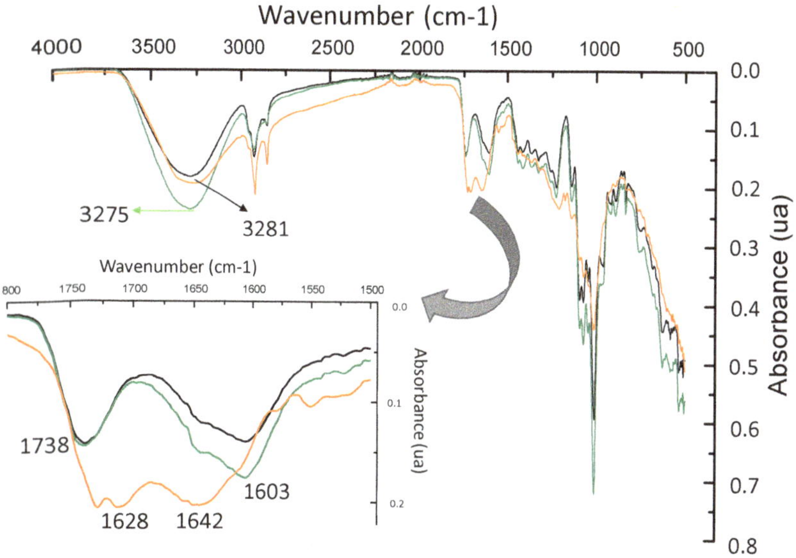

Figure 4. FTIR spectra of PC (------) and PC-5E (------) films and CBSE-2 (------).

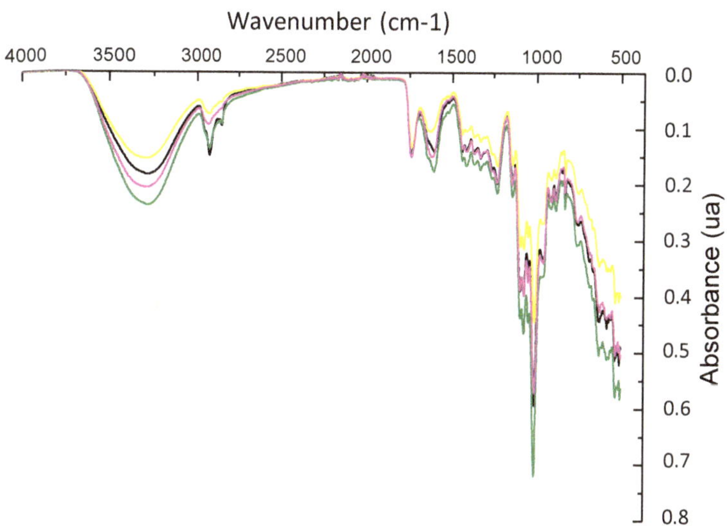

Figure 5. FTIR spectra of PC (------), PC-5E (------), PC-5E-1ZnO (------) and PC-5E-3ZnO (------) films.

3.6. X-ray Diffraction Analysis

The XRD pattern of ZnO/Zn-NPs is shown in Figure 6. Different characteristic diffraction peaks at 2θ = 31.8, 34.5, 36.3, 47.5, 56.6, 62.9, 66.5, 68.3, 69.1, 72.8 and 77.1° were observed. These peaks were ascribed, respectively, to the (1 0 0), (0 0 2), (1 0 1), (1 0 2), (1 1 0), (1 0 3), (2 0 2), (1 1 2), (2 0 1), (0 0 4) and (2 0 2) planes of hexagonal wurtzite structure ZnO according to JCPDS Card No. 79-0206. Similar planes were reported for ZnO-NPs by other authors [88–90]. Besides, these nanoparticles showed a small but noticeable peak at 46°, indicating the presence of interstitial zinc in the ZnO lattices [91].

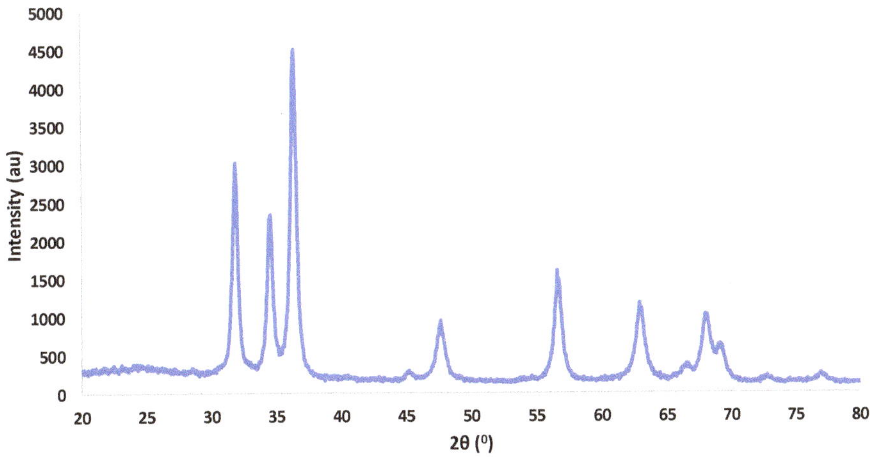

Figure 6. XRD pattern of ZnO/Zn-NPs.

Regarding pectin-based films, no specific diffraction peaks were observed for the neat pectin film (Figure 7), giving a clear indication of its amorphous structure as already reported by other authors [19,26,43]. The diffraction pattern of the studied nanoparticles was overcome by that of pectin at low ZnO/Zn-NP content [28], resulting in a completely amorphous structure for pectin-based films with 1 wt% of ZnO/Zn-NPs. However, when the concentration of ZnO/Zn-NPs was increased up to 3 wt%, two diffraction peaks between 30 and 35° were detected in films (circled area in Figure 7), corresponding to the major diffraction peaks observed in nanoparticles and suggesting the good incorporation of ZnO/Zn-NPs into the pectin matrix.

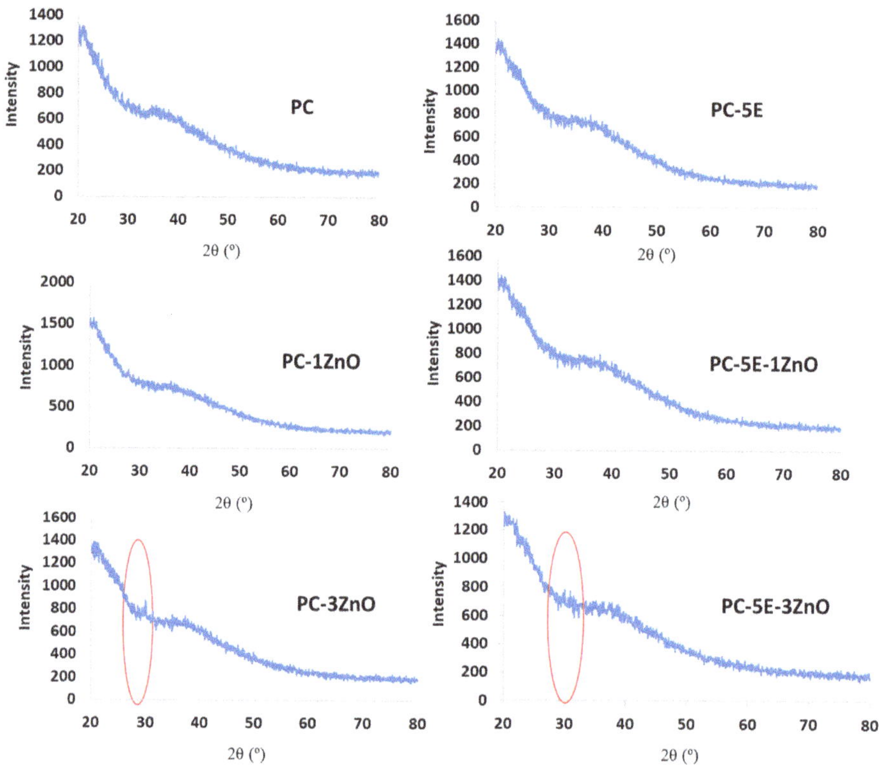

Figure 7. XRD patterns of pectin-based films.

3.7. Morphological Analysis

The microstructure of active pectin-based films containing CBSE-2 and ZnO/Zn-NPs was studied by SEM to get insight into the incorporation of the active compounds, their organisation along the biopolymer film matrix and their possible influence on the films' final properties. The cross-section micrographs of all pectin-based films are shown in Figure 8. A uniform and homogeneous surface was observed in all samples, demonstrating the good incorporation of CBSE-2 and ZnO/Zn-NPs into the pectin matrix. In addition, PC5E-1ZnO and PC5E-3ZnO films showed an apparent improvement in the nanoparticle dispersion through the pectin matrix when compared to film samples without the incorporation of CBSE-2. This behaviour was related to the presence of some antioxidant compounds in CBSE-2, increasing the possibility of the formation of hydrogen bonds between active compounds, nanoparticles and the pectin matrix [55]. The presence of both active additives in pectin-based films also placed ZnO/Zn-NPs mainly on the surface of the pectin films, particularly at the highest nanoparticle concentration. This behaviour may be due to the higher affinity between pectin and CBSE-2, displacing the nanoparticles to the material surface [70].

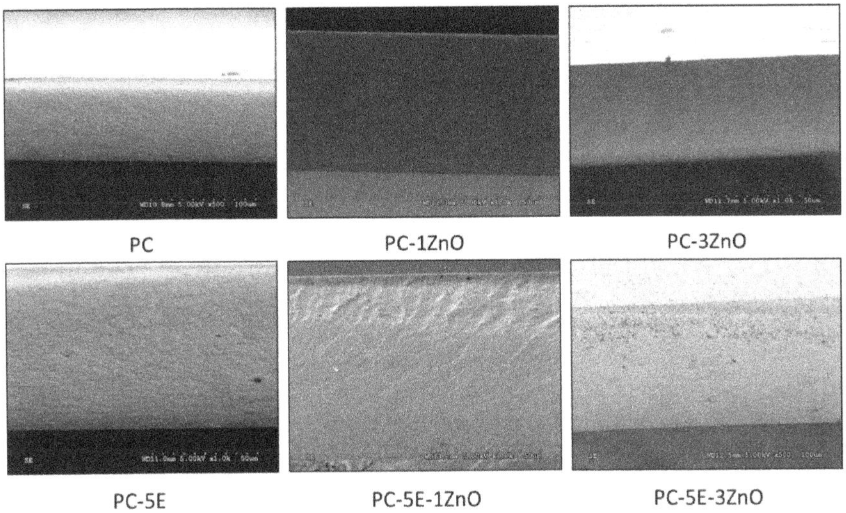

Figure 8. SEM micrographs of cross-sections of pectin-based films.

3.8. Photocatalytic Properties

Methylene blue (MB) is a cationic-type model dye which is commonly used in medicine and paper and textile industries [92]. The mechanism of the photocatalytic degradation of MB with ZnO-NPs has been reported by other authors. When ZnO is exposed to irradiation, electron–hole pairs are created and afterwards superoxide radicals are formed by the reaction of electrons with oxygen and hydroxyl radicals generated by the reaction of holes with hydroxide ions/water, which results in the rupture of organic bonds and the degradation of the MB molecules [93–95]. In the case of CBSE-2, antioxidant compounds, such as epicatechin or caffeic acid, might produce hydroxyl radicals under irradiation able to degrade the MB molecules [96].

The results obtained for MB degradation with all pectin-based formulations are shown in Table 1. As can be seen, the neat pectin film did not produce any photocatalytic activity. However, a significant increase ($p < 0.05$) in photodegradation efficiency was obtained in formulations with nanoparticles, showing values of around 90% at 3 wt% of ZnO/Zn-NP loading (PC-3ZnO and PC-5E-3ZnO nanocomposites). The positive effect of ZnO/Zn-NPs in the photodegradation of MB in aqueous solution has already been demonstrated [91]. The presence of interstitial zinc in these nanoparticles enhanced their photocatalytic activity compared to ZnO-NPs by an increase in the lifetime of the electron–hole pairs, with this effect being more pronounced with increasing ZnO/Zn-NP content [94]. Similar results were obtained in starch/polyvinyl alcohol/TiO_2-NP composite films after 90 min of testing [8].

4. Conclusions

CBSE-2 and ZnO/Zn-NPs were successfully added to pectin to obtain biodegradable films with no major changes in moisture content, structural, transparency and morphological properties. These additives showed high compatibility with the pectin matrix and their incorporation greatly improved oxygen, thermal and UV barrier properties, which are very important for the intended application of these biomaterials in food packaging. In particular, the barrier to oxygen was enhanced by 50%, while the screen to UV radiation reached 98%, which are essential properties for the packaging of food highly sensitive to oxidative degradation. Besides, high photocatalytic activity was also demonstrated in the nanocomposite films with the addition of ZnO/Zn-NPs, which was related to the presence of interstitial zinc in these nanoparticles. In conclusion, the pectin-based bionanocomposite films obtained in this work have shown great potential to be used in different fields, particularly in food packaging. Further specific studies on the developed biomaterials, such as mechanical performance and water absorption measurements, should be carried out to finally assess their suitability for this application. Moreover, the processes and materials proposed in this work could also contribute to the circular bioeconomy concept in the agriculture and food processing sectors by valorising cocoa bean shell residues and using biopolymers to obtain sustainable functional materials for food packaging and preservation.

Author Contributions: Conceptualization and methodology, A.C.M., A.J. and M.C.G.; formal analysis, investigation and writing—original draft preparation, A.C.M.; resources, writing—review and editing, supervision and funding acquisition, A.J. and M.C.G. All authors have read and agreed to the published version of the manuscript.

Funding: The authors would like to thank the Spanish Ministry of Science, Innovation and Universities (MAT2017-84909-C2-1-R) and Generalitat Valenciana (IDIFEDER/2018/007) for financially supporting this research.

Conflicts of Interest: The authors declare no conflict of interest.

References

1. Sivakanthan, S.; Rajendran, S.; Gamage, A.; Madhujith, T.; Mani, S. Antioxidant and antimicrobial applications of biopolymers: A review. *Food Res. Int.* **2020**, *136*, 109327. [CrossRef]
2. Gao, H.-X.; He, Z.; Sun, Q.; He, Q.; Zeng, W.-C. A functional polysaccharide film forming by pectin, chitosan, and tea polyphenols. *Carbohydr. Polym.* **2019**, *215*, 1–7. [CrossRef] [PubMed]
3. Mellinas, C.; Valdés, A.; Ramos, M.; Burgos, N.; Del Carmen Garrigós, M.; Jiménez, A. Active edible films: Current state and future trends. *J. Appl. Polym. Sci.* **2016**, *133*. [CrossRef]
4. Mendes, J.F.; Norcino, L.B.; Martins, H.H.A.; Manrich, A.; Otoni, C.G.; Carvalho, E.E.N.; Piccoli, R.H.; Oliveira, J.E.; Pinheiro, A.C.M.; Mattoso, L.H.C. Correlating emulsion characteristics with the properties of active starch films loaded with lemongrass essential oil. *Food Hydrocoll.* **2020**, *100*, 105428. [CrossRef]
5. Kumar, A.; Rao, K.M.; Han, S.S. Application of xanthan gum as polysaccharide in tissue engineering: A review. *Carbohydr. Polym.* **2018**, *180*, 128–144. [CrossRef] [PubMed]
6. Cazón, P.; Vázquez, M.; Velazquez, G. Cellulose-glycerol-polyvinyl alcohol composite films for food packaging: Evaluation of water adsorption, mechanical properties, light-barrier properties and transparency. *Carbohydr. Polym.* **2018**, *195*, 432–443. [CrossRef]
7. Kalkan, S.; Otağ, M.R.; Engin, M.S. Physicochemical and bioactive properties of edible methylcellulose films containing Rheum ribes L. extract. *Food Chem.* **2020**, *307*, 125524. [CrossRef]
8. Lin, D.; Huang, Y.; Liu, Y.; Luo, T.; Xing, B.; Yang, Y.; Yang, Z.; Wu, Z.; Chen, H.; Zhang, Q.; et al. Physico-mechanical and structural characteristics of starch/polyvinyl alcohol/nano-titania photocatalytic antimicrobial composite films. *LWT* **2018**, *96*, 704–712. [CrossRef]
9. Singh, T.P.; Chatli, M.K.; Sahoo, J. Development of chitosan based edible films: Process optimization using response surface methodology. *J. Food Sci. Technol.* **2015**, *52*, 2530–2543. [CrossRef]
10. Sani, I.K.; Pirsa, S.; Tağı, Ş. Preparation of chitosan/zinc oxide/Melissa officinalis essential oil nano-composite film and evaluation of physical, mechanical and antimicrobial properties by response surface method. *Polym. Test.* **2019**, *79*, 106004. [CrossRef]
11. Bernhardt, D.C.; Pérez, C.D.; Fissore, E.N.; De'Nobili, M.D.; Rojas, A.M. Pectin-based composite film: Effect of corn husk fiber concentration on their properties. *Carbohydr. Polym.* **2017**, *164*, 13–22. [CrossRef] [PubMed]
12. Almasi, H.; Azizi, S.; Amjadi, S. Development and characterization of pectin films activated by nanoemulsion and Pickering emulsion stabilized marjoram (*Origanum majorana* L.) essential oil. *Food Hydrocoll.* **2020**, *99*, 105338. [CrossRef]
13. Mellinas, C.; Ramos, M.; Jiménez, A.; Garrigós, M.C. Recent Trends in the Use of Pectin from Agro-Waste Residues as a Natural-Based Biopolymer for Food Packaging Applications. *Materials* **2020**, *13*, 673. [CrossRef] [PubMed]
14. Marenda, F.R.B.; Mattioda, F.; Demiate, I.M.; de Francisco, A.; de Oliveira Petkowicz, C.L.; Canteri, M.H.G.; de Mello Castanho Amboni, R.D. Advances in Studies Using Vegetable Wastes to Obtain Pectic Substances: A Review. *J. Polym. Environ.* **2019**, *27*, 549–560. [CrossRef]

15. Cataldo, V.A.; Cavallaro, G.; Lazzara, G.; Milioto, S.; Parisi, F. Coffee grounds as filler for pectin: Green composites with competitive performances dependent on the UV irradiation. *Carbohydr. Polym.* **2017**, *170*, 198–205. [CrossRef]
16. Espitia, P.J.; Du, W.X.; de Jesús Avena-Bustillos, R.; Soares, N.D.; McHugh, T.H. Edible films from pectin: Physical-mechanical and antimicrobial properties—A review. *Food Hydrocoll.* **2014**, *35*, 287–296. [CrossRef]
17. Kalathaki, I.; Alba, K.; Muhamedsalih, H.; Kontogiorgos, V. Fabrication and characterisation of metal-doped pectin films. *Food Hydrocoll.* **2019**, *92*, 259–266. [CrossRef]
18. Noreen, A.; Nazli, Z.-H.; Akram, J.; Rasul, I.; Mansha, A.; Yaqoob, N.; Iqbal, R.; Tabasum, S.; Zuber, M.; Zia, K.M. Pectins functionalized biomaterials; a new viable approach for biomedical applications: A review. *Int. J. Biol. Macromol.* **2017**, *101*, 254–272. [CrossRef]
19. Nisar, T.; Wang, Z.-C.; Yang, X.; Tian, Y.; Iqbal, M.; Guo, Y. Characterization of citrus pectin films integrated with clove bud essential oil: Physical, thermal, barrier, antioxidant and antibacterial properties. *Int. J. Biol. Macromol.* **2018**, *106*, 670–680. [CrossRef] [PubMed]
20. Giosafatto, C.V.L.; Di Pierro, P.; Gunning, P.; Mackie, A.; Porta, R.; Mariniello, L. Characterization of Citrus pectin edible films containing transglutaminase-modified phaseolin. *Carbohydr. Polym.* **2014**, *106*, 200–208. [CrossRef] [PubMed]
21. Baron, R.D.; Pérez, L.L.; Salcedo, J.M.; Córdoba, L.P.; do Amaral Sobral, P.J. Production and characterization of films based on blends of chitosan from blue crab (*Callinectes sapidus*) waste and pectin from Orange (*Citrus sinensis Osbeck*) peel. *Int. J. Biol. Macromol.* **2017**, *98*, 676–683. [CrossRef] [PubMed]
22. Akhter, R.; Masoodi, F.A.; Wani, T.A.; Rather, S.A. Functional characterization of biopolymer based composite film: Incorporation of natural essential oils and antimicrobial agents. *Int. J. Biol. Macromol.* **2019**, *137*, 1245–1255. [CrossRef]
23. Šešlija, S.; Nešić, A.; Ružić, J.; Kalagasidis Krušić, M.; Veličković, S.; Avolio, R.; Santagata, G.; Malinconico, M. Edible blend films of pectin and poly(ethylene glycol): Preparation and physico-chemical evaluation. *Food Hydrocoll.* **2018**, *77*, 494–501. [CrossRef]
24. Porta, R.; Di Pierro, P.; Sabbah, M.; Regalado-Gonzales, C.; Mariniello, L.; Kadivar, M.; Arabestani, A. Blend films of pectin and bitter vetch (Vicia ervilia) proteins: Properties and effect of transglutaminase. *Innov. Food Sci. Emerg. Technol.* **2016**, *36*, 245–251. [CrossRef]
25. Eghbal, N.; Degraeve, P.; Oulahal, N.; Yarmand, M.S.; Mousavi, M.E.; Gharsallaoui, A. Low methoxyl pectin/sodium caseinate interactions and composite film formation at neutral pH. *Food Hydrocoll.* **2017**, *69*, 132–140. [CrossRef]
26. Kodoth, A.K.; Ghate, V.M.; Lewis, S.A.; Prakash, B.; Badalamoole, V. Pectin-based silver nanocomposite film for transdermal delivery of Donepezil. *Int. J. Biol. Macromol.* **2019**, *134*, 269–279. [CrossRef]
27. Lorevice, M.V.; Otoni, C.G.; de Moura, M.R.; Mattoso, L.H.C. Chitosan nanoparticles on the improvement of thermal, barrier, and mechanical properties of high- and low-methyl pectin films. *Food Hydrocoll.* **2016**, *52*, 732–740. [CrossRef]
28. Chaichi, M.; Hashemi, M.; Badii, F.; Mohammadi, A. Preparation and characterization of a novel bionanocomposite edible film based on pectin and crystalline nanocellulose. *Carbohydr. Polym.* **2017**, *157*, 167–175. [CrossRef]
29. Nešić, A.; Gordić, M.; Davidović, S.; Radovanović, Ž.; Nedeljković, J.; Smirnova, I.; Gurikov, P. Pectin-based nanocomposite aerogels for potential insulated food packaging application. *Carbohydr. Polym.* **2018**, *195*, 128–135. [CrossRef]
30. Borzi, F.; Torrieri, E.; Wrona, M.; Nerín, C. Polyamide modified with green tea extract for fresh minced meat active packaging applications. *Food Chem.* **2019**, *300*, 125242. [CrossRef]
31. Heras-Mozos, R.; Muriel-Galet, V.; López-Carballo, G.; Catalá, R.; Hernández-Muñoz, P.; Gavara, R. Development and optimization of antifungal packaging for sliced pan loaf based on garlic as active agent and bread aroma as aroma corrector. *Int. J. Food Microbiol.* **2019**, *290*, 42–48. [CrossRef] [PubMed]
32. Moustafa, H.; Youssef, A.M.; Darwish, N.A.; Abou-Kandil, A.I. Eco-friendly polymer composites for green packaging: Future vision and challenges. *Compos. Part B Eng.* **2019**, *172*, 16–25. [CrossRef]
33. Valdés, A.; Mellinas, A.C.; Ramos, M.; Garrigós, M.C.; Jiménez, A. Natural additives and agricultural wastes in biopolymer formulations for food packaging. *Front. Chem.* **2014**, *2*, 6. [CrossRef]
34. Panak Balentić, J.; Ačkar, Đ.; Jokić, S.; Jozinović, A.; Babić, J.; Miličević, B.; Šubarić, D.; Pavlović, N. Cocoa Shell: A By-Product with Great Potential for Wide Application. *Molecules* **2018**, *23*, 1404. [CrossRef]

35. Okiyama, D.C.G.; Navarro, S.L.B.; Rodrigues, C.E.C. Cocoa shell and its compounds: Applications in the food industry. *Trends Food Sci. Technol.* **2017**, *63*, 103–112. [CrossRef]
36. Sirelkhatim, A.; Mahmud, S.; Seeni, A.; Kaus, N.H.M.; Ann, L.C.; Bakhori, S.K.M.; Hasan, H.; Mohamad, D. Review on Zinc Oxide Nanoparticles: Antibacterial Activity and Toxicity Mechanism. *Nano-Micro Lett.* **2015**, *7*, 219–242. [CrossRef] [PubMed]
37. Mocanu, A.; Isopencu, G.; Busuioc, C.; Popa, O.M.; Dietrich, P.; Socaciu-Siebert, L. Bacterial cellulose films with ZnO nanoparticles and propolis extracts: Synergistic antimicrobial effect. *Sci. Rep.* **2019**, *9*, 17687. [CrossRef]
38. EFSA Panel on Food Contact Materials, Enzymes, Flavourings and Processing Aids (CEF). Safety assessment of the substance zinc oxide, nanoparticles, for use in food contact materials. *EFSA J.* **2016**, *14*, 4408.
39. Indumathi, M.P.; Rajarajeswari, G.R. Mahua oil-based polyurethane/chitosan/nano ZnO composite films for biodegradable food packaging applications. *Int. J. Biol. Macromol.* **2019**, *124*, 163–174.
40. Chaudhary, P.; Fatima, F.; Kumar, A. Relevance of Nanomaterials in Food Packaging and its Advanced Future Prospects. *J. Inorg. Organomet. Polym. Mater.* **2020**, *1*, 3. [CrossRef]
41. Soubhagya, A.S.; Moorthi, A.; Prabaharan, M. Preparation and characterization of chitosan/pectin/ZnO porous films for wound healing. *Int. J. Biol. Macromol.* **2020**, *157*, 135–145. [CrossRef] [PubMed]
42. Ngo, T.M.P.; Dang, T.M.Q.; Tran, T.X.; Rachtanapun, P. Effects of Zinc Oxide Nanoparticles on the Properties of Pectin/Alginate Edible Films. *Int. J. Polym. Sci.* **2018**, *2018*, 5645797. [CrossRef]
43. Kodoth, A.K.; Ghate, V.M.; Lewis, S.A.; Badalamoole, V. Application of pectin-zinc oxide hybrid nanocomposite in the delivery of a hydrophilic drug and a study of its isotherm, kinetics and release mechanism. *Int. J. Biol. Macromol.* **2018**, *115*, 418–430. [CrossRef]
44. Kodoth, A.K.; Badalamoole, V. Pectin Based Graft Copolymer–ZnO Hybrid Nanocomposite for the Adsorptive Removal of Crystal Violet. *J. Polym. Environ.* **2019**, *27*, 2040–2053. [CrossRef]
45. Mellinas, A.C.; Jiménez, A.; Garrigós, M.C. Optimization of microwave-assisted extraction of cocoa bean shell waste and evaluation of its antioxidant, physicochemical and functional properties. *LWT* **2020**, *127*, 109361. [CrossRef]
46. Younis, H.G.R.; Zhao, G. Physicochemical properties of the edible films from the blends of high methoxyl apple pectin and chitosan. *Int. J. Biol. Macromol.* **2019**, *131*, 1057–1066. [CrossRef] [PubMed]
47. Dash, K.K.; Ali, N.A.; Das, D.; Mohanta, D. Thorough evaluation of sweet potato starch and lemon-waste pectin based-edible films with nano-titania inclusions for food packaging applications. *Int. J. Biol. Macromol.* **2019**, *139*, 449–458. [CrossRef] [PubMed]
48. Shahmohammadi Jebel, F.; Almasi, H. Morphological, physical, antimicrobial and release properties of ZnO nanoparticles-loaded bacterial cellulose films. *Carbohydr. Polym.* **2016**, *149*, 8–19. [CrossRef] [PubMed]
49. Lei, Y.; Wu, H.; Jiao, C.; Jiang, Y.; Liu, R.; Xiao, D.; Lu, J.; Zhang, Z.; Shen, G.; Li, S. Investigation of the structural and physical properties, antioxidant and antimicrobial activity of pectin-konjac glucomannan composite edible films incorporated with tea polyphenol. *Food Hydrocoll.* **2019**, *94*, 128–135. [CrossRef]
50. Wang, W.; Ma, X.; Jiang, P.; Hu, L.; Zhi, Z.; Chen, J.; Ding, T.; Ye, X.; Liu, D. Characterization of pectin from grapefruit peel: A comparison of ultrasound-assisted and conventional heating extractions. *Food Hydrocoll.* **2016**, *61*, 730–739. [CrossRef]
51. Jiang, W.; Qi, J.-R.; Liao, J.; Wan, Z.; Liang, W.; Huang, J.; Cao, Y.; Xiao, J.; Yang, X.-Q. Structural characterization of pectin-bismuth complexes and their aggregation in acidic conditions. *Int. J. Biol. Macromol.* **2020**, *154*, 788–794. [CrossRef]
52. Wang, C.; Chang, T.; Dong, S.; Zhang, D.; Ma, C.; Chen, S.; Li, H. Biopolymer films based on chitosan/potato protein/linseed oil/ZnO NPs to maintain the storage quality of raw meat. *Food Chem.* **2020**, *332*, 127375. [CrossRef]
53. Kanmani, P.; Rhim, J.-W. Properties and characterization of bionanocomposite films prepared with various biopolymers and ZnO nanoparticles. *Carbohydr. Polym.* **2014**, *106*, 190–199. [CrossRef] [PubMed]
54. Vishnu Chandar, J.; Shanmugan, S.; Mutharasu, D.; Azlan, A.A. Impact of ZnO Nanoparticles on Thermal Properties of Poly(3-hydroxybutyrate-co-10 mol % 3-hydroxyhexanoate) Copolymer. *Polym. Sci. Ser. A* **2019**, *61*, 504–513. [CrossRef]
55. Ramos, M.; Fortunati, E.; Peltzer, M.; Dominici, F.; Jiménez, A.; del Carmen Garrigós, M.; Kenny, J.M. Influence of thymol and silver nanoparticles on the degradation of poly(lactic acid) based nanocomposites: Thermal and morphological properties. *Polym. Degrad. Stab.* **2014**, *108*, 158–165. [CrossRef]

56. Shankar, S.; Tanomrod, N.; Rawdkuen, S.; Rhim, J.-W. Preparation of pectin/silver nanoparticles composite films with UV-light barrier and properties. *Int. J. Biol. Macromol.* **2016**, *92*, 842–849. [CrossRef] [PubMed]
57. Shankar, S.; Wang, L.-F.; Rhim, J.-W. Incorporation of zinc oxide nanoparticles improved the mechanical, water vapor barrier, UV-light barrier, and antibacterial properties of PLA-based nanocomposite films. *Mater. Sci. Eng. C* **2018**, *93*, 289–298. [CrossRef]
58. Ramos, M.; Jiménez, A.; Peltzer, M.; Garrigós, M.C. Characterization and antimicrobial activity studies of polypropylene films with carvacrol and thymol for active packaging. *J. Food Eng.* **2012**, *109*, 513–519. [CrossRef]
59. Soren, S.; Kumar, S.; Mishra, S.; Jena, P.K.; Verma, S.K.; Parhi, P. Evaluation of antibacterial and antioxidant potential of the zinc oxide nanoparticles synthesized by aqueous and polyol method. *Microb. Pathog.* **2018**, *119*, 145–151. [CrossRef]
60. Shabaani, M.; Rahaiee, S.; Zare, M.; Jafari, S.M. Green synthesis of ZnO nanoparticles using loquat seed extract; Biological functions and photocatalytic degradation properties. *LWT* **2020**, *134*, 110133. [CrossRef]
61. Janani, N.; Zare, E.N.; Salimi, F.; Makvandi, P. Antibacterial tragacanth gum-based nanocomposite films carrying ascorbic acid antioxidant for bioactive food packaging. *Carbohydr. Polym.* **2020**, *247*, 116678. [CrossRef] [PubMed]
62. Arrieta, M.P.; Peponi, L. Polyurethane based on PLA and PCL incorporated with catechin: Structural, thermal and mechanical characterization. *Eur. Polym. J.* **2017**, *89*, 174–184. [CrossRef]
63. Correa, J.P.; Molina, V.; Sanchez, M.; Kainz, C.; Eisenberg, P.; Massani, M.B. Improving ham shelf life with a polyhydroxybutyrate/polycaprolactone biodegradable film activated with nisin. *Food Packag. Shelf Life* **2017**, *11*, 31–39. [CrossRef]
64. Nanni, A.; Messori, M. A comparative study of different winemaking by-products derived additives on oxidation stability, mechanical and thermal proprieties of polypropylene. *Polym. Degrad. Stab.* **2018**, *149*, 9–18. [CrossRef]
65. Valdés, A.; Mellinas, A.C.; Ramos, M.; Burgos, N.; Jiménez, A.; Garrigós, M.C. Use of herbs, spices and their bioactive compounds in active food packaging. *RSC Adv.* **2015**, *5*, 40324–40335. [CrossRef]
66. Jha, P. Effect of plasticizer and antimicrobial agents on functional properties of bionanocomposite films based on corn starch-chitosan for food packaging applications. *Int. J. Biol. Macromol.* **2020**, *160*, 571–582. [CrossRef]
67. Maryam Adilah, Z.A.; Jamilah, B.; Nur Hanani, Z.A. Functional and antioxidant properties of protein-based films incorporated with mango kernel extract for active packaging. *Food Hydrocoll.* **2018**, *74*, 207–218. [CrossRef]
68. Dairi, N.; Ferfera-Harrar, H.; Ramos, M.; Garrigós, M.C. Cellulose acetate/AgNPs-organoclay and/or thymol nano-biocomposite films with combined antimicrobial/antioxidant properties for active food packaging use. *Int. J. Biol. Macromol.* **2019**, *121*, 508–523. [CrossRef]
69. Ramos, M.; Jiménez, A.; Peltzer, M.; Garrigós, M.C. Development of novel nano-biocomposite antioxidant films based on poly (lactic acid) and thymol for active packaging. *Food Chem.* **2014**, *162*, 149–155. [CrossRef]
70. Noshirvani, N.; Ghanbarzadeh, B.; Rezaei Mokarram, R.; Hashemi, M. Novel active packaging based on carboxymethyl cellulose-chitosan-ZnO NPs nanocomposite for increasing the shelf life of bread. *Food Packag. Shelf Life* **2017**, *11*, 106–114. [CrossRef]
71. Vartiainen, J.; Tammelin, T.; Pere, J.; Tapper, U.; Harlin, A. Biohybrid barrier films from fluidized pectin and nanoclay. *Carbohydr. Polym.* **2010**, *82*, 989–996. [CrossRef]
72. Farahnaky, A.; Sharifi, S.; Imani, B.; Dorodmand, M.M.; Majzoobi, M. Physicochemical and mechanical properties of pectin-carbon nanotubes films produced by chemical bonding. *Food Packag. Shelf Life* **2018**, *16*, 8–14. [CrossRef]
73. Rhim, J.-W.; Park, H.-M.; Ha, C.-S. Bio-nanocomposites for food packaging applications. *Prog. Polym. Sci.* **2013**, *38*, 1629–1652. [CrossRef]
74. Fortunati, E.; Armentano, I.; Zhou, Q.; Puglia, D.; Terenzi, A.; Berglund, L.A.; Kenny, J.M. Microstructure and nonisothermal cold crystallization of PLA composites based on silver nanoparticles and nanocrystalline cellulose. *Polym. Degrad. Stab.* **2012**, *97*, 2027–2036. [CrossRef]
75. Espitia, P.J.; Soares, N.D.; Teófilo, R.F.; dos Reis Coimbra, J.S.; Vitor, D.M.; Batista, R.A.; Ferreira, S.O.; de Andrade, N.J.; Medeiros, E.A. Physical–mechanical and antimicrobial properties of nanocomposite films with pediocin and ZnO nanoparticles. *Carbohydr. Polym.* **2013**, *94*, 199–208. [CrossRef] [PubMed]

76. Pirsa, S. Biodegradable film based on pectin/Nano-clay/methylene blue: Structural and physical properties and sensing ability for measurement of vitamin C. *Int. J. Biol. Macromol.* **2020**, *163*, 666–675. [CrossRef]
77. Adilah, A.N.; Jamilah, B.; Noranizan, M.A.; Hanani, Z.A.N. Utilization of mango peel extracts on the biodegradable films for active packaging. *Food Packag. Shelf Life* **2018**, *16*, 1–7. [CrossRef]
78. Mir, S.A.; Dar, B.N.; Wani, A.A.; Shah, M.A. Effect of plant extracts on the techno-functional properties of biodegradable packaging films. *Trends Food Sci. Technol.* **2018**, *80*, 141–154. [CrossRef]
79. Luzi, F.; Di Michele, A.; Torre, L.; Puglia, D. Active role of ZnO nanorods in thermomechanical and barrier performance of poly(vinyl alcohol-co-ethylene) formulations for flexible packaging. *Polymers* **2019**, *11*, 922. [CrossRef]
80. Mendes, J.F.; Martins, J.T.; Manrich, A.; Sena Neto, A.R.; Pinheiro, A.C.M.; Mattoso, L.H.C.; Martins, M.A. Development and physical-chemical properties of pectin film reinforced with spent coffee grounds by continuous casting. *Carbohydr. Polym.* **2019**, *210*, 92–99. [CrossRef] [PubMed]
81. Nešić, A.; Ružić, J.; Gordić, M.; Ostojić, S.; Micić, D.; Onjia, A. Pectin-polyvinylpyrrolidone films: A sustainable approach to the development of biobased packaging materials. *Compos. Part B Eng.* **2017**, *110*, 56–61. [CrossRef]
82. Viana, R.M.; Sá, N.M.; Barros, M.O.; de Fátima Borges, M.; Azeredo, H.M. Nanofibrillated bacterial cellulose and pectin edible films added with fruit purees. *Carbohydr. Polym.* **2018**, *196*, 27–32. [CrossRef] [PubMed]
83. Pereira, P.H.F.; Oliveira, T.Í.S.; Rosa, M.F.; Cavalcante, F.L.; Moates, G.K.; Wellner, N.; Waldron, K.W.; Azeredo, H.M.C. Pectin extraction from pomegranate peels with citric acid. *Int. J. Biol. Macromol.* **2016**, *88*, 373–379. [CrossRef] [PubMed]
84. Mzoughi, Z.; Abdelhamid, A.; Rihouey, C.; Le Cerf, D.; Bouraoui, A.; Majdoub, H. Optimized extraction of pectin-like polysaccharide from Suaeda fruticosa leaves: Characterization, antioxidant, anti-inflammatory and analgesic activities. *Carbohydr. Polym.* **2018**, *185*, 127–137. [CrossRef]
85. Kpodo, F.M.; Agbenorhevi, J.K.; Alba, K.; Bingham, R.J.; Oduro, I.N.; Morris, G.A.; Kontogiorgos, V. Pectin isolation and characterization from six okra genotypes. *Food Hydrocoll.* **2017**, *72*, 323–330. [CrossRef]
86. Kazemi, M.; Khodaiyan, F.; Labbafi, M.; Saeid Hosseini, S.; Hojjati, M. Pistachio green hull pectin: Optimization of microwave-assisted extraction and evaluation of its physicochemical, structural and functional properties. *Food Chem.* **2019**, *271*, 663–672. [CrossRef]
87. Sheikh, M.; Asghari, M.; Afsari, M. Effect of tiny amount of zinc oxide on morphological and thermal properties of nanocomposite PEBA thin films. *Alex. Eng. J.* **2018**, *57*, 3661–3669. [CrossRef]
88. Vijayakumar, S.; Krishnakumar, C.; Arulmozhi, P.; Mahadevan, S.; Parameswari, N. Biosynthesis, characterization and antimicrobial activities of zinc oxide nanoparticles from leaf extract of Glycosmis pentaphylla (Retz.) DC. *Microb. Pathog.* **2018**, *116*, 44–48. [CrossRef]
89. Arciniegas-Grijalba, P.A.; Patiño-Portela, M.C.; Mosquera-Sánchez, L.P.; Guerra Sierra, B.E.; Muñoz-Florez, J.E.; Erazo-Castillo, L.A.; Rodríguez-Páez, J.E. ZnO-based nanofungicides: Synthesis, characterization and their effect on the coffee fungi *Mycena citricolor* and *Colletotrichum* sp. *Mater. Sci. Eng. C* **2019**, *98*, 808–825. [CrossRef]
90. Sharma, S.; Kumar, K.; Thakur, N.; Chauhan, S.; Chauhan, M.S. The effect of shape and size of ZnO nanoparticles on their antimicrobial and photocatalytic activities: A green approach. *Bull. Mater. Sci.* **2020**, *43*, 20. [CrossRef]
91. Lin, S.-T.; Thirumavalavan, M.; Jiang, T.-Y.; Lee, J.-F. Synthesis of ZnO/Zn nano photocatalyst using modified polysaccharides for photodegradation of dyes. *Carbohydr. Polym.* **2014**, *105*, 1–9. [CrossRef] [PubMed]
92. Pai, S.; Sridevi, H.; Varadavenkatesan, T.; Vinayagam, R.; Selvaraj, R. Photocatalytic zinc oxide nanoparticles synthesis using Peltophorum pterocarpum leaf extract and their characterization. *Optik* **2019**, *185*, 248–255. [CrossRef]
93. Raja, A.; Ashokkumar, S.; Pavithra Marthandam, R.; Jayachandiran, J.; Khatiwada, C.P.; Kaviyarasu, K.; Ganapathi Raman, R.; Swaminathan, M. Eco-friendly preparation of zinc oxide nanoparticles using Tabernaemontana divaricata and its photocatalytic and antimicrobial activity. *J. Photochem. Photobiol. B Biol.* **2018**, *181*, 53–58. [CrossRef]
94. Rafaie, H.A.; Nor, R.M.; Azmina, M.S.; Ramli, N.I.T.; Mohamed, R. Decoration of ZnO microstructures with Ag nanoparticles enhanced the catalytic photodegradation of methylene blue dye. *J. Environ. Chem. Eng.* **2017**, *5*, 3963–3972. [CrossRef]

95. Ali, J.; Irshad, R.; Li, B.; Tahir, K.; Ahmad, A.; Shakeel, M.; Khan, N.U.; Khan, Z.U.H. Synthesis and characterization of phytochemical fabricated zinc oxide nanoparticles with enhanced antibacterial and catalytic applications. *J. Photochem. Photobiol. B Biol.* **2018**, *183*, 349–356. [CrossRef] [PubMed]
96. Noda, Y.; Kaneyuki, T.; Mori, A.; Packer, L. Antioxidant Activities of Pomegranate Fruit Extract and Its Anthocyanidins: Delphinidin, Cyanidin, and Pelargonidin. *J. Agric. Food Chem.* **2002**, *50*, 166–171. [CrossRef] [PubMed]

foods

MDPI

Article

Formulation of a Bio-Packaging Based on Pure Cellulose Coupled with Cellulose Acetate Treated with Active Coating: Evaluation of Shelf Life of Pasta Ready to Eat

Valeria Bugatti [1,2], Gianluca Viscusi [1] and Giuliana Gorrasi [1,*]

1 Department of Industrial Engineering, University of Salerno, via Giovanni Paolo II 132, 84084 Fisciano, Italy; vbugatti@unisa.it (V.B.); gviscusi@unisa.it (G.V.)
2 Nice Filler s.r.l., via Loggia dei Pisani 25, 80133 Napoli, Italy
* Correspondence: ggorrasi@unisa.it

Received: 1 September 2020; Accepted: 4 October 2020; Published: 7 October 2020

check for updates

Abstract: An active packaging based on pure cellulose coupled with cellulose acetate coated with layered double hydroxide (LDH), hosting 4-hydroxybenzoate (listed in EC-Directive 10/2011) as an antimicrobial agent, was formulated and realized. The release of 4-hydroxybenzoate ionically bonded to the LDH layers was much slower than the molecule freely dispersed into the coating. The capability of the active packaging to inhibit *Pseudomonas*, *Escherichia coli*, *Salmonella* and Lactic Bacteria was evaluated, as well as the global migration with three different food simulant (i.e., acetic acid at 3% (*v*/*v*), ethanol at 50% (*v*/*v*) and vegetable oil) that demonstrated, in compliance with the migration limits of the EU regulation, the suitability of the prepared packaging to be employed as food contact material. Ready to east cooked tomato pasta was packaged into the active trays and in uncoated, as control, up to 30 days at 4 °C. Organoleptic characteristics, mold evolution, total mesophilic aerobic counts (TBC), *Enterobacteriaceae*, Lactic Bacteria and *Pseudomonas*, and in colony forming unit per gram (CFU/g), showed a significant activity of 4-hydroxybenzoate in increasing the shelf life of the pasta ready to eat.

Keywords: bio-packaging; pasta; layered double hydroxide; active coating

1. Introduction

In the global food industry, takeaway food is one of the most growing sectors. It is linked with increased continuous demand for packaging, the majority of which is single use, to satisfy hygienic issues and food safety. Food packaging materials are almost 26% of all plastics produced worldwide, and about 90% becomes waste after only one use [1]. Furthermore, single-use plastics are non-recyclable materials [2], and this represents a true environmental problem. As demand for packaging increases, the development of better options for food packaging that is decoupled from the use of limited resources, becomes ever more important. Based on circular economy principles, food packaging materials should be reusable, recyclable, or compostable by design and fit within solutions that carry this out in practice and on an industrial scale [3]. Pasta is the most known and appreciated Italian food worldwide. Its consumption has largely increased in the past decade, from 7 to 12 million tons/year, due to the largely offered pasta meals in fast food and international markets. Ready to eat pasta, in contrast to frozen one, generally is more convenient for reduced preparation times and by selecting takeaway packaging. The most common ready to eat pasta is generally composed of pasta and sauce (the most commonly used is tomato sauce). Two mixed physical phases with very different characteristics make pasta with tomato sauce ready to eat an extremely dynamic system subjected to several modifications during the storage [4,5].

In order to extend the shelf life of the packaged pasta as much as possible and to control microbial growth, an active packaging solution is the most promising. It has been demonstrated that the simple blend of antimicrobials into the food formulation is not the best choice because they are often neutralized in reactions with the food during storage [5]. Active packaging solutions are a valid alternative to overcome such limitations since they allow the controlled release of antimicrobial agents during storage and maintain their critical concentration necessary to inhibit the microbial growth [6,7]. Several strategies to achieve active molecules' control release and modeling the transport phenomena have been reported [8–13]. Due to the importance of the topic, studies on the development of food packaging (composites and bio-composites) films with controlled release properties are increasing more and more [14–29]. The present paper reports the preparation of active trays based on compostable pure cellulose coupled with biodegradable cellulose acetate, as support for an active coating. The coating was realized using a food-grade resin in which was dispersed 4-hydroxybenzoate (listed in EC-Directive 10/2011) as an antimicrobial agent. Such active molecules have been previously ionically bonded to layered double hydroxides (LDH), an anionic natural clay, as a nano-carrier. The structural and thermal characterization of the active filler, as well as the study of the inhibition controlled release of 4-hydroxybenzoate was conducted. The in-vitro antimicrobial of the active trays was evaluated against *Pseudomonas, Escherichia coli, Salmonella* and Lactic Bacteria were evaluated, and the global migration with three different food simulants. Fresh cooked pasta with tomato, ready to eat, was packaged into the coated trays and uncoated, as control. The evaluation of the organoleptic characteristics, the evolution of *molds, TBC,* and *Enterobacteriaceae,* Lactic Bacteria and *Pseudomonas,* in CFU/g, was followed up 30 days of storage at 4 °C, highlighting a significant effect of LDH-(4-hydroxybenzoate) in improving the shelf life of the pasta meal.

2. Materials and Methods

2.1. Materials

Cellulose acetate (AC311025) used for the paper trays was supplied by Emilplast srl. (Modena, Italy) in film form, 25 μm thick. Cellulose acetate film was coupled to pure virgin cellulose cardboard in the form of trays for food. The active filler, having the trade name of N6B6® is based on an LDH intercalated with antimicrobial 4-hydroxybenzoate anion, listed in EC-Directive 10/2011. It was produced by Nicefiller Ltd., a startup of the University of Salerno (Italy). The synthesis was conducted accordingly to a previously reported procedure [30]. The active molecule in the nano-hybrid N6B6® is equal to 36 weight percentage (wt%).

The coating was done using a water-based resin for food packaging (Colorgraf 359509, solid content 40 ± 2%, viscosity 40 s at 20 °C) purchased from Colorgraf spa (Lainate, Italy). Its constituents are in accordance with the EC-Directive 2002/72, including amendments. The food-grade resin and the active filler at 10 wt% were mixed using high energy ball milling at ambient temperature, for 30 min at 450 rpm and coated on cellulose acetate by using an automatic coater. The cellulose acetate treated by coating with the active filler has been coupled to pure virgin cellulose cardboard to make paper trays for food (named active tray). Control trays based on the same component (i.e., cellulose coupled with cellulose acetate with no coating) were prepared and used as control (named uncoated tray). Freshly cooked pasta with tomato sauce was packaged (≅100 g) either in the active trays or in the uncoated and closed in an untreated cellulose acetate flowpack.

2.2. Methods

2.2.1. Analysis of Active Filler and Packaging

X-ray diffraction (XRD) patterns were taken in reflection with an automatic Bruker diffractometer D8 (Karlsruhe, Germany), using nickel-filtered Cu Kα radiation (Kα = 1.54050 Å) and operating at 40 kV and 40 mA, with a step scan of 0.05° of 2ϑ and 3 sec of counting time.

Thermogravimetric analyses (TGA) were carried out in air atmosphere with a Mettler TC-10 thermo-balance (Mettler-Toledo GmbH, Greifensee, Switzerland) from 30 to 700 °C, in airflow, at a heating rate of 10 °C/min.

The release kinetics of the active molecule 4-hydroxybenzoate, which is a chromophore, were followed using a Shimadzu UV-2401 PC spectrometer (Shimadzu, Kyoto, Japan). The considered band was 246 nm, highlighted by the calibration line in physiological solution. The tests were performed using 4 cm^2 square samples of packaging material placed into 25 mL physiological solution (NaCl 0.9%) and stirred, at room temperature, at 100 rpm in an orbital shaker (VDRL MOD. 711+, Asal S.r.l., Milan, Italy) for 30 days. The release medium (25 mL physiological solution) was withdrawn at fixed time intervals and replenished with fresh medium. The physiological solution withdrawn at every time, containing the active molecule, is analyzed at UV, and the absorbance corresponding to 246 nm is measured. From the calibration line, the concentration and, consequently, the percentage released is determined, which is reported as a function of time considering the packaging containing the active molecule anchored to the hydrotalcite compared with the packaging containing the free molecule.

The in-vitro effect of inhibition against *Pseudomonas, Escherichia coli, Salmonella,* and Lactic Bacteria by the active trays were analyzed following the directive ISO 22196:2011: such method evaluates the antibacterial activity of treated plastics, surfaces, and other non-porous materials. The test microorganism was prepared by growth in a liquid culture medium. The suspension of the test microorganism was standardized by dilution in a nutritive broth, then control and test surfaces were inoculated with microorganisms, in triplicate, and the microbial inoculum was covered with a sterile film. In order to provide a comparison, all microbiological assay runs were performed with the necessary parallel controls for the whole time of the experiments. Microbial concentrations were determined at "time zero" by elution, followed by dilution and plating. A control was run to verify that the neutralization/elution method effectively neutralized the antimicrobial agents in the antimicrobial surface being tested. The inoculated, covered control and antimicrobial test surfaces were allowed to incubate undisturbed in a humid environment for 24 h. After incubation, microbial concentrations were determined. The reduction of microorganisms relative to initial concentrations and the control surface was calculated. By including the proper controls and being able to make these reduction calculations, such assay allowed to evaluate whether the treated tray is bacteriostatic, having the ability to inhibit the growth of microorganisms. The procedural steps consisted of preparing the microbial suspension to be used as an inoculum; inoculation of microorganisms was in triplicate, both in untreated and in treated specimens, and covering with a sterile film. Incubation took place at 35 °C for 24 h in humid conditions; the microbial count present on the surface of the different treated and control specimens were determined after incubation by washing with a neutralizer (10 mL) and serial dilutions for the plate count with PCA medium. Microorganisms used were those provided by the official method ISO 22196: 2011. The size of the inoculum was between 2.5 × 105 and 10 × 105 CFU/mL. The effectiveness was measured by comparing the survival of the bacteria placed in contact with the treated and untreated materials. The various bacteria were grown in the nutrient broth (PBS). An aliquot of this culture was placed in contact with at least three treated and untreated samples. The samples were divided into portions of 50 × 50 mm size, inoculated with bacterial culture, and covered with a sterile film of 40 × 40 mm size. The colony-forming units were enumerated with a microbiological counting technique, and the antibacterial activity R was calculated from the results of the microbial count, which represented the elimination capacity on a logarithmic basis over a period of 24 h of the bacteria being in contact with the treated material. R = (Ut − U_0) − (At − U_0). The higher the R, the more the treated material has the ability to kill bacteria. The units of measurement of U_0, Ut, and At are Log (CFU/cm^2).

Overall migration tests were performed on the paper active trays according to the following procedure: film specimens with 1 dm^2 of surface area (10 cm × 10 cm, 0.10 mm thickness) were put into contact with 100 mL simulant (preconditioned at 40 °C) in a borosilicate glass tube closed with a screw cap internally layered with Teflon®. The obtained surface/volume ratio was 10 dm^2/L. Migration tests after contact for 10 days at 40 °C were performed using as simulants D2 (Vegetable oil), D1 (Ethanol at 50%) and B (Acetic acid at 3%). The overall migration limits applies to the sum of all substances that can migrate from the food contact material to the food simulant. The overall migration test was performed on different aliquots from the same contact sample. The overall migration results were calculated by using 6 dm^2/kg food (6 dm^2/L simulant) as a conventional EU surface/volume ratio. A known aliquot of the simulant from the contact solution was transferred into a weighted quartz capsule and evaporated to dryness until constant weight. From the differences between the weights, the overall migration was derived in accordance with EN 1186 Migration Testing for Food Contact Materials. The data were averaged on five samples.

Standard deviations of Figures 3–5 were calculated using the classical equation:

$$\sqrt{\frac{\Sigma(x-x_a)^2}{(n-1)}}$$

where x represents the value obtained from each experiment, x_a the medium value, and n the number of experiments conducted.

2.2.2. Analysis of Packaged Pasta

Organoleptic properties (appearance, color, texture, smell, and taste) have been analyzed according to the regulation UNI–U590B2560. Food sensory evaluation ensured that products and packaging were free from defects and/or contamination, and also measured and interpreted human responses based upon sight, smell, touch, taste, and hearing. This is an important aspect of product development and marketing since it offers insight into consumer behavior and quality assurance. For the sensory analyses, the pasta samples were cooked with water and salt for 20 min, and afterward, tomato sauce was added. The sensory analysis was conducted with ten non-trained academic tasters from the University of Salerno. The samples (around 25 g) were offered to the tasters on a disposable plastic plate, and each sample was identified by three random numbers. A water bottle was offered with the samples to remove any residual flavor from the mouth. Appearance, aroma, flavor, texture, and overall impression were evaluated using a 9-point hedonic scale with a minimum of 1 (extremely disliked) and a maximum of 9 (extremely liked) [31,32]

Microbiological tests on freshly cooked pasta, either packaged in active trays or in trays with no treatment, were conducted with the aim to measure: total mesophilic aerobic count (TBC) (ISO 4833:2003) [33], molds [34] Enterobacteriacae (AFNOR AES 10/07-01/08), Lactic Bacteria (ISO 15214:1998) [35], and Pseudomonas (ISO 13720:2010). Samples were divided into small pieces, placed in the trays with and without active filler and closed in an untreated cellulose acetate flowpack (165 × 135 × 45 mm). A total of 24 trays (3 replicates for 2 packages) for four storage times 0, 10, 20, and 30 days were stored at 4 °C for 30 days. At each storage time, samples were analyzed to follow the incidence of visible molds on the pasta, by analyzing the organoleptic properties. Mold evolution on pasta was evaluated as follows: ten grams of pasta sample were, in triplicate, aseptically homogenized in 90 mL of sterile 0.9% NaCl solution in a stomacher (Lab-Blender 400, PBI International, Milano, Italy), and decimally diluted in 0.1 % (*w*/*v*) sterile peptone saline solution (0.9 % NaCl) before plating on the selective media. In particular, molds were detected on potato dextrose agar (PDA) supplemented with 100 mg/L of chloramphenicol and incubated at 25 °C for 3 5 days (ISO 21527 1:2008). Total mesophilic aerobic counts (TBC) were enumerated on plate count agar (PCA) supplemented with 100 mg/L of cycloheximide after incubation at 30 °C for 24 h (ISO 4833:2003). Pseudomonas were grown on Pseudomonas agar base (PSA, amended with Pseudomonas CFC selective supplement, Oxoid) at 30 °C for 24 h (ISO 13720:2010); Lactic acid bacteria population (LABs) were counted on De Man, Rogosa, and Sharpe agar (MRS) supplemented with 100 mg/L of cycloheximide after incubation at 30 °C under anaerobic conditions for 48 h (ISO 15214:1998). Enterobacteriaceae were evaluated by RAPID'Enterobacteriaceae medium, high-performance medium for the enumeration of Enterobacteriaceae in food and environmental samples at 30 °C for 24 h (according to AFNOR AES 10/07-01/08). Microbiological values reported as (Log(CFU/g) were the average of three replicates.

Statistical analyses were conducted using Origin Lab. Results were expressed as the mean value ± standard deviation (SD). ANOVA and Tukey's tests were performed to compare the obtained results at a significance level of $\alpha < 5\%$. The curve fitting for the Gompertz equation (Equation (1)) was evaluated by non-linear regression. The coefficient of determination (R^2) was evaluated to assess the goodness of the fitting.

3. Results and Discussion

3.1. Analysis of Active Filler and Trays

Figure 1 shows the LDH in nitrate form (a), with the peak characteristic of the basal spacing at $2\vartheta = 10.2°$, corresponding to an interlayer distance d = 8.6 Å. The intercalation of the 4-hydroxybenzoate anion into LDH galleries occurred successfully (Figure 1b), and this was demonstrated by a shift of the peak toward lower angle, $2\vartheta = 5.6°$, that corresponded to an increased interlayer distance (d = 14.86 Å) having the 4-hydroxybenzoate anion a steric hindrance higher than nitrate.

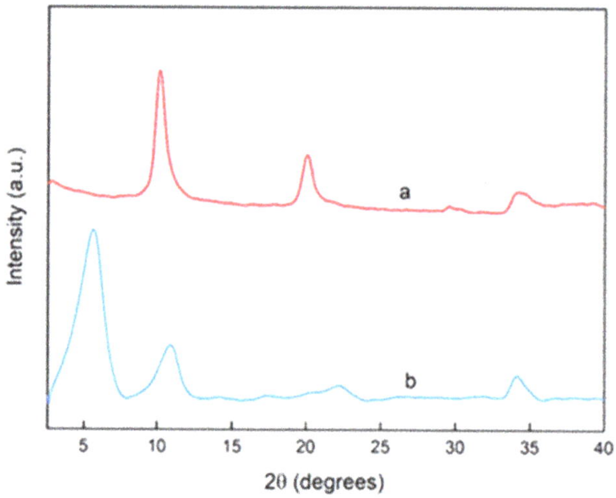

Figure 1. XRD of pristine LDH in nitrate form (**a**) and LDH hosting 4-hydroxybenzoate (**b**).

Figure 2 reports the TGA analysis on of 4-hydroxybenzoic acid (a), LDH-NO_3 (b), and LDH-(4-hydroxybenzoate) (c) in airflow. LDH with intercalated nitrate anion shows three main degradation steps [36]: the loss of absorbed water between the LDH galleries at about 150 °C, the thermal decomposition of nitrate anions at around 250 °C, the dehydroxylation of the LDH layers after 400 °C. It is evident that the 4-hydroxybenzoic acid degrades in one main step, centered at about 238 °C. The thermal stability of 4-hydroxybenzoate intercalated between LDH's sheets results improved; in fact, the main decomposition of the nano-hybrid was centered at about 434 °C. Such behavior has already been found for several organic molecules intercalated into LDH layers [37], allows us to hypothesize a protecting effect from the inorganic LDH nano-carrier also for 4-hydroxybenzoate.

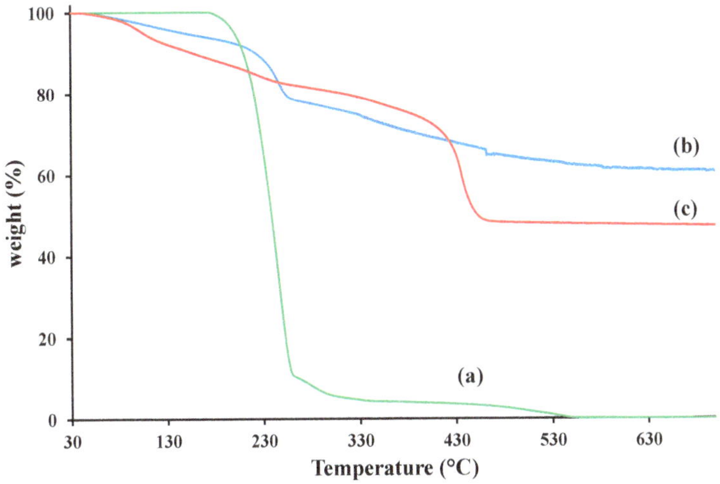

Figure 2. TGA analysis of 4-hydroxybenzoic acid (**a**), LDH-NO_3 (**b**), and LDH-(4-hydroxybenzoate) (**c**).

The release of 4-hydroxybenzoate (wt%) of the active molecule was evaluated as a function of time (days) from the active tray. A sample in which the active molecule was simply dispersed into the food-grade resin at the same percentage (i.e., 3.6%) and coated on the cellulose acetate was also prepared, following the same experimental conditions reported in Section 2.1.

The release of 4-hydroxybenzoate freely dispersed into the coating medium was very fast and completed in almost 3 days (Figure 3). The release mode of the active molecule anchored to the LDH layers showed a first fast step, corresponding to the release from the surface of the material, followed by a second step related to the de-intercalation due to ionic exchange with chloride in physiological solution. It is evident that in the whole investigated range of time, the 4-hydroxybenzoate inside the active tray showed a slower release compared to the same molecule freely dispersed into the coating. In addition, in the considered range of contact time (i.e., 30 days) the molecule bonded to the nano-carrier was not completely released. This result was in agreement with XRD analysis that demonstrated the successful intercalation of the sorbate between the LDH galleries.

Pseudomonas, *Escherichia coli*, *Salmonella*, and *Lactobacillus* strains were used to test the bacterial inhibition capability from the active tray filled with LDH-(4-hydroxybenzoate). It is evident from Table 1 that the prepared active trays exerted a significant antibacterial activity with respect to all considered strains. The degree of inhibition of the 4-hydroxybenzoate molecule present in the active trays towards the bacteria analyzed ranged from 1 to 5 orders of magnitude higher than the untreated ones, highlighting a significant inhibition towards *Pseudomonas* and *Salmonella*; with respect to *Escherichia Coli* and *Lactobacillus*, even if slightly lower, there was still evidence of a bacteriostatic activity in the treated tray. The importance of the capability from the active packaging to have such important bacteria inhibition, common in food spoilage, allowed us to use that for storage of cooked tomato pasta. Although a certain cause and effect correlation for the mechanisms of action of parabens has not yet been established, several studies showed that such compounds could be active at the cytoplasmic membrane and capable of inhibiting both membrane transport and the electron transport system [38].

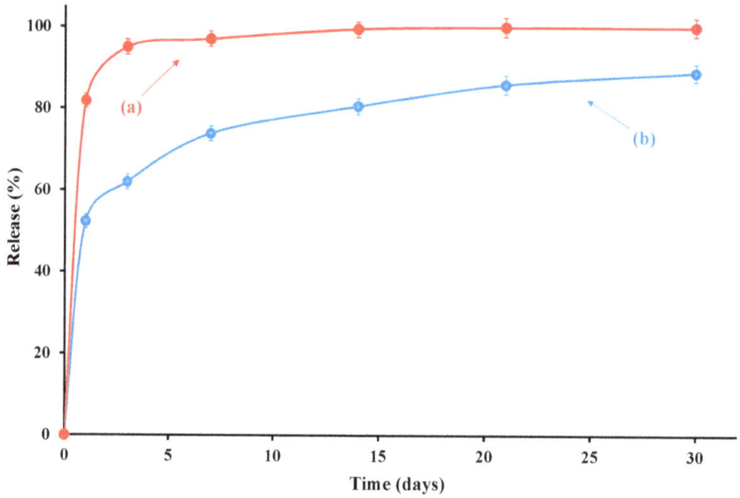

Figure 3. Release of 4-hydroxybenzoate (wt%) from: (**a**) cellulose acetate coated with 3.6 wt% of 4-hydroxybenzoate simply dispersed into the food-grade resin, and (**b**) cellulose acetate coated with active molecule anchored to LDH.

Table 1. Inhibition of *Pseudomonas, Escherichia coli, Salmonella,* and *Lactobacillus* from the active trays, following the directive ISO 22196:2011.

Bacterial Strain	***Pseudomonas aeruginosa* ATCC 15442**	***Escherichia coli* ATCC 8739**	***Lactobacillus sakei* ATCC 15521**	***Salmonella enterica typhimurium* ATCC 14028**
Sample size (mm × mm)	50 × 50	40 × 40	50 × 50	50 × 50
Sample thickness (mm)	1.0	0.070	0.1	1.0
Inoculum volume (mL)	0.4	0.4	0.4	0.4
Number of bacteria available in the inoculum	410.000	25.000	35.000	120.000
Uo—Counts bacteria recovered from untreated specimens after inoculation (Log units)	4.4 ± 0.0026 (Log CFU/cm^2)	4.3 ± 0.0024 (Log CFU/cm^2)	3.4 ± 0.0021 (Log CFU/cm^2)	3.9 ± 0.0022 (Log CFU/cm^2)
Ut—Count of bacteria recovered from non-treated samples after 24 h from inoculation (Log units)	5.3 ± 0.0032 (Log CFU/cm^2)	5.7 ± 0.0034 (Log CFU/cm^2)	2.5 ± 0.0014 (Log CFU/cm^2)	4.3 ± 0.0025 (Log CFU/cm^2)
At—Count of bacteria recovered from treated samples after 24 h from inoculation (Log units)	n.d. *	4.6 ± 0.0026 (Log CFU/cm^2)	0.8 ± 0.0011 (Log CFU/cm^2)	1.6 ± 0.0013 (Log CFU/cm^2)
Antibacterial activity R = (Ut − Uo) − (At − Uo) (ISO 22196:2011)	>5.3	1.1	1.6	2.6

* Not detectable.

In order to demonstrate that the prepared active packaging is suitable for food contact, we performed overall migration tests on the active trays. Table 2 reports the overall migration, evaluated on the active trays, in different food simulants, according to UNI EN 1186-1:2003 and UNI EN 1186-9: 2003 simulant D2 (Vegetable oil), D1 (Ethanol at 50%), and B (Acetic acid at 3%). The experimental results, in compliance with the migration limits, demonstrate the suitability of the considered material for food contact. The prepared packaging, although new, can be easily used for food contact, due to the respected security level to preserve consumer health.

Table 2. Overall migration from the active trays, according to UNI EN 1186-1:2003 and UNI EN 1186-9: 2003.

Simulant	B—Acetic Acid at 3% (*v*/*v*)	D1—Ethanol at 50% (*v*/*v*)	D2—Vegetable Oil	Limits
Temperature of the test (°C)	40	40	40	
Contact time (days)	10	10	10	
Overall migration (mg/dm^2)	7.41 ± 0.03	9.05 ± 0.08	7.72 ± 0.05	10

3.2. Evaluation of Shelf Life on Packaged Pasta Ready to Eat

Cooked pasta with tomato sauce (≅100 g), ready to eat, was packaged into the active trays and uncoated. Organoleptic properties, *Moulds*, *Total mesophilic aerobic counts* (TBC), *Enterobacteriaceae*, Lactic Bacteria and *Pseudomonas* were evaluated up to 30 days of storage at 4 °C. Table 3 below shows the organoleptic properties after 30 days of storage.

Table 3. Organoleptic analysis at 30 days in uncoated and active trays evaluated according to regulation M.I. 1923A Rev0 2012.

Characteristic Observed	Uncoated Trays	Active Trays
appearance	Water condensation on the external packaging. Presence of visible mold	No water condensation on the external packaging. The appearance of the food typically regular, even if with water droplets on the entire surface of the sample
color	Not regular, with a slightly darker note.	Typical regular.
texture	not regular.	typical regular.
smell	not acceptable due to the presence of degradative notes	typical regular
taste	not acceptable due to the presence of degradative notes	acceptable with the absence of significant degradative notes
visible moulds (%)	≅80	≅5

The incidence molds on the surface of the cooked pasta, was evaluated by counting the number of pieces of pasta with visible mold on each bag for each replicate in the untreated packaging and in the active packages, expressed as a percentage of infected pieces in respect to the total. It was evident that the pasta packaged in uncoated trays and the active ones showed several differences in terms of appearance, smell, taste, and color. In addition, the presence of visible molds on the surface was significantly different ($p < 0.05$): pasta stored in an uncoated tray was covered by ≅80% of mold on the surface, while pasta packaged in the active tray shows only ≅5% of visible molds. The presence of the active molecule on the surface of the bio-packaging has an antimicrobial and anti-mold effect; the anti-mold and antimicrobial activity of the molecule were known, but the hydrotalcite system modified with the organic molecule means that the release of the molecule occurred faster in the first hours and then more slowly over 30 days to ensure bacteriostatic activity on molds and bacteria both in the first few hours to block the proliferation of microbes, but also during the whole investigated storage time to ensure the maintenance of quality, food safety and freshness of the pasta for longer times. Figures 4 and 5 report the mold and TBC evolution, respectively, as a function of the time (days), evaluated on pasta packaged either in the active tray or uncoated. The results were presented as the mean ± SD (standard deviation).

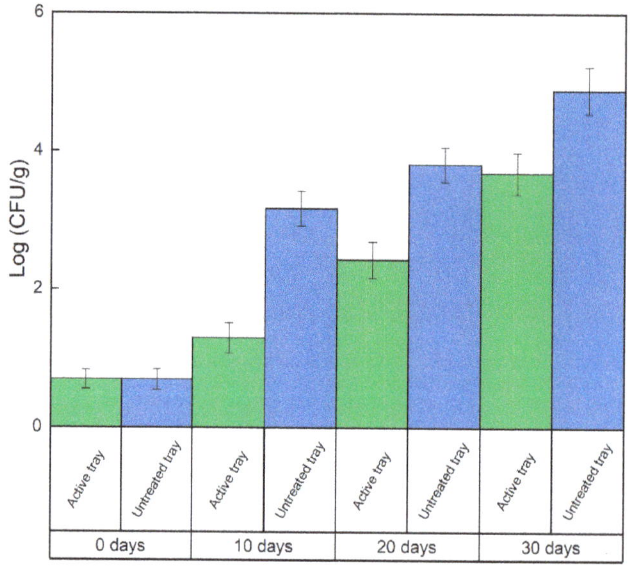

Figure 4. Mold evolution, Log (CFU/g) on packaged pasta @4 °C as a function of time.

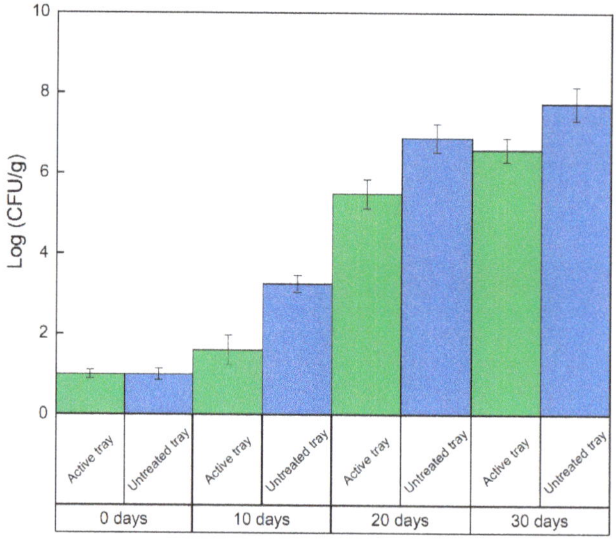

Figure 5. TBC evolution, Log (CFU/g) on packaged pasta @4 °C as a function of time.

The microbiological analysis of molds and TBC present in both the pasta packed in the uncoated tray and in pasta packed in the tray treated with the active filler was carried out in triplicate at different times; time zero, after 10 days, 20 days, and 30 days. At all times of analysis, it could be observed that the Log values (CFU/g) in the active tray were lower than those of the uncoated tray, reaching a difference of almost 1–2 logarithmic orders of magnitude. Such results, which were significantly different ($p < 0.05$), highlighted that the active trays containing the LDH-(4-hydroxybenzoate) possessed a strong inhibitory power towards molds and TBC on the considered packaged food. The Gompertz equation was applied to estimate the shelf life (SL) of ready-to-eat pasta by fitting Equation (1) [39]:

$$\log(CFU) = K + A * \exp\left\{-exp\left\{\left[(\mu_{max} * 2.7182) * \frac{\lambda - t}{A}\right] + 1\right\}\right\} \quad (1)$$

where K (log(CFU/g)) is the initial level of the bacterial count, μ_{max} is the maximum growth rate, λ is the lag phase (days), A is the maximum bacteria growth achieved at the stationary phase, and t is the time (days). After estimating Gompertz's equation parameters, shelf life was evaluated by using Equation (2):

$$S.L. = \lambda - \frac{A * \left\{ln\left[-ln\left(\frac{\log(1 * 10^2) - K}{A}\right)\right] - 1\right\}}{\mu_{max} * 2.7182} \quad (2)$$

where $1 * 10^2$ is the acceptability limit for the mold evolution of cooked pasta. All analyses were carried out in triplicate. The coefficient of determination (R^2) is reported in Table 4. The media and standard deviations were calculated. The results concerning the mold evolution on packaged pasta are reported in Figure 6.

Table 4. Gompertz's equation parameters evaluated by using Equations (1) and (2).

	Untreated Tray	Active Tray
K	0.70 ± 0.07^a	0.68 ± 0.04^a
A	1.62 ± 0.18^a	1.35 ± 0.20^b
μ_{max} (days $^{-1}$)	0.11 ± 0.02^a	0.06 ± 0.01^b
λ (days)	7.07 ± 1.16^a	14.78 ± 2.05^b
Shelf life (days)	20.70 ± 1.21^a	54.47 ± 1.35^b
R^2	0.989	0.990

Means having different superscript lowercase letters for a parameter were significantly different ($p < 0.05$) through the Tukey test.

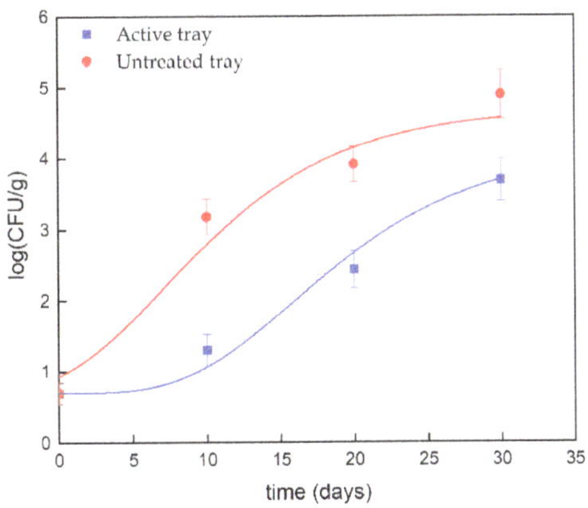

Figure 6. Evolution of molds on packaged pasta during storage time (scatters) and fitting curves (solid lines) obtained from Equation (1).

The evaluated parameters from Equation (2) are reported in Table 4. The statistical analysis showed a significant difference ($p < 0.05$).

Table 4 clearly evidences the differences between the two types of packaging, suggesting that 4-hydroxybenzoate affected the maximum cell growth rate and the cell growth rate in the stationary phase. Moreover, it is worth noting a noticeable increase in lag time, which doubled for the active tray (from 7 days to 14 days) and a halving of μ_{max} (0.11 and 0.06 days^{-1} for the untreated and the active tray, respectively). As shown in the above table, the active tray slowed the growth of the molds during cooked pasta storage, allowing to extend the shelf life up to about 55 days, noticeably prolonged in comparison with the shelf life of pasta stored in untreated tray (about 21 days). Table 5 shows the microbiological analysis respect to *Enterobacteriaceae,* Lactic Bacteria, and *Pseudomonas* evaluated on the cooked pasta stored either in the active and uncoated trays. It is worth noting that the starting bacteria counts, for all considered strains, are very low. The very low values of bacteria counts, in all cases, remained below 10 CFU/g up 20 days of storage in all packaging. At 30 days it was visible as an increasing 2–3 order of magnitudes for the pasta packaged into untreated trays, and still a value of such strains below 10 CFU/g for the pasta stored in the active trays. The statistical analysis on active and untreated trays showed a significant difference ($p < 0.05$) after 30 days. It follows that the active packaging, was able to prevent the considered bacteria proliferation for this long time.

Table 5. Evaluation of *Enterobacteriaceae,* Lactic Bacteria and *Pseudomonas* in the uncoated and active trays.

Time (Days)	Sample	Enterobacteriaceae (CFU/g)	Lactic Bacteria (CFU/g)	*Pseudomonas* spp. (CFU/g)
0	Untreated tray	$<10^{a}$	$<10^{a}$	$<10^{a}$
	Active tray	$<10^{a}$	$<10^{a}$	$<10^{a}$
10	Untreated tray	$<10^{a}$	$<10^{a}$	$<10^{a}$
	Active tray	$<10^{a}$	$<10^{a}$	$<10^{a}$
20	Untreated tray	$<10^{a}$	$<10^{a}$	$<10^{a}$
	Active tray	$<10^{a}$	$<10^{a}$	$<10^{a}$
30	Untreated tray	$\cong 10^{3\,a}$	$\cong 10^{3\,a}$	$\cong 10^{2\,a}$
	Active tray	$<10^{b}$	$<10^{b}$	$<10^{b}$

Means having different superscript lowercase letters for a parameter are significantly different ($p < 0.05$) through the Tukey test.

Figure 7 shows, as a visual example, images of pasta stored for two months at 4 °C in both prepared packaging.

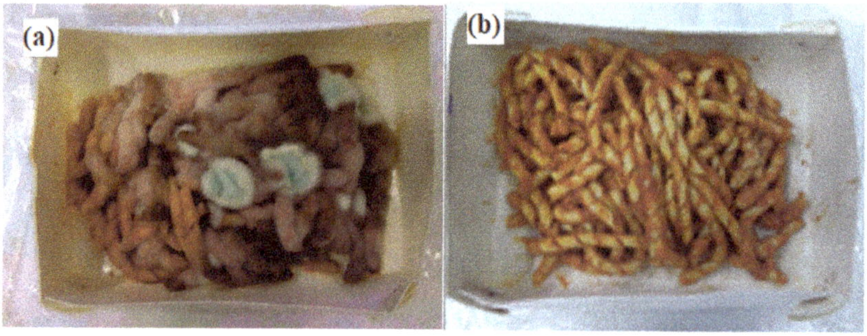

Figure 7. Pictures taken on cooked pasta with tomato sauce stored for two months at 4 °C in: (**a**) an uncoated tray; (**b**) and active tray.

After 2 months at 4 °C, molds covered almost the total surface of the pasta stored in the uncoated tray. A very different picture (few molds ≅ 5%) was seen for the pasta packaged in the active packaging, demonstrating the inhibitory effect of fillers towards the proliferation of molds, even after two months of storage.

4. Conclusions

Cellulose-based trays coupled with cellulose acetate, coated with an active filler, were prepared and tested as packaging for cooked pasta with tomato sauce, ready to eat. The active coating was based on a food-grade resin filled with layered double hydroxide (LDH) nanofiller hosting 4-hydroxybenzoate, as an antimicrobial, listed in EC-Directive 10/2011/EC of 14 January, 2011. The in-vitro bacterial inhibition on the active trays against *Pseudomonas, Escherichia coli, Salmonella,* and Lactic Bacteria was analyzed, and a significant antibacterial activity was evidenced, in particular respect to *Pseudomonas* and *Salmonella.* Global migration tests on the active trays using acetic acid at 3% (*v*/*v*), ethanol at 50% (*v*/*v*) and vegetable oil, as food simulants, resulted in compliance with the migration limits imposed from EU regulation, demonstrating the suitability of the prepared material for food contact. Ready to eat, freshly cooked pasta with tomato sauce was packaged in the active and uncoated trays, as control, up to 30 days at 4 °C. Organoleptic analyses, mold count, Total mesophilic aerobic counts (TBC), *Enterobacteriaceae,* Lactic Bacteria, and *Pseudomonas* count showed an important antimicrobial activity of 4-hydroxybenzoate for increasing the shelf life of the packaged pasta using a cheap, safe, and easily industrially scalable process.

Author Contributions: V.B. and G.G. conceived the paper. V.B. and G.V. collected data. All authors equally interpreted the data. G.G. wrote the paper. All authors have read and agreed to the published version of the manuscript.

Funding: Project Prin 2017 "MultIFunctional poLymer cOmposites based on groWn matERials (MIFLOWER)" (grant number: 2017B7MMJ5_001) from the Italian Ministry of Education University and Research.

Conflicts of Interest: The authors declare no conflict of interest.

References

1. Preventing Plastic Waste in Europe—European Environment Agency. Available online: https://www.eea.europa.eu/publications/preventing-plastic-waste-in-europe (accessed on 1 September 2020).
2. *Proposal for a Directive of the European Parliament and of the Council on the Reduction of the Impact of Certain Plastic Products on the Environment (Text with EEA Relevance)*; European Commission: Brussel, Belgium, 2018; Volume 340.
3. A European Strategy for Plastics in a Circular Economy. Available online: https://ec.europa.eu/environment/circular-economy/pdf/plastics-strategy.pdf (accessed on 1 September 2020).
4. Carini, E.; Curti, E.; Littardi, P.; Luzzini, M.; Vittadini, E. Water dynamics of ready to eat shelf stable pasta meals during storage. *Innov. Food Sci. Emerg. Technol.* **2013**, *17*, 163–168. [CrossRef]
5. Carini, E.; Curti, E.; Cassotta, F.; Najm, N.E.O.; Vittadini, E. Physico-chemical properties of ready to eat, shelf-stable pasta during storage. *Food Chem.* **2014**, *144*, 74–79. [CrossRef]
6. Suppakul, P.; Miltz, J.; Sonneveld, K.; Bigger, S.W. Active packaging technologies with an emphasis on antimicrobial packaging and its applications. *J. Food Sci.* **2003**, *68*, 408–420. [CrossRef]
7. Appendini, P.; Hotchkiss, J.H. Review of antimicrobial food packaging. *Innov. Food Sci. Emerg. Technol.* **2002**, *3*, 113–126. [CrossRef]
8. Langer, R.S.; Peppas, N.A. Present and future applications of biomaterials in controlled drug delivery systems. *Biomaterials* **1981**, *2*, 201–214. [CrossRef]
9. Mallapragada, S.K.; Peppas, N.A. Crystal dissolution-controlled release systems: I. Physical characteristics and modeling analysis. *J. Control Release* **1997**, *45*, 87–94. [CrossRef]
10. Collins, R. Mathematical modelling of controlled release from implanted drug- impregnated monoliths. *Pharm. Sci. Technol. Today* **1998**, *1*, 269–276. [CrossRef]

11. Siepmann, J.; Peppas, N.A. Modeling of drug release from delivery systems based on hydroxypropyl methylcellulose (HPMC). *Adv. Drug Deliv. Rev.* **2001**, *48*, 139–157. [CrossRef]
12. Siepmann, J.; Lecomte, F.; Bodmeier, R. Diffusion-controlled drug delivery systems: Calculation of the required composition to achieve desired release profiles. *J. Control Release* **1999**, *60*, 379–389. [CrossRef]
13. Smith, K.L.; Herbig, S.M. Controlled Release. In *Membrane Handbook*; Springer US: New York, NY, USA, 1992; pp. 915–935.
14. Lopez-Rubio, A.; Gavara, R.; Lagaron, J.M. Bioactive packaging: Turning foods into healthier foods through biomaterials. *Trends Food Sci. Technol.* **2006**, *17*, 567–575. [CrossRef]
15. Buonocore, G.G.; Conte, A.; Corbo, M.R.; Sinigaglia, M.; Del Nobile, M.A. Mono- and multilayer active films containing lysozyme as antimicrobial agent. *Innov. Food Sci. Emerg. Technol.* **2005**, *6*, 459–464. [CrossRef]
16. Bugatti, V.; Sorrentino, A.; Gorrasi, G. Encapsulation of Lysozyme into halloysite nanotubes and dispersion in PLA: Structural and physical properties and controlled release analysis. *Eur. Polym. J.* **2017**, *93*, 495–506. [CrossRef]
17. Bugatti, V.; Viscusi, G.; Naddeo, C.; Gorrasi, G. Nanocomposites based on PCL and halloysite nanotubes filled with lysozyme: Effect of draw ratio on the physical properties and release analysis. *Nanomaterials* **2017**, *7*, 213. [CrossRef]
18. Bugatti, V.; Vertuccio, L.; Zuppardi, F.; Vittoria, V.; Gorrasi, G. Pet and active coating based on a ldh nanofiller hosting p-hydroxybenzoate and food-grade zeolites: Evaluation of antimicrobial activity of packaging and shelf life of red meat. *Nanomaterials* **2019**, *9*, 1727. [CrossRef] [PubMed]
19. Bugatti, V.; Brachi, P.; Viscusi, G.; Gorrasi, G. Valorization of tomato processing residues through the production of active bio-composites for packaging applications. *Front. Mater.* **2019**, *6*. [CrossRef]
20. Gorrasi, G.; Bugatti, V.; Vertuccio, L.; Vittoria, V.; Pace, B.; Cefola, M.; Quintieri, L.; Bernardo, P.; Clarizia, G. Active packaging for table grapes: Evaluation of antimicrobial performances of packaging for shelf life of the grapes under thermal stress. *Food Packag. Shelf Life* **2020**, *25*, 100545. [CrossRef]
21. Gemili, S.; Yemenicioğlu, A.; Altinkaya, S.A. Development of cellulose acetate based antimicrobial food packaging materials for controlled release of lysozyme. *J. Food Eng.* **2009**, *90*, 453–462. [CrossRef]
22. Buonocore, G.G.; Nobile, M.A.; Panizza, A.; Bove, S.; Battaglia, G.; Nicolais, L. Modeling the Lysozyme Release Kinetics from Antimicrobial Films Intended for Food Packaging Applications. *J. Food Sci.* **2003**, *68*, 1365–1370. [CrossRef]
23. Buonocore, G.G.; Sinigaglia, M.; Corbo, M.R.; Bevilacqua, A.; La Notte, E.; Nobile, M.A.D. Controlled Release of Antimicrobial Compounds from Highly Swellable Polymers. *J. Food Prot.* **2004**, *67*, 1190–1194. [CrossRef]
24. Bugatti, V.; Gorrasi, G.; Montanari, F.; Nocchetti, M.; Tammaro, L.; Vittoria, V. Modified layered double hydroxides in polycaprolactone as a tunable delivery system: In vitro release of antimicrobial benzoate derivatives. *Appl. Clay Sci.* **2011**, *52*, 34–40. [CrossRef]
25. Gorrasi, G. Dispersion of halloysite loaded with natural antimicrobials into pectins: Characterization and controlled release analysis. *Carbohydr. Polym.* **2015**, *127*, 47–53. [CrossRef] [PubMed]
26. Gorrasi, G.; Bugatti, V.; Tammaro, L.; Vertuccio, L.; Vigliotta, G.; Vittoria, V. Active coating for storage of Mozzarella cheese packaged under thermal abuse. *Food Control* **2016**, *64*, 10–16. [CrossRef]
27. Gorrasi, G.; Bugatti, V. Edible bio-nano-hybrid coatings for food protection based on pectins and LDH-salicylate: Preparation and analysis of physical properties. *LWT—Food Sci. Technol.* **2016**, *69*, 139–145. [CrossRef]
28. Gorrasi, G.; Vertuccio, L. Evaluation of zein/halloysite nano-containers as reservoirs of active molecules for packaging applications: Preparation and analysis of physical properties. *J. Cereal Sci.* **2016**, *70*, 66–71. [CrossRef]
29. Gorrasi, G.; Attanasio, G.; Izzo, L.; Sorrentino, A. Controlled release mechanisms of sodium benzoate from a biodegradable polymer and halloysite nanotube composite. *Polym. Int.* **2017**, *66*, 690–698. [CrossRef]
30. Frunza, M.; Lisa, G.; Popa, M.I.; Miron, N.D.; Nistor, D.I. Thermogravimetric analysis of layered double hydroxides with chloramphenicol and salicylate in the interlayer space. *J. Therm. Anal. Calorim.* **2008**, *93*, 373–379. [CrossRef]
31. Cardello, H.M.A.B.; Da Silva, M.A.P.A.; Damasio, M.H. Measurement of the relative sweetness of stevia extract, aspartame and cyclamate/saccharin blend as compared to sucrose at different concentrations. *Plant Foods Hum. Nutr.* **1999**, *54*, 119–129. [CrossRef]

32. Cuquel, F.L.; de Oliveira, C.F.S.; Lavoranti, O.J. Perfil sensorial de onze cultivares de pêssegos. *Cienc. Tecnol. Aliment.* **2012**, *32*, 70–75. [CrossRef]
33. ISO 4833:2003(en), Microbiology of Food and Animal Feeding Stuffs—Horizontal Method for the Enumeration of Microorganisms—Colony-count Technique at 30 degrees C. Available online: https://www.iso.org/obp/ui/#iso:std:iso:4833:ed-3:v1:en (accessed on 27 April 2020).
34. ISO—ISO 21527-1:2008—Microbiology of Food and Animal Feeding Stuffs—Horizontal Method for the Enumeration of Yeasts and Moulds—Part 1: Colony Count Technique in Products with Water Activity Greater than 0,95. Available online: https://www.iso.org/standard/38275.html (accessed on 27 April 2020).
35. ISO—ISO 15214:1998—Microbiology of Food and Animal Feeding Stuffs—Horizontal Method for the Enumeration of Mesophilic Lactic Acid Bacteria—Colony-Count Technique at 30 Degrees C. Available online: https://www.iso.org/standard/26853.html (accessed on 27 April 2020).
36. Bugatti, V.; Vertuccio, L.; Zara, S.; Fancello, F.; Scanu, B.; Gorrasi, G. Green pesticides based on cinnamate anion incorporated in layered double hydroxides and dispersed in pectin matrix. *Carbohydr. Polym.* **2019**, *209*, 356–362. [CrossRef]
37. Gorrasi, G.; Bugatti, V. Mechanical dispersion of layered double hydroxides hosting active molecules in polyethylene: Analysis of structure and physical properties. *Appl. Clay Sci.* **2016**, *132–133*, 2–6. [CrossRef]
38. Eklund, T. Inhibition of Growth and Uptake Processes in Bacteria by Some Chemical Food Preservatives. *J. Appl. Bacteriol.* **1980**, *48*, 423–432. [CrossRef] [PubMed]
39. Corbo, M.R.; Del Nobile, M.A.; Sinigaglia, M. A novel approach for calculating shelf life of minimally processed vegetables. *Int. J. Food Microbiol.* **2006**, *106*, 69–73. [CrossRef] [PubMed]

foods

MDPI

Article

Lid Films of Poly(3-hydroxybutyrate-*co*-3-hydroxyvalerate)/Microfibrillated Cellulose Composites for Fatty Food Preservation

Eva Hernández-García, Amparo Chiralt, Maria Vargas * and Sergio Torres-Giner

Research Institute of Food Engineering for Development (IIAD), Universitat Politècnica de València (UPV), 46022 Valencia, Spain

* Correspondence: mavarco@tal.upv.es

Abstract: The present work evaluates the food packaging performance of previously developed films of poly(3-hydroxybutyrate-*co*-3-hydroxyvalerate) (PHBV) reinforced with atomized microfibrillated cellulose (MFC) compatibilized by a reactive melt-mixing process. To this end, the novel green composite films were originally applied herein as lids in aluminum trays to preserve two dissimilar types of fatty foods, namely minced pork meat and sunflower oil. Results indicated that the PHBV/MFC films effectively preserved the physicochemical and microbiological quality of pork meat for one week of storage at 5 °C. In particular, the compatibilized green composite lid film yielded the lowest weight loss and highest oxidative stability, showing values of 0.935% and 0.78 malonaldehyde (MDA)/kg. Moreover, none of the packaged meat samples exceeded the acceptable Total Aerobial Count (TAC) level of 5 logs colony-forming units (CFU)/g due to the improved barrier properties of the lids. Furthermore, the green composite films successfully prevented sunflower oil oxidation in accelerated oxidative storage conditions for 21 days. Similarly, the compatibilized PHBV/MFC lid film led to the lowest peroxide value (PV) and conjugated diene and triene contents, with respective values of 19.5 meq O_2/kg and 2.50 and 1.44 g/100 mL. Finally, the migration of the newly developed PHBV-based films was assessed using two food simulants, proving to be safe since their overall migration levels were in the 1–3 mg/dm^2 range and, thus, below the maximum level established by legislation.

Keywords: PHBV; cellulose; sustainable packaging; food quality and shelf life; migration

Citation: Hernández-García, E.; Chiralt, A.; Vargas, M.; Torres-Giner, S. Lid Films of Poly(3-hydroxybutyrate-*co*-3-hydroxyvalerate)/Microfibrillated Cellulose Composites for Fatty Food Preservation. *Foods* **2023**, *12*, 375. https://doi.org/10.3390/foods12020375

Academic Editor: Elsa M. Gonçalves

Received: 12 December 2022
Revised: 5 January 2023
Accepted: 10 January 2023
Published: 13 January 2023

1. Introduction

Food preservation aims to extend shelf life and provide safer products to consumers using varied materials and technologies [1]. In this sense, packaging solutions based on plastics have been extensively used due to their transparency, flexibility, ease of processing, reduced cost, low weight, and high versatility [2]. In addition, packaging materials can provide other functions to the food product, such as active and bioactive properties [3]. Nowadays, the development of sustainable materials and designs represents a fundamental challenge in the food packaging industry [4,5]. However, most of the conventional materials currently used for food packaging are based on non-biodegradable petrochemical polymers that cause high environmental impacts related to their disposal.

A possible packaging alternative approach is the use of biopolymers, which are macromolecules that are either obtained from natural sources or biodegradable or show both features [6–10]. Packaging materials based on biopolymers can be feasibly composted in industrial facilities and, in some cases, in domestic composts and natural environments [11,12]. Among biopolymers, polyhydroxyalkanoates (PHAs), such as poly(3-hydroxybutyrate) (PHB), are semi-crystalline aliphatic polyesters obtained from microorganisms that rapidly biodegrade in composting facilities and even in the environment [13].

PHB is a homopolyester that shows physical properties similar to those of polypropylene (PP) and polyethylene terephthalate (PET) [14]. PHB shows high crystallinity and low thermal stability, yielding rigid and brittle materials, which limits its use in food packaging applications [15]. However, the copolymer of 3-hydroxybutyrate (3HB) with 3-hydroxyvalerate (3HV), that is, poly(3-hydroxybutyrate-*co*-3-hydroxyvalerate) (PHBV), shows higher ductility and reduced crystallinity, opening up new possibilities in food packaging [16].

Furthermore, a newly designed strategy for improving the performance of PHBV-based packaging materials within the framework of the Circular Economy is its reinforcement with renewable and biodegradable micro- or nanostructured fillers, such as microfibrillated cellulose (MFC) [17]. In this sense, an innovative and scalable technique was developed in a previous study of the research group to prepare green composite films based on PHBV and atomized MFC [18]. Atomization, also called the spraying method, is a process in which emulsions or other multiphase systems are broken into small drops of liquid by high-speed fluids (typically compressed gas) or fluids with centrifugal force, and then solidified into powder. The atomization process of MFC suspensions resulted in the formation of ultrathin cellulose structures, having a rod-like morphology and sizing approximately 3 ± 1 μm, which were composed of fibers with thicknesses in the nano-range. The atomized MFC particles were thereafter melt-mixed with PHBV and subsequently processed into films by thermo-compression. Incorporating the nanostructured fibers of cellulose successfully improved the physical properties of the biopolyester, mainly in terms of tensile and gas barrier properties, without reducing thermal stability or significantly altering the optical properties. Moreover, these green composite formulations were compatibilized with different reactive additives to enhance the filler-to-matrix interfacial adhesion and, hence, their performance for food packaging applications. Among all the range of formulations tested, optimal results were attained for PHBV containing intermediate MFC content of 5 wt% and compatibilized with the combination of triglycidyl isocyanurate (TGIC) with dicumyl peroxide (DCP) by a reactive melt-mixing process.

Thus, the present study focuses on applying the previously prepared compatibilized green composite films as rigid lids for food packaging applications. To this end, two types of fatty foods were selected: minced pork meat and sunflower oil. On the one hand, pork meat is a high-protein food of high biological value. It also contains varying amounts of lipids, including saturated fatty acids, which usually range from 2.5 to 11.9 g/100 g of meat and can be much higher depending on the level of trimming and other factors [19]. Minced pork meat is known to deteriorate during storage due to microbial spoilage, surface dehydration, and myoglobin and lipid oxidation, which could decrease its nutritional value and cause changes in appearance and production of off-flavors and odors [20,21]. On the other hand, sunflower oil is rich in unsaturated fatty acids, mainly oleic and linoleic [22]. These are associated with oxidation that can modify the organoleptic properties of the oil and cause losses in nutritional value and quality [23]. Therefore, in both foods, lipid oxidation should be prevented or at least delayed [24], and the PHBV/MFC composites can be regarded as excellent candidates to accomplish this objective due to their improved barrier against oxygen. Therefore, green composites of PHBV with 5 wt% MFC, with and without compatibilizers, were prepared in the form of films and applied as lids in trays containing the fatty foods. The quality and shelf-life of the two packaged foodstuffs were evaluated throughout storage and compared with films of neat PHBV and a commercial high-barrier multilayer habitually used in the food packaging industry. Finally, migration studies were performed using food simulants to corroborate the safety and potential use of green composite films in food packaging.

2. Materials and Methods

2.1. Materials

Commercial PHBV with food-grade status was supplied as ENMAT Y1000P in the form of pellets by Tianan Biologic Materials (Ningbo, China). The 3HV fraction in the copolyester is ~2 mol%, and the molecular weight (M_W) is ~2.8 × 10^5 g/mol. The aqueous suspension of nanocellulose, Exilva F01-V grade, was provided by Borregaard ChemCell (Barcelona, Spain) in the form of a gel with a solid content of 10 wt%. The compatibilizers TGIC, reference 379506, with molecular weight (M_W) of 297.26 g/mol, and DCP, reference 329541, with M_W of 270.37 g/mol and 98% purity, were purchased from Sigma-Aldrich S.A. (Madrid, Spain). Commercial high-barrier multilayer film, based on polyamide 6 (PA6) and poly(ethylene-*co*-vinyl alcohol) (EVOH), with a total thickness of 140 µm, was supplied by WK THOMAS (Barcelona, Spain).

Minced pork meat (Consum S. Coop. V., Valencia, Spain) and sunflower oil (maximum acidity 0.2°, average values per 100 mL: energy value: 3397 kJ/826 kcal; fats: 92 g (7.3 g saturated, 50 g monounsaturated, and 34.3 g polyunsaturated); 0 g carbohydrates; 0 g proteins; 0 g salt) (Mercadona, Valencia) were purchased from local supermarkets. Trichloroacetic acid (TCA) (99.5% purity), 2-thio-barbituric acid (TBA) (>98% purity) reagent, glacial acetic acid (99.7% purity), sodium thiosulfate ($Na_2S_2O_3$) (99.5% purity), and isooctane (99.5% purity) were supplied by Panreac Química, SA (Castellar del Vallés, Spain). Iodine (99.5% purity) and 1-decanol (≥99.0% purity) were obtained by Acros Organics® (Geel, Belgium) and Sigma-Aldrich (Madrid, Spain), respectively. Potassium iodide (≥99.0% purity) were supplied by Sigma-Aldrich (Madrid, Spain). Microbiological media: buffered peptone water, Violet Bile Red agar (VRB), and Plate Count agar (PCA) were provided by Scharlab S.A. (Barcelona, Spain), whereas Man, Rogosa, and Sharpe agar (MRS) were supplied by Lankem-Labbox (Barcelona, Spain).

2.2. Preparation of PHBV-Based Green Composites

The green composite films were prepared using PHBV and the MFC powder, which was obtained by atomization in a laboratory with spray-drying equipment (Minispray Dryer Büchi B-290 Advanced, BUCHI Ibérica S.L.U., Barcelona, Spain) using compressed air in previously optimized conditions [18]. The PHBV pellets were first dried in a vacuum oven (vacuum TEM-TJP Selecta, S.A., Barcelona, Spain) at 60 °C and 0.2 bar for four hours. Afterward, the dried pellets and MFC powder were placed in a dissector containing phosphorus pentoxide (P_2O_5, 99% purity, reference 214701, Sigma-Aldrich, S.A.) at 25 °C for one week to remove the remaining water. Then the PHBV pellets were melt-mixed with MFC at 5 wt% in an internal mixer (HAAKE™ PolyLab™ QC, Thermo Fisher Scientific, Herzogenaurach, Germany) at 170 °C and 50 rpm. The mixing time was set at 5 min after analyzing the stability of neat PHBV during melt-mixing. The processed amount of each composition was approximately 50 g. PHBV samples with MFC and compatibilizers (TGIC and DCP) were used to produce the green composites (PHBV/MFC + TGIC + DCP). Samples without MFC with compatibilizers (PHBV + TGIC + DCP), with MFC without compatibilizers (PHBV/MFC), and neat PHVB were also prepared as control materials. Table 1 summarized the set of formulations prepared.

Table 1. Summary of film compositions according to the weight content (wt%) of poly(3-hydroxybutyrate-*co*-3-hydroxyvalerate) (PHBV) and microfibrillated cellulose (MFC) in which triglycidyl isocyanurate (TGIC) and dicumyl peroxide (DCP) were added as parts per hundred resins (phr) of green composite.

Sample	PHBV (wt%)	MFC (wt%)	TGIC (phr)	DCP (phr)
PHBV	100	0	0	0
PHBV/MFC	95	5	0	0
PHBV + TGIC + DCP	100	0	1	0.25
PHBV/MFC + TGIC + DCP	95	5	1	0.25

The samples obtained from the internal mixer were cold-milled to have the product in powder, an easier to handle format for subsequent pressing. To this end, each dough was ground at two pulses of 30 s in a milling machine (Model M20, IKA, Staufen, Germany). The powder-like material from the milling process was stored in a desiccator at 25 °C and

0% relative humidity (RH) with silica gel for one week. It was necessary to use 2.5 g of PHBV powder to obtain a film sizing 10 cm × 10 cm with a thickness of approximately 130 μm. Compression molding was carried out in a hydraulic press (Model LP20, Labtech Engineering, Bangpoo, Thailand) by placing the samples between two sheets and inside a square frame of Teflon. The samples were preheated for 3 min at 200 °C without pressure, then for 4 min pressed at 200 °C and 100 bar, and finally cooled for 3 min until 80 °C. The resultant films were cut into round disks of approximately 7 cm in diameter, separated by absorbent laboratory paper, and stored in a desiccator with P_2O_5 (0% RH) at 25 °C. All the film samples remained in the desiccator for a minimum of 15 days to eliminate the remaining humidity and reduce the effect of physical aging of PHBV before using them as packaging lids.

2.3. *Evaluation of Packaged Foods*

All devices and work surfaces were first disinfected with 96% ethanol (Panreac S.A, Barcelona, Spain). All the packaging films were sterilized by exposure to ultraviolet (UV) light for one hour in a laminar flow cabinet (Bio II Advance, Telstar, Terrasa, Spain). The films were used as lids in aluminum cups sizing 6.6 cm in diameter and 2.4 cm in height (WPAL-050-100, Quima S.L, Valencia, Spain) in order to evaluate their effect on the food shelf life. The films were sealed manually using an extra instant adhesive glue (Pattex Crocodile, Ferreteria Moreno, Valencia, Spain). All food samples were packaged and handled inside the laminar flow cabinet.

2.3.1. Minced Pork Meat

Minced pork meat samples (10 g) were immediately transported after purchase to the laboratory facilities in a portable fridge at 5 °C and placed in the aluminum cups that were closed with lids made of the various films. All samples were stored for up to 7 days at 5 °C and 48% RH, and the packaged pork meat samples were analyzed in terms of weight loss, pH, lipid oxidation, color changes, and microbial counts.

After 3 and 7 days of storage, the food samples were weighed to determine, in triplicate, their weight loss using an analytical balance (ME36S, ±0.0001 g accuracy, Sartorius, Goettingen, Germany). The pH values were also determined using a digital pH meter by directly inserting the electrode probe (Mettler-Toledo GmbH, Schwerzenbach, Switzerland) into the pork meat sample. Five measurements were taken for each packaging treatment and time. The oxidative stability of pork meat was monitored using the thiobarbituric acid reactive substances (TBARS) at the beginning and end of the storage, following the methodology described by Siu & Draper [25]. To this end, each packaged pork meat sample was taken from the trays and placed in bags (Stomacher 440 Classic Strainer Bags, Worthing, UK) with 50 mL of distilled water and homogenized for 2 min using a homogenizer (Masticator Paddle blender, IUL Instruments, Barcelona, Spain). Then, 50 mL of 10% *v*/*v* TCA was added, and the homogenate was filtered with a vacuum pump using Whatman #1 filter paper (Whatman N°1, Whatman International Ltd., Kent, UK). After that, 8 mL of the clear filtrate was added to 2 mL of 0.06 M TBA reagent and incubated for 90 min at 80 °C. The absorbance was read at 532 nm, and the results were obtained in triplicate and expressed as mg of malonaldehyde (MDA) per kg of meat sample.

The CIE L*a*b* color coordinates of the packaged pork meat were obtained using the illuminant D65/10° observer from the reflection spectra of the sample surface using the MINOLTA colorimeter spectrum (model CM-5, Minolta Co., Tokyo, Japan). Measurements were performed by placing the colorimeter on six random points of the lid films that were used to preserve the meat samples in the trays. The lightness (L^*) and the color coordinates a^* and b^* in the CIE L*a*b* color space were determined. The lightness value, L^*, defines black at 0 and white at 100. The a^* axis represents green–red color (negative values toward green and positive toward red), whereas the b^* axis represents blue–yellow color (negative values toward blue and positive toward yellow). The chroma (C_{ab}^*) and hue (h_{ab}^*) were determined by using Equations (1) and (2), respectively. The total color difference at the different storage times with respect to the initial day was calculated using Equation (3). Three packaged pork meat samples with each packaging film were analyzed in duplicate after 3 and 7 days of storage.

$$C^*_{ab} = \sqrt{(a^*)^2 + (b^*)^2} \quad (1)$$

$$h^*_{ab} = arctg\frac{b^*}{a^*} \quad (2)$$

$$\Delta E = \sqrt{(\Delta L^*)^2 + (\Delta a^*)^2 + (\Delta b^*)^2} \quad (3)$$

Microbiological analyses of the packaged pork meat samples were carried out after the different storage times (0, 3, and 7 days). The packaged meat samples were aseptically taken from the container in the laminar flow cabinet and subsequently placed in sterile bags with 90 mL of peptone water (Scharlab S.A., Barcelona, Spain). The stomacher bags were homogenized for 3 min in the homogenizer, and the resultant homogenate was then 10-fold serially diluted using Tryptic Soy Broth (TSB) (Scharlab S.A., Barcelona, Spain) and used to enumerate total viable counts, total coliforms, and lactic acid bacteria. The total aerobic counts (TAC) and total coliforms were determined on PCA and VRB plates, respectively, after incubation at 37 °C for 48 h. The lactic acid bacteria (LAB) were enumerated using MRS plates after incubation at 30 °C for 72 h. After incubation, all the colonies were counted, and the results were expressed as colony-forming units per gram (CFU/g). All the microbial tests were performed in triplicate.

2.3.2. Sunflower Oil

Sunflower oil (50 mL) was placed in the aluminum trays and closed with lids made of the different films. The trays were stored for up to 21 days at 30 °C and 53% RH and exposed to a fluorescent light (model T8-G13, Leroy Merlin, Valencia, Spain) at an intensity of 1000–1500 lux, measured using a digital Luxometer (model RS Pro ILM1332A, RS Components, Madrid, Spain). The oxidative stability of the sunflower oil was measured in terms of the peroxide value (PV) and diene and triene content after 0, 4, 11, and 21 days of storage.

The titrimetric method was used to determine the PV of the samples using an automatic titrator (Titrando, Metrohm Ion Analysis, Switzerland) [26]. To this end, 1 g of sunflower oil was dissolved in 10 mL of solvent (glacial acetic acid:1-decanol volume ratio of 3:2, containing 10–15 mg/L of iodine) and mixed with 200 µL of an oversaturated potassium iodide solution (130 g potassium iodide dissolved in 100 mL of distilled water). The mixture was thoroughly sacked and kept in the dark for 1 min. Then, 50 mL of distilled water was added, and the solution was titrated with 0.01 M or 0.001 M $Na_2S_2O_3$, depending on the PV predicted. A blank control sample, without sunflower oil, was also prepared following the same procedure. All the analyses were performed in triplicate.

The conjugated dienes and trienes were also determined utilizing a spectrophotometric method [27]. The absorbance of the appropriately diluted samples in isooctane was measured in a spectrophotometer (Evolution 201 VisibleUV, ThermoScientific, Herzogenaurach, Germany) at wavelengths of 232 nm and 268 nm to determine the conjugated dienes and trienes, respectively.

2.4. Migration Tests

The overall migration levels of the PHBV-based films were evaluated in two food simulants, 10% v/v ethanol (simulant A) and olive oil (simulant D2), by immersing 0.5 dm^2 of film sample in 50 mL of simulant at the normalized conditions of 40 °C for ten days. Total migration in 10% ethanol was evaluated by monitoring sample weight loss, whereas in the case of the vegetable oil it was obtained by extraction of the oil absorbed in the film sample followed by quantification by Gas Chromatography with Flame-Ionization Detection (GC-FID, GC 2400, PerkinElmer España SL, Madrid, Spain), according to ISO

1186 [28] and [29] standards, respectively. Three replicates per sample in each simulant were tested and the migration results were averaged.

2.5. Statistical Analysis

The statistical analyses of the data were performed through an analysis of variance (ANOVA) using Statgraphics Centurion XVII-X64 software (Manugistics Corp., Rockville, MD, USA). One-way and multifactor ANOVA were used to analyze the influence of the kind of packaging and storage time on the properties of packaged food. Fisher's least significant difference (LSD) procedure was used at the 95% confidence level.

3. Results and Discussion

3.1. Physicochemical Properties of Packaged Minced Pork Meat

Figure 1 shows the visual aspect of the packaged minced pork meat according to the type of lid film based on PHBV. All the samples showed a similar appearance, presenting low transparency due to the inherently high crystallinity of the microbial copolyester.

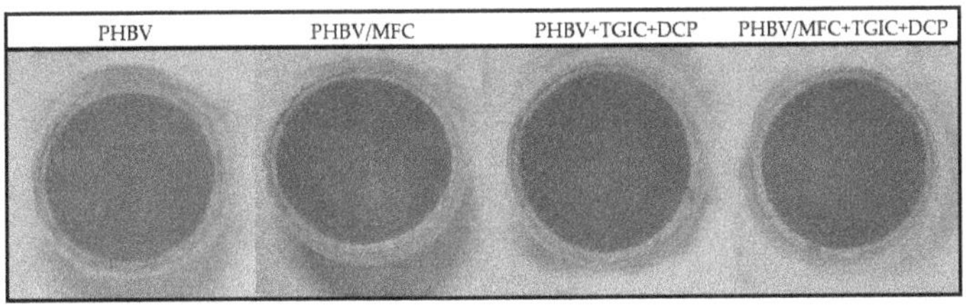

Figure 1. Images of the minced pork meat samples packaged in trays with the green composite lid films of poly(3-hydroxybutyrate-*co*-3-hydroxyvalerate) (PHBV) with microfibrillated cellulose (MFC) and compatibilized with triglycidyl isocyanurate (TGIC) and dicumyl peroxide (DCP).

In order to quantify the physicochemical properties of the packaged minced pork meat during storage, Figure 2 shows the changes in pH, weight loss, and TBARS according to the type of lid film. The figure note also includes the corresponding values of water vapor and oxygen permeability, since the physicochemical properties of the packaged foods are related to the barrier performance of the films. It can be seen in Figure 2a that the initial pH of the pork meat samples was 5.43 ± 0.11, which is in the range of the values previously reported for fresh pork meat samples [30]. As expected, the pH values of meat increased during storage due to the increased content of nitrogenous bases resulting from proteolysis caused by the activity of microorganisms [31]. This increase, slightly lower in the meat samples packaged in trays with the commercial high-barrier lid film, the so-called multilayer sample, was not affected significantly ($p > 0.05$) by any of the evaluated PHBV samples. The lower pH values observed in the pork meat samples packaged with the commercial multilayer film can be explained by its lower oxygen permeance (2.15×10^{-16} $m^3 \cdot m^{-2} \cdot Pa^{-1} \cdot s^{-1}$ [21]) compared to the permeance values of the developed green composite films, which are in the range from 1.5×10^{-15} to 7.5×10^{-16} $m^3 \cdot m^{-2} \cdot Pa^{-1} \cdot s^{-1}$ [18], and this then could potentially delay the activity of spoilage microorganisms. The latter observation is also in accordance with the lower oxidation level and reduced total microbial counts of the pork meat packaged in these samples, as described below. Similar results were obtained in pork meat slices packaged in polylactide (PLA) films, showing pH values in the 5–6 range during the first week of cold storage [21].

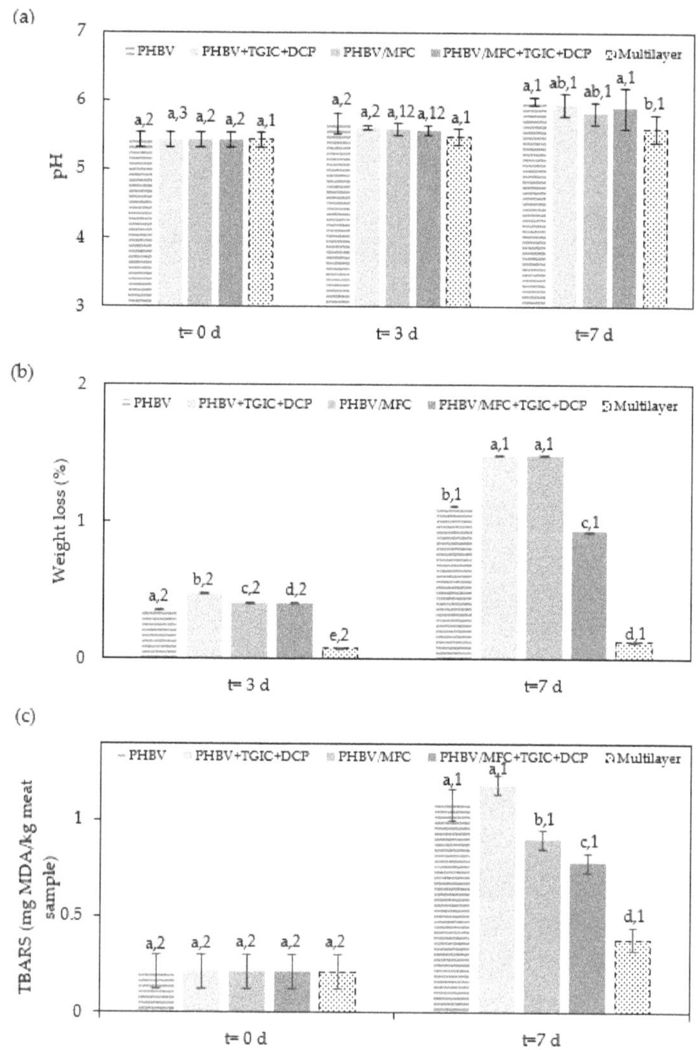

Figure 2. (**a**) Changes in pH, (**b**) weight loss, and (**c**) thiobarbituric acid reactive substances (TBARS) of the cold-stored minced pork meat samples packaged in trays with the green composite lid films of poly(3-hydroxybutyrate-*co*-3-hydroxyvalerate) (PHBV) with microfibrillated cellulose (MFC) and compatibilized with triglycidyl isocyanurate (TGIC) and dicumyl peroxide (DCP) and the multilayer commercial film. Oxygen permeability (OP, $m^3 \cdot m \cdot m^{-2} \cdot Pa^{-1} \cdot s^{-1}$) and water vapor permeability (WVP, $kg \cdot m \cdot m^{-2} \cdot Pa^{-1} \cdot s^{-1}$) data of each film reported in previous studies * [18] and ** [21]: PHBV (OP* = 2.09×10^{-19} and WVP* = 1.84×10^{-15}), PHBV + TGIC + DCP (OP* = 2.70×10^{-19} and WVP* = 5.69×10^{-15}), PHBV/MFC (OP* = 1.24×10^{-19} and WVP* = 3.29×10^{-15}), PHBV/MFC + TGIC + DCP (OP* = 0.98×10^{-19} and WVP* = 3.29×10^{-15}), and Multilayer (OP** = 2.74×10^{-20} and WVP** = 7.95×10^{-16}, assuming monolayer). Various superscripts (letters) indicate significant differences among samples for the same storage time (a–e) and superscript numbers due to storage time for the same sample (1–3) ($p < 0.05$).

The weight loss of the pork meat during storage is also reported in Figure 2b. The commercial multilayer lid film resulted in a significantly ($p < 0.05$) lower weight loss during storage compared to the pork meat samples packaged in trays with the green composite lid films. This weight variation can be ascribed to water evaporation of exudate that naturally occurs through the lid film, which is formed due to the leakage of intramuscular fluids from the cut surface [32]. Therefore, as seen in the figure note, this phenomenon was related to the different water vapor barrier properties of the evaluated packaging materials. Furthermore, one can observe that the weight loss was slightly lower, but still significant ($p < 0.05$), in the tray using the unfilled PHBV lid film due to the presence of celluloses which is known to increase water absorption and hence permeability of the biopolyester film [18]. Thus, after one week of storage, the meat sample packaged in the trays with compatibilized green composite resulted in the lowest weight loss (0.935%) among the PHBV-based lid films. This value was significantly lower ($p < 0.05$) than the weight loss values attained in the samples packaged using the neat PHBV film (1.117%), PHBV + TGIC + DCP film (1.488%), and PHBV/MFC film (1.489%). In any case, the weight losses attained were relatively low since these values were kept below 2%. For instance, pork meat fillets packaged in PLA films yielded values of weight loss of 3.4% after one week of storage, whereas these films were able to successfully preserve the food samples during this period under refrigeration conditions [21]. In another study, Konuk Takma and Kore [33] reported similar results for chicken breast meat packaged in PET films assembled with antimicrobial chitosan and alginate coatings, reaching values of 2–3% after five days of storage under cold storage conditions of 4 °C. In any case, one should also consider the effect of the packaging design and the fact that mass transfer is preferable in films to trays due to the higher area exposed to vapor and gas permeation.

Lipid oxidation in minced pork meat samples was evaluated by monitoring TBARS formation, which measures the amount of MDA produced by secondary products of polyunsaturated fatty acid peroxidation. The initial TBARS of the pork meat samples was 0.21 ± 0.09. As also shown in Figure 2c, the values obtained were consistent with the reported oxygen barrier properties of the polymer materials. Thus, the meat samples packaged in trays with the multilayer lid film showed the lowest oxidation level at the end of storage, that is, 0.38 MDA/kg, which was similar to that found in previous studies carried out with refrigerated pork meat slices [21]. It can also be observed that the TBARS values significantly ($p < 0.05$) decreased in the trays with the lids made of the MFC-containing PHBV films due to the higher oxygen barrier promoted by nanocelluloses. This result was particularly noticeable for the food sample packaged in the tray with the compatibilized green composite, having the lowest permeability among all the PHBV-based films tested herein. These samples yielded a value of 0.78 MDA/kg, which represents a reduction of approximately 30% when compared with the meat samples packaged with the lids of the neat PHBV films (1.08 MDA/kg) and PHBV + TGIC + DCP (1.18 MDA/kg). However, after one week of storage, all the pork meat samples packaged using PHBV films exceeded the threshold of 0.5 mg MDA/kg, which is habitually considered the limit of the detection of off-flavors in pork meat by consumers [34]. This result suggests that the shelf life of minced pork meat packaged in food trays with these lid films should not exceed 4–5 days when stored in refrigeration conditions.

Furthermore, Figure 3 shows the evolution of the chromatic parameters (L^*, C_{ab}^*, h_{ab}^*) and the total color difference (ΔE) of the minced pork meat with respect to the initial values (t = 0). Samples showed differences in chrome and hue at the beginning of storage as affected by the type of lid film used in the food tray. These differences can be mainly explained by the light scattering effect of each type of film on the meat sample according to the reported differences in their optical properties [18,21]. In fact, the average percentage of internal transmittance of the green composite samples ranged from 67% to 72% at 550 nm, whereas the commercial multilayer sample showed an average internal transmittance of 90% at this wavelength. One can further observe that the packaged meat samples L^* and C_{ab}^* values hardly varied during the seven days of storage, whereas the h_{ab}^* values significantly ($p < 0.05$) increased. In this regard, one should consider that lipid oxidation and myoglobin oxidation are known to lead to discoloration in meat, and the two processes are frequently linked, since the oxidation of one of these compounds produces chemical species that promote the oxidation of the other [35]. In addition, the above-reported changes in the pH values can be associated with variations in the reduced form of myoglobin (Mb) and the susceptibility of muscle pigments to oxygenation and oxidation in pork meat. In particular, meat yellowness increases due to an increase in the relative amounts of myoglobin's oxygenated and oxidized forms (MbO_2 and MetMb) at the expense of the reduced form [36]. All these factors explain the reported significant ($p < 0.05$) color changes of the meat samples during the storage time. Furthermore, it is also worth mentioning that the samples packaged in the green composite films developed a redder hue and maintained the original color of the fresh meat. Finally, the total color differences after one week of storage of all the samples with respect to the initial day did not positively exceed the usual tolerance limit for food products ($\Delta E < 5$ [37]), except for the samples packaged in the trays with the unfilled PHVB films, which rose to the 5-units level.

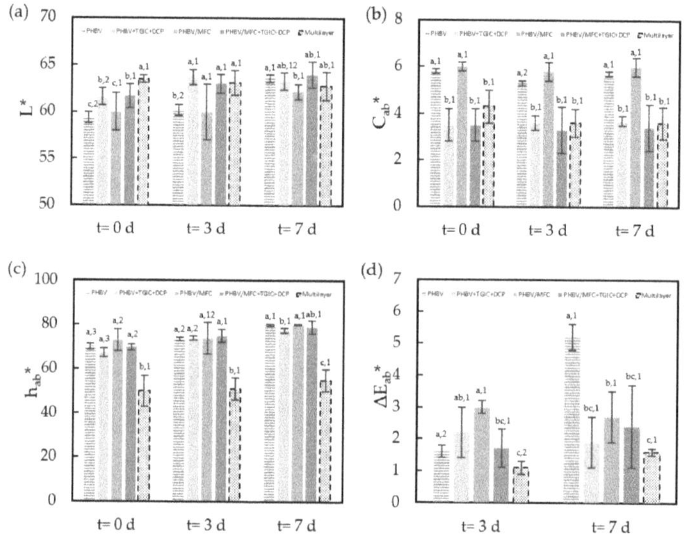

Figure 3. Development of color parameters in terms of (**a**) lightness (L^*), (**b**) chroma ($C_{ab}{}^*$), (**c**) hue ($h_{ab}{}^*$), and (**d**) total color difference ($\Delta E_{ab}{}^*$) of the cold-stored minced pork meat samples packaged in trays with the green composite lid films of poly(3-hydroxybutyrate-*co*-3-hydroxyvalerate) (PHBV) with microfibrillated cellulose (MFC) and compatibilized with triglycidyl isocyanurate (TGIC) and dicumyl peroxide (DCP) and the multilayer commercial film. Color parameters of each film reported in previous studies * [18] and ** [21]: L^* (PHBV (88.41 ± 0.11)*; PHBV + TGIC + DCP (89.39 ± 0.15)*; PHBV/MFC (88.40 ± 0.30)*; PHBV/MFC + TGIC + DCP (88.86 ± 0.09)*; Multilayer (84.60 ± 1.00)**), $C_{ab}{}^*$ (PHBV (10.24 ± 0.15)*; PHBV + TGIC + DCP (8.90 ± 0.40)*; PHBV/MFC (10.25 ± 0.99)*; PHBV/MFC + TGIC + DCP (9.40 ± 0.20)*; Multilayer (2.60 ± 0.40)**) and $h_{ab}{}^*$ (PHBV (95.25 ± 0.11)*; PHBV + TGIC + DCP (98.00 ± 0.40)*; PHBV/MFC (95.10 ± 0.30)*; PHBV/MFC + TGIC + DCP (98.50 ± 0.20)*; Multilayer (132.00 ± 4.00)**). Different superscripts (a–c) indicate significant differences among samples for the same storage time and different superscripts (1–3) indicate differences due to storage time for the same sample ($p < 0.05$).

3.2. Microbial Characteristics of Packaged Minced Pork Meat

Figure 4 shows the TAC, coliforms, and LAB counts of the packaged pork meat samples as a function of the storage time. As expected, the lowest microbial counts were observed in the meat samples packaged in trays with the commercial multilayer lid film due to its higher oxygen barrier. Similarly, the trays with the unfilled PHBV film yielded the meat samples with the highest bacterial growth, confirming the higher performance of the green composites. In many cases, differences among the meat samples packaged in the trays with PHBV-based lid films were only significant ($p < 0.05$) after three days of storage.

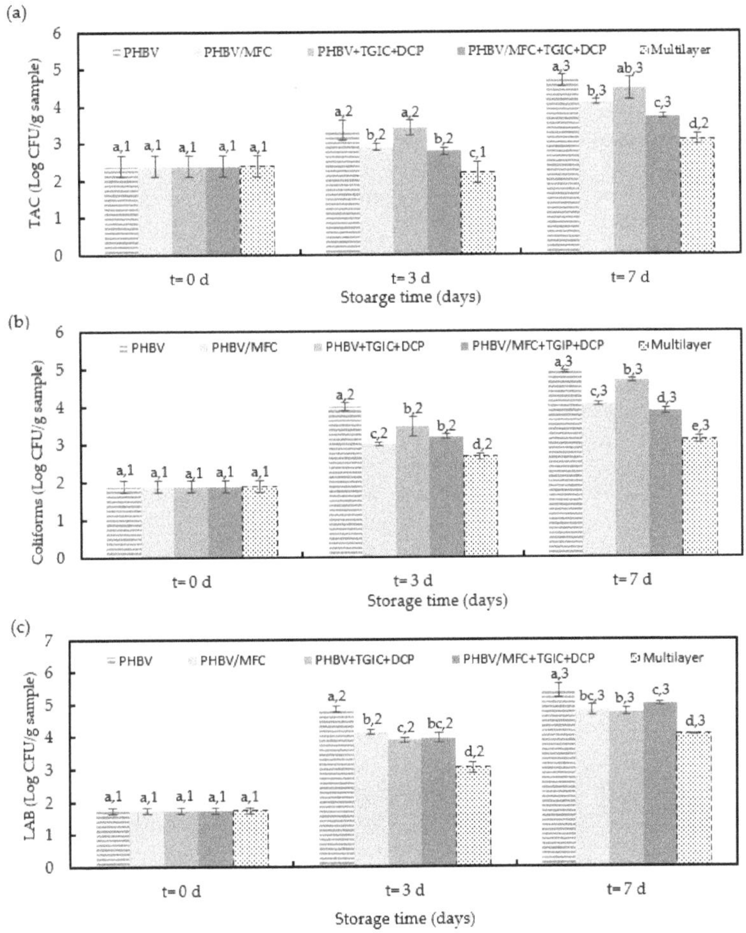

Figure 4. Total aerobic counts (TAC) (**a**), coliforms (**b**), and lactic acid bacteria (LAB) (**c**) of the cold-stored minced pork meat samples packaged in trays with the green composite lid films of poly(3-hydroxybutyrate-*co*-3-hydroxyvalerate) (PHBV) with microfibrillated cellulose (MFC) and compatibilized with triglycidyl isocyanurate (TGIC) and dicumyl peroxide (DCP) and the multilayer commercial film. Different superscripts (a–e) indicate significant differences among samples for the same storage time and different superscripts (1–3) indicate differences due to storage time for the same sample ($p < 0.05$).

In terms of TAC, which is the quantitative standard for identifying the conditions and degree of contamination of meat [38], it can be observed that counts increased during refrigerated storage from approximately 2.4 logs CFU/g to values in the 3.7–4.7 CFU/g range for the minced pork meat packaged in the trays with the green composite lid films. Moreover, the bacterial growth pattern in the minced pork meat samples was similar to all the packaged meat samples, showing low growth during the first days. Differences among the meat samples were not significant ($p > 0.05$). Moreover, among the trays with PHBV-based films, the lowest bacterial counts were attained for the meat samples packaged with compatibilized green composites. Although the meat samples packaged in the trays with the multilayer lid films showed lower values throughout the storage period, none of the packaged minced pork samples exceeded the acceptable TAC level of five logs CFU/g. The latter value corresponds to the maximum acceptable level of TAC in mechanically separated fresh pork meat established by the European Commission (EC) in Regulation Number 2073/2005 [39]. In this regard, unpackaged pork meat fillets tested in the same conditions reached a TAC value above 7 logs CFU/g [21]. Therefore, according to these microbiological criteria, all evaluated packaging materials accomplished the requested value of food safety and quality to preserve pork meat.

Concerning the total coliforms, microbial counts increased from approximately 1.9 logs CFU/g to values ranging from 3.88 to 4.93 log CFU/g in the pork meat samples packaged in the trays with green composite lid films. It can be observed that the use of lid films of PHBV with MFC yielded minced pork samples with slightly lower coliform counts but still significant ($p < 0.05$) compared with the meat samples packaged in trays with PHBV and PHBV + TGIC + DCP films, whereas the multilayer lid yielded to the lowest bacterial counts. These results correlate well with the pH values reported above since total coliform counts are related to the permeance of oxygen gas through the lid film that favors the activity of spoilage microorganisms. These results were similar to those obtained in other studies with refrigerated packaged beef meat in PLA/PVA multilayer films [40].

Finally, microbial counts for LAB, which are the dominant group of microorganisms isolated from meat products [41], increased in the minced meat samples from 1.7 log CFU/g to values in the 4–5.5 log CFU/g range after one week of storage. As also observed for the TAC and coliform counts, the growth pattern was similar in all the pork meat samples, though different counts were attained. In the case of the meat samples packaged with the multilayer film, LAB counts were significantly ($p < 0.05$) lower than the values attained in the samples packaged using PHBV-based films. Similarly, the newly developed MFC-containing PHBV films successfully reduced the LAB counts of the meat samples, though differences were lower than 2 logs and thus showed no significant differences ($p > 0.05$). Similar results were obtained in pork meat slices packaged in PLA films, showing LAB values increased from 1.3 logs CFU/g to values in the 4–5 log CFU/g range after one week of storage [21].

3.3. Oxidative Stability of Packaged Sunflower Oil

The visual aspect of the sunflower oil packaged in the trays with the various PHBV-based lid films can be seen in Figure 5. All the samples showed a similar appearance due to the lack of transparency of the PHBV film.

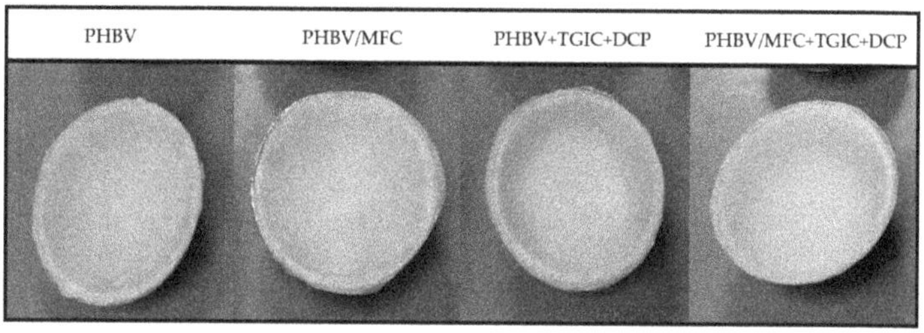

Figure 5. Images of the sunflower oil samples packaged in trays with the green composite lid films of poly(3-hydroxybutyrate-*co*-3-hydroxyvalerate) (PHBV) with microfibrillated cellulose (MFC) and compatibilized with triglycidyl isocyanurate (TGIC) and dicumyl peroxide (DCP).

Figure 6 shows the changes in PV of the sunflower oil samples during storage at 30 °C and 53% RH, with light to promote oil oxidation. The PV is associated with hydroperoxides derived from the primary oxidation of polyunsaturated fatty acids present in the sample. Thus, PV values are related to the initial oxidation stage [42]. It can be observed that the initial PV of sunflower oil was as low as 0.5 meq O_2/kg, which is in the range of the detection limit, and then it showed a continuous increase during storage. Similar to that observed for pork meat, the oil samples packaged in trays with the multilayer lid films reached the lowest PV values due to their high-oxygen-barrier properties. In these food samples, the maximum PV of 10 meq O_2/kg established in the Codex standard for fresh-named vegetable oils [43] was not exceeded throughout the storage period despite the accelerated oxidation conditions. In the case of the oil samples packaged with the PHBV-based lid films, after 21 days of storage, PV values reached levels between 19.5 and 24.8 meq O_2/kg, showing the samples packaged with compatibilized green composite films had the lowest value.

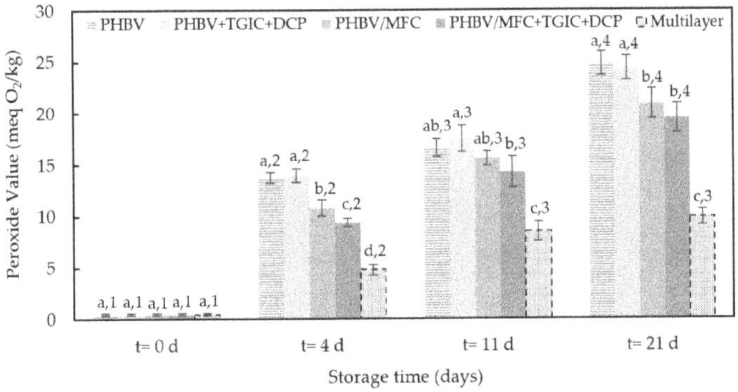

Figure 6. Peroxide values during accelerated storage conditions of the sunflower oil samples packaged in trays with the green composite lid films of poly(3-hydroxybutyrate-*co*-3-hydroxyvalerate) (PHBV) with microfibrillated cellulose (MFC) and compatibilized with triglycidyl isocyanurate (TGIC) and dicumyl peroxide (DCP) and the multilayer commercial film. Different superscripts (a–d) indicate significant differences among samples for the same storage time and different superscripts (1–4) indicate differences due to storage time for the same sample ($p < 0.05$).

Interestingly, and also in correlation to their higher oxygen barrier, the MFC-containing PHBV films induced significantly ($p < 0.05$) lower PV values in the packaged oil than the unfilled PHBV films, particularly in the case of the compatibilized green composite film. Moreover, these PV values are still below the 30–40 meq O_2/kg range, which is generally associated with a rancid taste [44]. Moreover, an open aluminum cup containing 50 mL of sunflower oil was considered to reach values of 27 meq O_2/kg, 74 meq O_2/kg, and 130 meq O_2/kg after 4, 11, and 21 days of storage, respectively. Similar results were obtained in other studies with the same oil and under the same accelerated conditions [26,45].

The changes in the conjugated dienes and trienes content throughout the accelerated oxidation conditions are shown in Figure 7. Conjugated dienes and trienes are formed by the rearrangement of the hydroperoxide double bounds; thus, conjugated dienes represent the primary degradation products of oil that can be used to confirm the PV content, while conjugated trienes are related to the secondary products of the oxidation [46]. All samples showed increased conjugated dienes, with a pattern similar to those previously observed during PV analysis. However, the conjugated trienes remained almost constant during the storage period for the oil samples packaged in the trays with the multilayer and PHBV/MFC film lids, despite being significantly ($p < 0.05$) higher for the latter. This observation indicates that sunflower oil oxidation did not reach the second stage of degradation in these food trays. Therefore, these films prevented the occurrence of primary degradation products. The latter observation is in accordance with results reported in previous studies carried out under the same accelerated conditions [26]. Conversely, a significant ($p < 0.05$) increase was noticed for the samples packaged using unfilled PHBV films during the last days of the assays, confirming the higher oxidation level reached in these food trays. In particular, after three weeks of storage, samples packaged with the compatibilized green composite film presented the lowest conjugated dienes and trienes, showing values of 2.50 and 1.44 g/100 mL, respectively, whereas for the oil packaged with the neat PHBV film these were 3.30 and 1.52 g/100 mL. These differences can be mainly explained by the higher oxygen barrier achieved in the PHBV films by the presence of MFC, particularly when the dispersion was improved by the use of the TGIC and DCP compatibilizers.

(a)

(b)

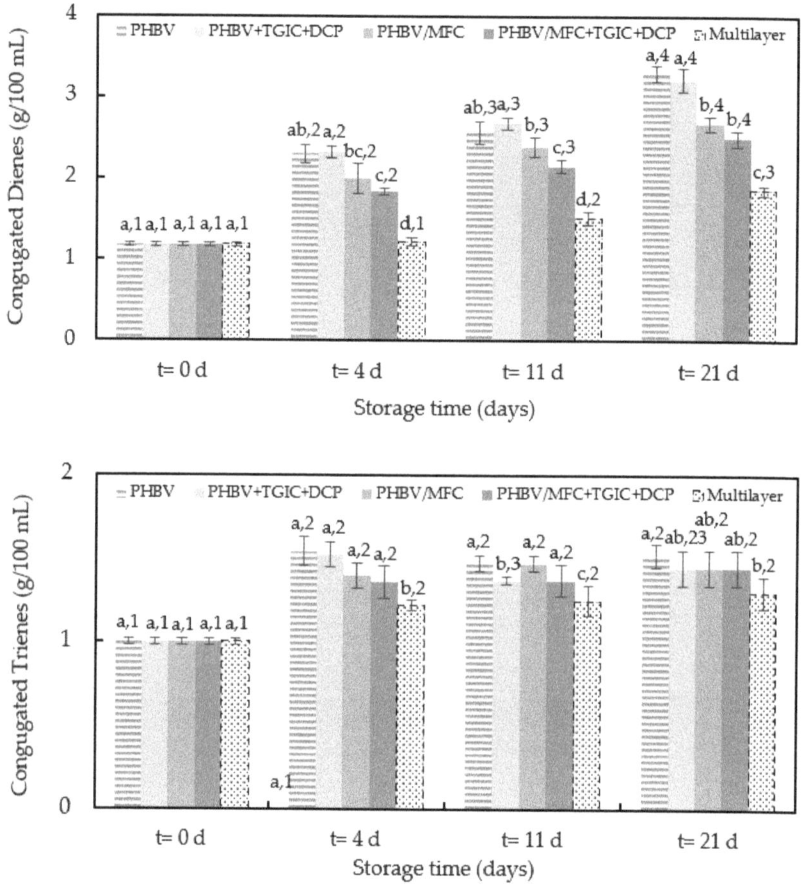

Figure 7. Conjugated dienes (**a**) and trienes (**b**) during accelerated storage conditions of the sunflower oil samples packaged in trays with the green composite lid films of poly(3-hydroxybutyrate-*co*-3-hydroxyvalerate) (PHBV) with microfibrillated cellulose (MFC) and compatibilized with triglycidyl isocyanurate (TGIC) and dicumyl peroxide (DCP) and the multilayer commercial film. Different superscripts (a–c) indicate significant differences among samples for the same storage time and different superscripts (1–4) indicate differences due to storage time for the same sample ($p < 0.05$).

3.4. Overall Migration in Food Simulants

The migration of the packaging constituents into the food is a prominent issue in food contact materials, especially for those containing nano and microfibers. Ethanol 10% v/v and olive oil, simulants A and D2, were chosen to simulate migration into minced pork meat and sunflower oil, respectively. The overall migration levels in the two evaluated foods, polar and non-polar simulants, are shown in Table 2.

Table 2. Overall migration levels of the green composite films of poly(3-hydroxybutyrate-*co*-3-hydroxyvalerate) (PHBV) with microfibrillated cellulose (MFC) and compatibilized with triglycidyl isocyanurate (TGIC) and dicumyl peroxide (DCP) into food simulants.

Film	Ethanol 10% v/v (mg/dm^2)	Vegetable Oil (mg/dm^2)
PHBV	1.7 ± 0.2 [a]	2.2 ± 0.1 [a]
PHBV/MFC	1.9 ± 0.5 [a]	2.6 ± 0.2 [b]
PHBV/MFC + TGIC + DCP	2.8 ± 0.4 [b]	2.7 ± 0.3 [b]

Different superscripts within the same column indicate significant differences among samples ([a,b]) ($p < 0.05$).

All the evaluated PHBV-based films showed overall migration levels of 1–3 mg/dm^2. Differences among the films were slight, though still significant ($p < 0.05$), which can be mainly related to sample heterogeneities but also to the potential release of MFC from the film surface. In any case, these values are well below 10 mg/dm^2, which is the maximum level established for plastic materials and articles intended to come into contact with food for both simulants under the evaluated exposure conditions, complying with the overall migration limit (OML) set out in Commission Regulation (EU) No. 10/2011 on plastic materials and articles intended to come into contact with food and its subsequent amendments [28,29]. Other studies have also evaluated the overall migration level of PHBV reinforced with nanocelluloses in different food simulants. For instance, Li et al. [47] evaluated the migration of PHBV with cellulose nanocrystal-grafted rosin (CNC-R) in both 10% v/v ethanol and isooctane. Authors reported respective values of 161.5 and 127.1 μg/kg for each food simulant, both below the maximum limit of 60 mg/kg, whereas the presence of the nanocrystals reduced the migration due to their nucleation effect and resultant increment of crystallinity.

Similarly, Zhang et al. [48] also reported that the incorporation of cellulose nanocrystals-silver nanohybrids (CNC-Ag) into PHBV reduced the migration levels in 10% v/v ethanol and isooctane by approximately 50%, and all were well below the maximum migration limits (60 mg/kg of simulant). Furthermore, the conditions selected in the overall migration tests cover the contact of the packaging material with the foodstuff for all long-storage periods at room or lower temperatures. It also covers packaging filling at high-temperature conditions and/or packaging subjected to heating, from 70 °C for 2 h to 100 °C for 15 min. Therefore, it can be concluded that all the films tested can be safely applied to food applications to preserve foodstuff in polar and non-polar media. In any case, this will have to be further confirmed by specific migration studies according to the intended use and particular legislation.

4. Conclusions

The negative impact on the environment of commercial multilayer food packaging materials makes it necessary to find novel biodegradable alternatives based on biopolymers derived from renewable resources, such as the microbial copolymer PHBV, following the Circular Bioeconomy principles. In previous studies, novel green composites of PHBV reinforced with MFC and compatibilized with TGIC and DCP by a reactive melt-mixing process were developed and characterized in terms of their physical properties, yielding very promising materials for food preservation applications. In this study, the optimal formulations of the previously prepared green composites were applied as lid films in aluminum food trays to ascertain their capacity to preserve two fatty foods: minced pork meat and sunflower oil. Results showed that the PHBV-based reinforced green composites effectively preserved the physicochemical and microbiological quality of minced pork meat samples during storage for one week at 5 °C. Regarding the prevention of oxidation of sunflower oil, all evaluated green composites also showed satisfactory performance in avoiding the formation of peroxide levels associated with a rancid taste and in preventing the occurrence of primary degradation products in accelerated storage conditions. For both fatty foods, the compatibilized green composite films accomplished higher values of food quality than the unfilled PHBV films, which was ascribed to the higher oxygen barrier attained in the copolyester by incorporating atomized nanocelluloses. Moreover, all PHBV-based films proved to be safe for their packaging application to preserve meat and oil products since their overall migration levels did not exceed in either of the two simulants (10% v/v ethanol and vegetable oil) the maximum level established by legislation. However, specific migration studies should be further conducted to confirm their safe use in food preservation applications according to particular legislation.

Author Contributions: Conceptualization, S.T.-G.; methodology, E.H.-G.; formal analysis, E.H.-G.; investigation, E.H.-G.; data curation, E.H.-G.; writing—original draft preparation, E.H.-G. and M.V.; writing—review and editing, E.H.-G., A.C. and S.T.-G.; supervision, M.V., S.T.-G. and A.C.; project administration, A.C. and S.T.-G.; funding acquisition, A.C. and S.T.-G. All authors have read and agreed to the published version of the manuscript.

Funding: This research was funded by the Spanish Ministry of Science and Innovation (MICI), grant number PID2021-128749OB-C33.

Data Availability Statement: Data is contained within the article and available on request.

Acknowledgments: E.H.-G. and S.T.-G. acknowledge MICI for her predoctoral research grant (BES 2017-082040) and his Ramón y Cajal contract (RYC2019-027784-I). The authors are also grateful to the Packaging Technologies Department of AINIA for the technical support provided to determine the overall migration tests.

Conflicts of Interest: The authors declare no conflict of interest.

References

1. Robertson, G.L. Packaging and Food and Beverage Shelf Life. In *The Stability and Shelf Life of Food*, 2nd ed.; Subramaniam, P., Ed.; Woodhead Publishing: Sawston, UK, 2016; pp. 77–106.
2. Torres-Giner, S.; Gil, L.; Pascual-Ramírez, L.; Garde-Belza, J.A. Packaging: Food waste reduction. In *Encyclopedia of Polymer Applications*; Mishra, M., Ed.; CRC Press: Boca Raton, FL, USA, 2019.
3. Radusin, T.; Torres-Giner, S.; Stupar, A.; Ristic, I.; Miletic, A.; Novakovic, A.; Lagaron, J.M. Preparation, characterization and antimicrobial properties of electrospun polylactide films containing *Allium ursinum* L. extract. *Food Packag. Shelf Life* **2019**, *21*, 100357. [CrossRef]
4. Garcia-Garcia, D.; Quiles-Carrillo, L.; Balart, R.; Torres-Giner, S.; Arrieta, M.P. Innovative solutions and challenges to increase the use of Poly(3-hydroxybutyrate) in food packaging and disposables. *Eur. Polym. J.* **2022**, *178*, 111505. [CrossRef]
5. Hernández-García, E.; Vargas, M.; Chiralt, A. Active Starch-Polyester Bilayer Films with Surface-Incorporated Ferulic Acid. *Membranes* **2022**, *12*, 976. [CrossRef]
6. Hottle, T.A.; Bilec, M.M.; Landis, A.E. Biopolymer production and end of life comparisons using life cycle assessment. *Resour. Conserv. Recycl.* **2017**, *122*, 295–306. [CrossRef]
7. Hernández-García, E.; Vargas, M.; Chiralt, A. Thermoprocessed starch-polyester bilayer films as affected by the addition of gellan or xanthan gum. *Food Hydrocoll.* **2021**, *113*, 106509. [CrossRef]
8. Hernández-García, E.; Vargas, M.; Chiralt, A.; González-Martínez, C. Biodegradation of PLA-PHBV Blend Films as Affected by the Incorporation of Different Phenolic Acids. *Foods* **2022**, *11*, 243. [CrossRef]
9. Hernández-García, E.; Vargas, M.; Chiralt, A. Starch-polyester bilayer films with phenolic acids for pork meat preservation. *Food Chem.* **2022**, *385*, 132650. [CrossRef]
10. Hernández-García, E.; Vargas, M.; Chiralt, A. Effect of active phenolic acids on properties of PLA-PHBV blend films. *Food Packag. Shelf Life* **2022**, *33*, 100824. [CrossRef]
11. Torres-Giner, S.; Figueroa-Lopez, K.J.; Melendez-Rodriguez, B.; Prieto, C.; Pardo-Figuerez, M.; Lagaron, J.M. Emerging Trends in Biopolymers for Food Packaging. In *Sustainable Food Packaging Technology*; Athanassiou, A., Ed.; WILEY-VCH GmbH: Weinheim, Germany, 2021; pp. 1–33. [CrossRef]
12. Hernández-García, E.; Vargas, M.; González-Martínez, C.; Chiralt, A. Biodegradable Antimicrobial Films for Food Packaging: Effect of Antimicrobials on Degradation. *Foods* **2021**, *10*, 1256. [CrossRef]
13. Arcos-Hernandez, M.V.; Laycock, B.; Pratt, S.; Donose, B.C.; Nikolić, M.A.L.; Luckman, P.; Werker, A.; Lant, P.A. Biodegradation in a soil environment of activated sludge derived polyhydroxyalkanoate (PHBV). *Polym. Degrad. Stat.* **2012**, *97*, 2301–2312. [CrossRef]
14. Bucci, D.Z.; Tavares, L.B.B.; Sell, I. Biodegradation and physical evaluation of PHB packaging. *Polym. Test.* **2007**, *26*, 908–915. [CrossRef]

15. Torres-Giner, S.; Montanes, N.; Fombuena, V.; Boronat, T.; Sanchez-Nacher, L. Preparation and characterization of compression-molded green composite sheets made of poly(3-hydroxybutyrate) reinforced with long pita fibers. *Adv. Polym. Technol.* **2018**, *37*, 1305–1315. [CrossRef]
16. Laycock, B.; Arcos-Hernandez, M.V.; Langford, A.; Buchanan, J.; Halley, P.J.; Werker, A.; Lant, S.; Pratt, P.A. Thermal properties and crystallization behavior of fractionated blocky and random polyhydroxyalkanoate copolymers from mixed microbial cultures. *J. Appl. Polym. Sci.* **2014**, *131*, 40836. [CrossRef]
17. Tajeddin, B. Cellulose-Based Polymers for Packaging Applications. In *Lignocellulosic Polymer Composites*; Wiley: Hoboken, NJ, USA, 2014; pp. 477–498. [CrossRef]
18. Freitas, P.A.V.; Barrrasa, H.; Vargas, F.; Rivera, D.; Vargas, M.; Torres-Giner, S. Atomization of Microfibrillated Cellulose and Its Incorporation into Poly(3-hydroxybutyrate-*co*-3-hydroxyvalerate) by Reactive Extrusion. *Appl. Sci.* **2022**, *12*, 2111. [CrossRef]
19. Cormier, A.; Vautour, G.; Allard, J. Canada-Wide Survey of the Nutritional Composition of Six Retail Pork Cuts. *J. Food Compost. Anal.* **1996**, *9*, 255–268. [CrossRef]
20. Mazzola, N.; Sarantópoulos, C.I.G.L. Packaging Design Alternatives for Meat Products. In *Food Processing*; Marc, R.A., Díaz, A.V., Izquierdo, G.D.P., Eds.; IntechOpen: London, UK, 2019. [CrossRef]
21. Hernández-García, E.; Vargas, M.; Torres-Giner, S. Quality and Shelf-Life Stability of Pork Meat Fillets Packaged in Multilayer Polylactide Films. *Foods* **2022**, *11*, 426. [CrossRef]
22. Khan, S.; Choudhary, S.; Pandey, A.; Khan, M.K.; Thomas, G. Sunflower oil: Efficient oil source for human consumption. *Emergent Life Sci. Res.* **2015**, *1*, 1–3.
23. Kucuk, M.; Caner, C. Effect of packaging materials and storage conditions on sunflower oil quality. *J. Food Lipids* **2005**, *12*, 222–231. [CrossRef]
24. Cenci-goga, B.T.; Lulietto, M.F.; Sechi, P.; Borgogni, E.; Karama, M.; Grispoldi, L. New Trends in Meat Packaging. *Microbiol. Res.* **2020**, *11*, 56–67. [CrossRef]
25. Siu, G.; Draper, H.H. A survey of malonaldehyde content of retail meats and fish. *J. Food Sci.* **1978**, *43*, 1147–1149. [CrossRef]
26. Talon, E.; Vargas, M.; Chiralt, A.; Gonzalez-Martínez, C. Antioxidant starchbased films with encapsulated eugenol. *Appl. Sunflower Oil Preserv. Lwt.* **2019**, *113*, 108290. [CrossRef]
27. European Commission. Regulation EEC/2568/91 on the characteristics of olive and olive pomace oils and their analytical methods. *Off. J. Eur. Communities L* **1991**, *248*, 1–83.
28. *UNE-EN-ISO 1186-2*; Asociación Española de Normalización. Materiales y Artículos en Contacto con Productos Alimenticios. Plásticos. Parte 2: Métodos de Ensayo para la Migración Global en Aceite de Oliva por Inmersión Total. Asociación Española de Normalización: Madrid, Spain, 2002.
29. *UNE-EN-ISO 1186-3*; Asociación Española de Normalización. Materiales y Artículos en Contacto con Productos Alimenticios. Plásticos. Parte 3: Métodos de Ensayo para la Migración Global en Simuladores de Alimentos Acuosos por Inmersión Total. Asociación Española de Normalización: Madrid, Spain, 2002.
30. Wang, G.; Liu, Y.; Yong, H.; Zong, S.; Jin, C.; Liu, J. Effect of ferulic acid-grafted-chitosan coating on the quality of pork during refrigerated storage. *Foods* **2021**, *10*, 1374. [CrossRef] [PubMed]
31. Daniloski, D.; Petkoska, A.T.; Galić, K.; Ščetar, M.; Kurek, M.; Vaskoska, R.; Nedelkoska, D.N. The effect of barrier properties of polymeric films on the shelf-life of vacuum packaged fresh pork meat. *Meat Sci.* **2019**, *158*, 107880. [CrossRef] [PubMed]
32. Stella, S.; Garavaglia, D.; Francini, S.; Viganò, V.; Bernardi, C.; Tirloni, E.A. Evaluation of the weight loss of raw beef cuts vacuum packaged with two different techniques. *Ital. J. Food Saf.* **2019**, *8*, 8111. [PubMed]
33. Konuk Takma, D.; Korel, F. Active packaging films as a carrier of black cumin essential oil: Development and effect on quality and shelf-life of chicken breast meat. *Food Packag. Shelf Life* **2019**, *19*, 210–217. [CrossRef]
34. Sheard, P.R.; Enser, M.; Wood, J.D.; Nute, G.R.; Gill, B.P.; Richardson, R.I. Shelf life and quality of pork and pork products with raised n-3 PUFA. *Meat Sci.* **2000**, *55*, 213–221. [CrossRef]
35. Faustman, C.; Sun, Q.; Mancini, R.; Suman, S.P. Myoglobin and lipid oxidation interactions: Mechanistic bases and control. *Meat Sci.* **2010**, *86*, 86–94. [CrossRef]
36. Karamucki, T.; Jakubowska, M.; Rybarczyk, A.; Gardzielewska, J. The influence of myoglobin on the colour of minced pork loin. *Meat Sci.* **2013**, *94*, 234–238. [CrossRef]
37. Hutchings, J.B. Instrumental specification. In *Food Colour and Appearance*; Springer: New York, NY, USA, 1999; pp. 199–237.
38. Huang, L.; Zhao, J.; Chen, Q.; Zhang, Y. Rapid detection of total viable count (TVC) in pork meat by hyperspectral imaging. *Food Res. Int.* **2013**, *54*, 821–828. [CrossRef]
39. European Commission. Regulation (EC) No 2073/2005 of 15 November 2005 on microbiological criteria for foodstuffs. *Off. J. Eur. Union.* **2005**, *50*, 1–26. Available online: http://data.europa.eu/eli/reg/2005/2073/oj (accessed on 1 September 2022).
40. Andrade, L.; González-Martínez, C.; Chiralt, A. Antimicrobial PLA-PVA multilayer films containing phenolic compounds. *Food Chem* **2021**, *375*, 131861. [CrossRef]
41. Qin, Y.-Y.; Yang, J.-Y.; Lu, H.-B.; Wang, S.-S.; Yang, J.; Yang, X.-C.; Cao, J.-X. Effect of chitosan film incorporated with tea polyphenol on quality and shelf life of pork meat patties. *Int. J. Biol. Macromol.* **2013**, *61*, 312–316. [CrossRef]
42. Sadeghi, E.; Karami, F.; Etminan, A. The effect of Ferulago angulata (Schlecht) Boiss essential oil on stabilization of sunflower oil during accelerated storage. *J. Food Process. Preserv.* **2017**, *41*, e12745. [CrossRef]
43. Codex-Alimentarius. Codex-Standards for fats and oils from vegetable sources. *Codex-Stan* **1999**, *210*, 210–1999.

44. Godswill, A.C.; Amagwula, I.O.; Victory, I.S.; Gonzaga, A.I. Effects of repeated deep frying on refractive index and peroxide value of selected vegetable oils. *Int. J. Adv. Acad.* **2018**, *4*, 106–119. [CrossRef]
45. Freitas, P.A.V.; González-Martínez, C.; Chiralt, A. Antioxidant starch composite films containing rice straw extract and cellulose fibres. *Food Chem.* **2023**, *400*, 134073. [CrossRef] [PubMed]
46. Mohdaly, A.A.A.; Sarhan, M.A.; Mahmoud, A.; Ramadan, M.F.; Smetanska, I. Antioxidant efficacy of potato peels and sugar beet pulp extracts in vegetable oils protection. *Food Chem.* **2010**, *123*, 1019–1026. [CrossRef]
47. Li, F.; Abdalkarim, S.Y.H.; Yu, H.Y.; Zhu, J.; Zhou, Y.; Guan, Y. Bifunctional reinforcement of green biopolymer packaging nanocomposites with natural cellulose nanocrystral-rosin hybrids. *Appl. Bio. Mater* **2020**, *3*, 1944–1954. [CrossRef]
48. Zhang, H.; Yu, H.Y.; Wang, C.; Yao, J. Effect of silver contents in cellulose nanocrystal/silver nanohybrids on PHBV crystallization and property improvements. *Carbohydr. Polym.* **2022**, *173*, 7–16. [CrossRef] [PubMed]

foods

Article

Quality and Shelf-Life Stability of Pork Meat Fillets Packaged in Multilayer Polylactide Films

Eva Hernández-García , María Vargas and Sergio Torres-Giner *

Research Institute of Food Engineering for Development (IIAD), Universitat Politècnica de Valencia (UPV), 46022 Valencia, Spain; evhergal@upvnet.upv.es (E.H.-G.); mavarco@tal.upv.es (M.V.)
* Correspondence: storresginer@upv.es

Abstract: In the present study, the effectiveness of a multilayer film of polylactide (PLA), fully bio-based and compostable, was ascertained to develop a novel sustainable packaging solution for the preservation of fresh pork meat. To this end, the multilayer PLA films were first characterized in terms of their thermal characteristics, structure, mechanical performance, permeance to water and aroma vapors and oxygen, and optical properties and, for the first time, compared with two commercial high-barrier multilayer packaging films. Thereafter, the multilayers were thermosealed to package fillets of fresh pork meat and the physicochemical changes, lipid oxidation levels, and microbiological counts were monitored in the food samples during storage under refrigeration conditions. Results showed that the meat fillets packaged in PLA developed a redder color and showed certain indications of dehydration and oxidation, being more noticeably after 11 days of storage, due to the higher water vapor and oxygen permeance values of the biopolymer multilayer. However, the pH changes and bacterial growth in the cold-stored fresh pork meat samples were minimal and very similar in the three tested multilayer films, successfully accomplishing the requirements of the food quality and safety standards at the end of storage.

Keywords: PLA; multilayer films; sustainable packaging; food quality and shelf life; meat preservation

check for updates

Citation: Hernández-García, E.; Vargas, M.; Torres-Giner, S. Quality and Shelf-Life Stability of Pork Meat Fillets Packaged in Multilayer Polylactide Films. *Foods* **2022**, *11*, 426. https://doi.org/10.3390/foods11030426

Academic Editor: Barbara Speranza

Received: 24 December 2021
Accepted: 26 January 2022
Published: 1 February 2022

1. Introduction

Food preservation aims to extend shelf life and provide safer products to consumers using different materials and technologies [1]. Advances in packaging materials have played an essential role in food preservation [2]. In the last decade, plastics, that is, polymers and additives, have extensively contributed to food preservation due to their balanced properties (e.g., transparency, flexibility, low cost, ease of processing, low weight, etc.) and high versatility based on the wide variety of formulations for specific product requirements [3]. In addition, food packaging can work beyond conventional protective properties, providing other functions to the food product, such as active and bioactive properties, convenience, and communication [4].

Barrier developments in the food packaging area have greatly helped to reduce food waste since the quality of most food products deteriorates due to mass transfer phenomena, such as moisture adsorption/desorption, the invasion of oxygen, flavor loss as well as sorption of undesirable odors, or the migration of the components from the container to the food. However, barrier packaging developments are not based on a single polymer since, in most cases, it does not fully meet the permeability requirements necessary for all vapors and gases [5]. Thus, to achieve the high-barrier performance required for correct food preservation, in addition to minimizing material costs and achieving lidding film adhesion, the food packaging industry primarily employs structures composed of multiple polymer layers. The so-called multilayer structures usually consist of up to 12 or even more layers of different polymers and/or coatings, in either symmetrical or asymmetrical assemblies, specially designed to be impervious to the penetration and migration of gases and moisture [6]. Nevertheless, multilayer packaging materials have created severe problems related to the disposal of plastics after single short use due to the polymers used are non-biodegradable and their layers are extremely difficult to separate and recycle [7]. In fact, in developed countries, an important part of municipal solid waste consists of polymer materials derived from multilayer structures [8]. Therefore, the development of sustainable solutions and management of post-consumer plastics represents a fundamental challenge in the food packaging area.

Innovations in sustainable food packaging appear as a way to minimize the environmental problems of petrochemical polymers. In this scenario, biopolymers offer the advantages of being produced from renewable resources and/or reintegrated into the carbon cycle through biodegradation by microorganisms and enzymes in a natural environment or compost [9,10]. Among biopolymers, polylactide (PLA) is a linear aliphatic polyester obtained from carbohydrate-rich plants, such as corn or wheat. Starch is separated and converted to dextrose (D-glucose) through enzymatic hydrolysis and fermented to lactic acid [11]. PLA is, thereafter, industrially synthesized by ring-opening polymerization (ROP) of lactide, the cyclic diester formed from two molecules of lactic acid with loss of two molecules of water, to achieve high molecular weights (M_{WS}) without solvents [12]. The resultant PLA packaging articles can be biodisintegrated with the right temperature and humidity conditions in industrial composting facilities, offering a sustainable alternative to non-biodegradable petroleum derived polymers [13]. PLA exhibits similar mechanical strength and higher transparency than polyethylene terephthalate (PET) or polystyrene (PS) [14]. Thus, PLA is ideal for rigid and transparent film applications and it can also be used in food trays, bottles, candy wraps, and cups [15]. Currently, various types of foods are packaged in PLA-based materials, with a high range of physical properties, water activity, and pH, such as fresh vegetables and salads [16]. However, the use of PLA films has been scarcely explored in meat preservation.

Pork and beef meat products play an important role in worldwide nutrition with large increases in global consumption due to key micronutrients and protein content. However, fresh red meat is highly perishable due to its biological composition and its shelf life is dependent on pre-slaughter, processing, and post-processing conditions [17]. Meat deterioration could lead to the decrease in the nutritional value and changes in appearance and production of off-flavors and odors. Microbial spoilage, water exudates, color loss due to myoglobin oxidation, and lipid oxidation represent the first indicators of meat deterioration [18]. With the increased demand for high-quality, safe, and extended shelf life of fresh products, multilayer packaging is commonly used in the meat industry. For instance, fresh pork can be only preserved for six days upon hygienic and temperature control, whereas its shelf life can be extended for up to 3 weeks when stored in vacuum packaging (VP) or modified atmosphere packaging (MAP, 0.4% carbon monoxide, ~60% carbon dioxide, and ~40% nitrogen) [19]. Although the presence of oxygen can be beneficial during the retail display to provide a bright red color that consumers associate with freshness, completely anoxic environments optimize keeping quality.

The present study aims to evaluate the physical properties, in terms of thermal, mechanical, barrier, and optical properties of multilayer PLA films and their feasibility for preserving fresh pork meat. To this end, the quality and shelf life of the meat packaged in the PLA films were compared with that preserved in two commercial multilayer films with a high barrier that are habitually used for meat packaging.

2. Materials and Methods

2.1. Materials

A bioriented PLA (BOPLA) film of 20 μm and labeled as compostable according to the "OK bio-based" (S206) and DIN EN 13432 (7H0052) standards, respectively, was provided as NATIVIA® NTSS 20 by Taghleef Industries S.L. (Jaén, Spain). It is a multilayer film composed of three layers of PLA that is heat-sealable on both sides, where the external layers have been modified to confer a minimum sealing temperature (MST) of 85 °C. A high-barrier multilayer film based on polyamide 6 (PA6) and poly(ethylene-*co*-vinyl alcohol) (EVOH), with a total thickness of 140 μm, was supplied as WK140 by WK THOMAS (Barcelona, Spain). A multilayer film, also based on EVOH and with a total thickness of 100 μm, was obtained from Cryovac Inc. (Sealed Air Spain, Buñol, Spain) with commercial reference VST200P. These commercial multilayers were marked as PLA, M1, and M2, respectively. The commercial multilayers M1 and M2 have been designed by manufacturers for lamination applications in barrier packaging, including the preservation of meat. A commercial polyvinyl chloride (PVC) cling film (Bosque Verde, Mercadona S.A., Valencia, Spain), employed for food wrapping applications, was also used as a control to handle and protect the non-packaged samples.

Absolute ethanol (EtOH) and D-limonene were both supplied by Sigma-Aldrich S.A. (Madrid, Spain). Magnesium nitrate-6-hydrate [$Mg(NO_3)_2$], trichloroacetic acid (TCA), and thiobarbituric acid (TBA) reagent were all supplied by Panreac Química, SA (Castellar del Vallés, Spain).

The pork meat was purchased in the form of a single piece of 1 kg from the loin from a local supermarket (Consum S. Coop. V., Valencia, Spain). The food samples were immediately transported to the laboratory facilities in a portable fridge at 5 °C and processed into fillets using a professional slicer (Smarty 250 IX, Manconi, Italy) in a cabinet. Microbiological media, that is, buffered peptone water, Violet Bile Red agar (VRB), and Plate Count agar (PCA) were provided by Scharlab S.L. (Barcelona, Spain). Man Rogosa and Sharpe Agar (MRS) were obtained from Lankem-Labbox (Barcelona, Spain).

2.2. *Film Characterization*

2.2.1. Thickness and Conditioning

Prior to characterization, the whole thickness of all films was measured using an electronic digital micrometer (Comecta S.A., Barcelona, Spain), having ±0.001 mm accuracy. Measurements were performed at six random points and values were averaged. Thereafter, the samples were preconditioned at 25 °C and 53% relative humidity (RH) for 1 week.

2.2.2. Thermal Characterization

Differential scanning calorimetry (DSC) was carried out to determine the melting temperature (T_m) of the polymers present in the multilayers using a DSC823^e Star^e (Mettler-Toledo GmbH, Schwerzenbach, Switzerland). Films samples (~10 mg) were placed into aluminum pans, tightly sealed, and heated from −50 °C to 250 °C at 10 °C/min. As a reference, an empty aluminum pan was used.

Their thermal degradation was analyzed by thermogravimetric analysis (TGA) using a Star^e System analyzer (Mettler-Toledo GmbH). Film samples (~3 mg) were subjected to a heating program from 25 °C to 600 °C at a heating rate of 10 °C/min under a nitrogen atmosphere (10 mL/min). In each test, the corresponding mass curve was obtained as a function of temperature. All thermal tests were performed in triplicate.

2.2.3. Multilayer Structure Determination

The determination of the number of layers and each layer thickness was performed by image analysis using light optical microscopy (Olympus BX50 light microscope, Tokyo, Japan) using 50× magnification. The samples were cross-sectioned with a microtone (Microtone M240, Especialidades Médicas Myr, S.L., Tarragona, Spain) and stained with an iodopovine solution (10 g/100 mL, Viatris Inc., Madrid, Spain).

The microstructure and surface of the films were observed by Field Emission Scanning Electron Microscope (FESEM, Ultra 55, Zeiss, Oxford Instruments, Abingdon, UK). To this end, the samples were mounted on holders using double-sided carbon tape, coated with a platinum layer (EM MED020 sputter coater, Leica Biosystems, Barcelona, Spain), and observed using an accelerated voltage of 2 kV. The film samples were previously cryo-fractured with liquid nitrogen to obtain their cross-sections.

2.2.4. Mechanical Analysis

The mechanical properties of the multilayer films were evaluated using a universal mechanical testing press (Stable Micro System TA-XT plus, Haslemere, England), following the standard method ASTM D882 [20]. The samples, sizing 25 mm × 100 mm, were positioned in tension test clips (Model A/TG, Stable Micro Systems) and subjected to a tensile test at a speed of 50 mm/min until breaking. The force–distance curves obtained in the test were transformed into Henky stress–strain curves.

2.2.5. Permeance Measurements

Water vapor permeance of the multilayer films was determined according to ASTM E96/E96M gravimetric methodology [21] at 25 °C and 53% RH. Film samples (Ø = 3.5 cm) were placed and sealed in Payne permeability cups filled with 5 mL of distilled water (100% RH). Then, the cups with the films were placed into desiccators containing an $Mg(NO_3)_2$ over-saturated solution and weighed periodically (ME36S, ± 0.00001 g accuracy, Sartorius, Goettingen, Germany) for one week. The water vapor permeance was calculated considering the water vapor transmission rate (WVTR), which was determined from the slope of the weight loss vs. time and correcting for permeant partial pressure. For the permeance to limonene, the procedure was similar to that described previously for water vapor, placing 5 mL of D-limonene inside the Payne permeability cups and storing them under the controlled environmental conditions of 25 °C and 53% RH. Limonene permeation rate (LPR) was obtained from the steady-state permeation slopes of weight loss vs. time and corrected for permeant partial pressure. In both cases, cups with aluminum films were used as control samples to estimate and subtract the vapor loss through the sealing. All the vapor permeance measurements were performed in triplicate.

The oxygen permeance of the films was determined using an Ox-Tran equipment Model 1/50 (Mocon, Minneapolis, MN, USA) according to ASTM D3985-05 [22]. Tests conditions were also set at 25 °C and 53% RH and the exposure area during the test was 50 cm^2. The permeance values were derived from the oxygen transmission rate (OTR) measurements, which were corrected with the gas partial pressure and recorded in triplicate.

2.2.6. Optical Evaluation

The optical properties of the multilayer films were determined, in triplicate, by measuring the reflection spectrum of the films in the wavelength range from 400 to 700 nm using a MINOLTA spectro-colorimeter (model CM-5, Minolta Co., Tokyo, Japan). Transparency was measured by the internal transmittance (Ti) at 700 nm, applying the Kubelka–Munk theory of multiple scattering to determine the film reflection spectra (R) using the black (R_0) and white (R_g) backgrounds, whereas the CIE $L^*a^*b^*$ (CIELAB) color coordinates were determined using Equations (1)–(3), considering the illuminant D65 and 10° observer:

$$T_i = \sqrt{(a + R_0)^2 - b^2} \tag{1}$$

$$a = \frac{1}{2}\left[R + \left(\frac{R_0 - R + R_g}{R_0 \times R_g}\right)\right] \tag{2}$$

$$b = \sqrt{a^2 - 1} \tag{3}$$

The chromatic parameters hue angle ($h_{ab}{}^*$), chroma ($C_{ab}{}^*$), and opacity (O) were obtained from Equations (4)–(6), respectively:

$$h_{ab}{}^* = arctg\left(\frac{b^*}{a^*}\right) \tag{4}$$

$$C_{ab}{}^* = \sqrt{a^{*2} + b^{*2}} \tag{5}$$

$$O = A_{500} \times L \quad (6)$$

where A_{500} and L correspond to the absorbance at 500 nm and the film thickness, respectively.

2.3. Pork Meat Characterization

2.3.1. Preparation of Pork Meat Samples

Prior to preparing the samples, all utensils and work surfaces were disinfected to avoid cross-contamination with 96% ethanol (Panreac SA, Barcelona, Spain). The as-received multilayer films were also sterilized by exposure to ultraviolet (UV) light for 30 min in a laminar flow cabinet (Bio II Advance, Telstar, Terrassa, Spain). Then, the pork meat was processed into 10 g slices using a professional slicer (Smarty 250 IX, Manconi, Italy) and immediately placed inside the multilayer films. To this end, the films were first cut into samples, sizing 10 cm × 10 cm, and heat-sealed with a vacuum sealer machine (Vacio Press Elite, SAECO, Gaggio Montano, Italy), placing the meat fillets in between. Some pork meat fillets were also wrapped in PVC cling films to be used as the control samples (without packaging).

All samples were stored under refrigerated conditions, at 5 °C and 48% RH, for up to 15 days to carry out the evaluation. These conditions were selected according to previous studies that evaluated the performance of biodegradable packaging films for meat preservation applications [23,24].

2.3.2. Physico-Chemical Evaluation

The pH values were determined using a digital pH meter by direct insertion of the electrode probe (Mettler-Toledo GmbH) into the pork meat samples. The following measurements were performed at day 0 (unpackaged) and at days 3, 7, 11, and 15. Five measurements were taken, in duplicate, for each sample.

Weight loss of the samples was quantified as a function of storage time with a scale (ME36S, ±0.00001 g accuracy) from Sartorius. The optical properties of the packaged pork meat were also characterized by measuring the reflection spectrum of the packaged samples at a wavelength of 400 to 700 nm at six random points on the sample surface, using the MINOLTA colorimeter spectrum, with a standard white plate as a background. Color coordinates (CIE $L^*a^*b^*$), hue (h_{ab}^*), chroma (C_{ab}^*), and total color difference (ΔE, Equation (7)) were measured during storage using the illuminant D65/10° observer.

$$\Delta E = \sqrt{(\Delta L^*)^2 + (\Delta a^*)^2 + (\Delta b^*)^2} \quad (7)$$

The degree of lipid oxidation was evaluated by quantifying the thiobarbituric acid reactive species (TBARS) according to the method described by Siu and Draper [25]. For this, at the beginning and end of storage, 10 g of each sample were placed in bags (Stomacher 440 Classic Strainer Bags, Worthing, UK) with 50 mL of distilled water and homogenized for 2 min using a homogenizer (Masticator Paddle blender, IUL Instruments, Barcelona, Spain). Then, 50 mL of 10 (vol/vol%) TCA was added and the homogenate was filtered with a vacuum pump using Whatman #1 filter paper (Whatman N°1, Whatman International Ltd., Kent, UK). Thereafter, 8 mL of the clear filtrate was added to 2 mL of 0.06 M TBA reagent and incubated for 90 min at 80 °C. The absorbance was read at 532 nm and the results were expressed as mg of malonaldehyde (MDA) per kg of meat.

2.3.3. Microbial Analysis

Pork samples were analyzed for bacterial growth at the selected different storage times, that is, 0, 3, 7, 11, and 15 days. A total of 10 g of meat sample packaged in each of the multilayer films was aseptically taken using sterile tweezers in the laminar flow cabinet and, subsequently, placed in sterile bags (Stomacher 440 Classic Strainer Bags) with 90 mL of peptone water (Scharlab S.L.). The Stomacher bags were homogenized for 3 min using a homogenizer (Masticator Paddle blender, IUL Instruments). Thereafter, serial decimal dilutions were prepared and plated on PCA, VRB, and MRS agars for the total aerobic counts (TAC), total coliforms, and lactic acid bacteria (LAB) counts, respectively. The plates used for the total coliform and total aerobic counts were incubated at 37 °C for 48 h, while the conditions for those used for the LAB counts were 30 °C for 72 h. After incubation, all the colonies were counted, and the results were expressed as colony-forming units per gram (CFU/g). All the microbial tests were performed in triplicate.

2.4. *Statistical Analysis*

Results were subjected to analysis of variance (ANOVA) using Statgraphics Centurion XVII-64 software (Manugistics Corp., Rockville, MD, USA). To this end, significant differences were assumed with a significance level greater than 95% ($p < 0.05$).

3. Results and Discussion

3.1. *Multilayer Structure*

Prior to evaluating the performance of the multilayers, their structures were ascertained by determining the thermal properties and observing the film samples in the microscope. In Figure 1, the DSC and TGA curves of the three multilayer films are gathered. One can observe in Figure 1a, showing the DSC thermograms, that the PLA multilayer showed a glass transition temperature (T_g) at ~67 °C, having a low-intensity endothermic peak due to physical aging or the presence of a plasticizer with a low-melting profile, followed by a T_m value of 148.9 °C. These thermal transition values are characteristic of PLA, confirming the absence of other semicrystalline polymers in the film sample. The M1 multilayer showed three endothermic peaks, centered at 109.8 °C, 182.4 °C, and 220.3 °C, which can be respectively ascribed to the melting process of low-density polyethylene (LDPE), poly(ethylene-*co*-vinyl alcohol) with 32 mol% ethylene content (EVOH32), and PA6. In the case of M2, this multilayer film sample also yielded multiple endothermic peaks, particularly observed at 56.4 °C, 85.8 °C, 127.3 °C, and 156.4 °C. The first two melting peaks suggest the presence of a metallocene polyethylene wax with a very low melting point and a copolymer of ethylvinylacetate (EVA). The latter copolymer combines molecules of a hydrophilic and a hydrophobic part, which leads to an excellent tie-layer behavior [26]. The medium melting point can be ascribed to high-density polyethylene (HDPE) or a propylene-ethylene copolymer (coPP). The following low-intensity exothermic peak can be related to the melting of poly(ethylene-*co*-vinyl alcohol) with 44 mol% or, more probably, 48 mol% ethylene contents (EVOH44 or EVOH48) [27]. Finally, the last and intense exothermic peak observed at ~200 °C can be due to the deacetylation of EVA. In particular, copolymers with high vinyl acetate contents (up to 50 wt%) can release acetic acid prior to thermal decomposing at higher temperatures [28].

Figure 1b shows the TGA curves of the different multilayer films to better ascertain their composition and corroborate the DSC results shown above. From these curves, the corresponding values of T_{onset} (initial degradation temperature) and T_{deg} (temperature of maximum degradation rate) were determined. In the case of PLA, one can observe that the film degraded in a single and rapid step, showing values of T_{onset} and T_{deg} of 310.6 °C and 360.3 °C, respectively. These values are similar to those reported previously for PLA, which releases lactic acid, oligomers of lactic acid (OLAs), acetaldehydes, carbon dioxide, carbon monoxide, and ketones during thermal decomposition [29]. A similar degradation profile, but with higher thermal stability, was observed in the case of the M1 film. In particular, this commercial multilayer sample was thermally stable up to 350.4 °C and then degraded in two steps with T_{deg} values of 444 °C and 546.4 °C. The presence of two degradation peaks has been reported for the thermal degradation of polyolefins, which is based on the decomposition of the C–C covalent bonds followed by the breakdown, at a lower decomposition rate, of the polymer chains by free radicals [30]. Finally, the M2 sample presented the typical thermal degradation profile of EVA, showing an initial mass loss from 180.2 °C that involves the formation and release of acetic acid, the so-called deacetylation, being more noticeable in the 325–410 °C range, followed by degradation of the ethylene blocks [28]. For EVOH, T_{deg} values of approximately 391 °C and 409 °C have been reported for EVOH32 and EVOH44, respectively, which are also in the decomposition range observed for the multilayers studied herein [31].

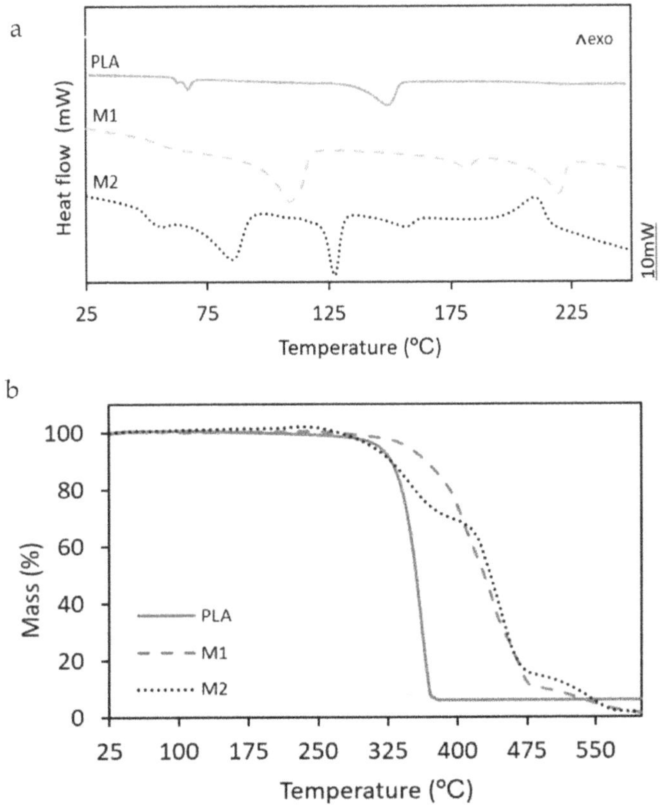

Figure 1. (**a**) Differential scanning calorimetry (DSC) thermograms and (**b**) thermogravimetric analysis (TGA) curves of the polylactide (PLA), M1, and M2 multilayer films.

Layer thickness and distribution of the commercial multilayers were determined by light optical microscopy after staining. Figure 2 shows the cross-section of the M1 and M2 multilayers, where one can observe the presence of different layers by having different color contrasts. In the M1 multilayer sample, up to 4 layers can be seen. The thicker one of approximately 90 µm, with lower color contrast, can be ascribed to LDPE. Then, three colored layers can be identified, having a different contrast. In the case of EVOH, the staining was more intense due to the higher number of hydroxyl groups, which correspond to the inner layer with a thickness of 5 µm. This layer was surrounded by two PA6 layers, having a thickness of approximately 20 µm. The morphological analysis of the M2 indicated the presence of 5 layers. The structural one, without color and of approximately 60 µm, that corresponds to the polyolefin. Then, the layers stained can be ascribed to two thin layers EVOH of 5 µm, the most intense ones, that adhered to two layers of EVA, of approximately 16 µm. The presence of EVA, as suggested above during the DSC and TGA evaluations, would contribute to improving layer adhesion and toughness in the multilayer, while EVOH is aimed to provide barrier to oxygen, odors, and flavors.

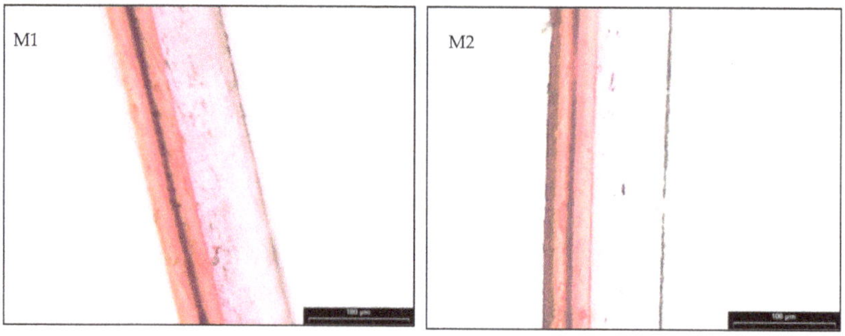

Figure 2. Stained cross-sections of the M1 (**left**) and M2 (**right**) multilayer films.

The structure of the multilayers was further analyzed and confirmed by FESEM. Figure 3 shows the cross-sectional (left images) and superficial micrographs (right images) of the film samples. One can see that the PLA film was composed of 3 layers, having the external thermosealable layers a mean thickness of approximately 1 μm. In the case of the M1, it can be discerned that this film sample was composed of 4 layers since the EVOH layer started to delaminate from the surrounding PA6 layers due to the cryo-fracture process. Finally, the presence of several layers can be inferred in the M2 film, where the thickest layer, shown at the top, would correspond to the polyolefin. Moreover, the top views revealed that all multilayer films exhibited homogeneous and smooth surfaces without any pores. Similar morphologies were previously reported for multilayer film structures based on PLA [32]. Table 1 summarizes the layer thickness obtained from the microscopy analysis.

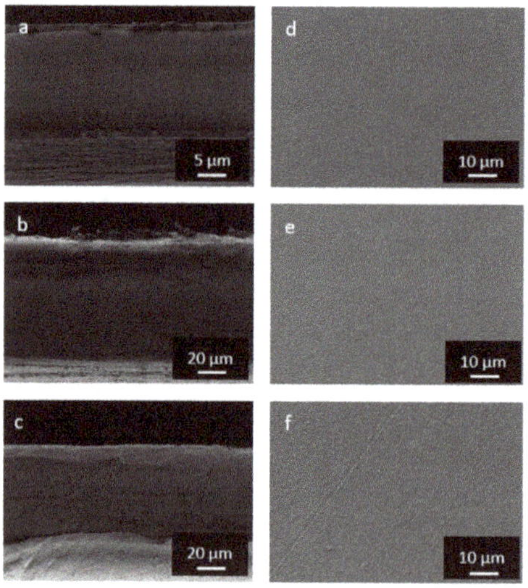

Figure 3. Field emission electron microscopy (FESEM) micrographs of the polylactide (PLA) (**a**,**d**), M1 (**b**,**e**), and M2 (**c**,**f**) multilayer films in their cross-sections (left) and top views (right).

Table 1. Layer thicknesses of the polylactide (PLA), M1, and M2 multilayer films.

Film	Thickness (μm)					
	Layer 1	Layer 2	Layer 3	Layer 4	Layer 5	Total
PLA	1	18	1	-	-	20
M1	21	5	22	90	-	138
M2	5	17	5	16	60	103

Therefore, from the structural analysis, one can consider that the PLA film was composed of a three-layer structure of the neat biopolymer, whereas the two commercial multilayers presented a structure of PA6/EVOH32/PA6/LDPE (M1) and EVOH48/EVA/EVOH48 /EVA/coPP (M2). Although EVOH is a non-chlorine barrier material for fresh meat packaging, copolymers with lower ethylene contents and, hence, higher oxygen barrier, are very sensitive to moisture conditions, especially at a relative humidity of above 80%. Therefore, as in the case of M1, this copolymer, EVOH32, was incorporated in multilayer structures protected between water-barrier polymers. For M2, EVOH48 was applied in the external layer due to its higher ethylene content. In both cases, sustainable alternatives are urged due to the increasing environmental concerns related to its poor recyclability.

3.2. Mechanical Properties of Multilayers

Figure 4 shows the tensile stress–strain curves of the multilayer films that allowed obtaining the parameters of elastic modulus (E), tensile stress at break (σ_b), and deformation at break (ϵ_b). One can observe that the three tested films presented a very dissimilar mechanical performance, suggesting different applications. On the one hand, the M1 multilayer presented the lowest mechanical resistance, yielding values of E and σ_b of 258 ± 24 and 42 ± 3 MPa, respectively, and ϵ_b of 67 ± 8%. These values correspond to a flexible film with high ductility. In contrast, the multilayer PLA film showed E and σ_b values of 2167 ± 209 and 89 ± 4 MPa, respectively, with a ϵ_b value of 5 ± 1%. This mechanical performance is related to strong but brittle materials. In the case of M2, this multilayer presented intermediate values but closer to those of the M1 multilayer, that is, E, σ_b, and ϵ_b values of 529 ± 27 MPa, 71 ± 9 MPa, and 56 ± 8%, respectively. Therefore, in terms of mechanical performance, the commercial multilayer films of PA6/EVOH32/PA6/LDPE and EVOH48/EVA/EVOH48/EVA/coPP presented characteristics of films suitable for flexible packaging applications, while the PLA multilayer film, being more rigid and less deformable, can be of more interest for rigid packaging uses such as trays, lids, or protective sheets. It is also worth indicating that the mechanical properties attained for the PLA films are similar to those reported for both injection-molded parts [33] and thermo-compressed films [34] made of neat PLA. Thus, it can be considered that the presence of the external thermosealable PLA layers did not alter the original mechanical characteristics of the biopolyester, and good adhesion between these with the inner PLA layer can be inferred due to they share the same composition.

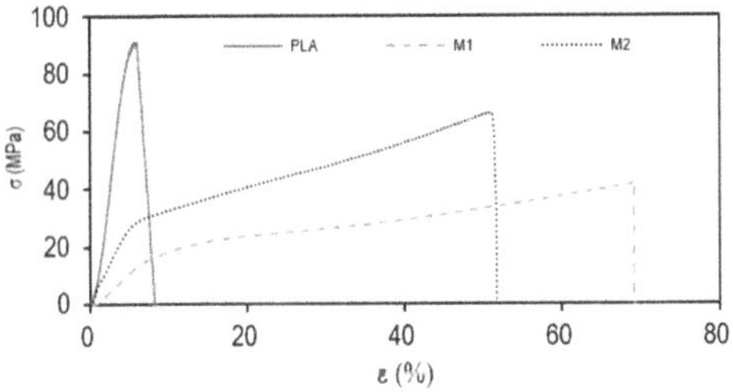

Figure 4. Typical stress (σ) versus deformation (ϵ) curves of the polylactide (PLA), M1, and M2 multilayer films.

3.3. Barrier Properties of Multilayers

Table 2 shows the thicknesses, measured by the micrometer, and barrier properties to water and limonene vapors and oxygen of the multilayer films. Barrier performance was expressed in terms of permeance since it is determined not only by the intrinsic permeability of each constituent material of the multilayer structure but also by thickness sample. As it can be seen in the table, all film samples showed significant differences ($p < 0.05$) and the M1 multilayer presented the highest barrier to water vapor and oxygen gas. In particular, the values of permeance to the vapors of water and limonene were 5.79×10^{-12} and 1.51×10^{-10} kg.m^{-2}.Pa^{-1}.s^{-1}, respectively, whereas the oxygen permeance showed a value of 2.15×10^{-16} m^3.m^{-2}.Pa^{-1}.s^{-1}. This high-barrier performance can be related to both the low permeability of its constituents and the higher film thickness, that is, nearly 140 µm. In particular, the LDPE polyolefin contributes to its high-water vapor barrier since it shows a very low water vapor permeability with a value of 1.2×10^{-15} kg.m.m^{-2}.Pa^{-1}.s^{-1} at 38 °C and 90% RH (standard tropical conditions) [35]. It is worth mentioning that the equivalent water vapor permeance of a 137-µm film made of neat LDPE results in a value of 8.76×10^{-12} kg.m^{-2}.Pa^{-1}.s^{-1}, which is in the order but nearly two times higher than that of the multilayer film sample tested herein. This difference could be attributed to the lower temperature used herein to determine the barrier properties, that is, 25 °C, or potential treatment of the film with a water-barrier coating, such as silicon oxide (SiOx) (1.3×10^{-17} kg.m.m^{-2}.Pa^{-1}.s^{-1} at 38 °C and 90% RH) [36]. In terms of the oxygen barrier, the presence of PA6 and, more importantly, EVOH32 can contribute to its very low permeance, particularly at low humidity conditions that are achieved when these polymers are protected by LDPE at the external layers. For instance, in the case of EVOH32, that is, the copolymer containing 32 mol% of ethylene, oxygen permeability varies from 0.77 to 91×10^{-21} m^3.m.m^{-2}.Pa^{-1}.s^{-1} for 0 and 75% RH, respectively, at 23 °C [35]. One can further observe that the M2 sample, that is, EVOH48/EVA/EVOH48/EVA/coPP multilayer, also presented a high-barrier performance, but still significantly lower ($p < 0.05$) than that of PA6/EVOH32/PA6/LDPE in terms of water vapor and oxygen. This multilayer showed permeance values of 2.32×10^{-11} kg.m^{-2}.Pa^{-1}.s^{-1} and 9.71×10^{-15} m^3.m^{-2}.Pa^{-1}.s^{-1} for water vapor and oxygen, respectively. In comparison to the other commercial multilayer, this lower permeance can result from its lower thickness, that is, 98 µm, and also, in the case of oxygen, due to its higher permeability since it is based on an EVOH copolymer with a lower vinyl alcohol content. For instance, EVOH44 is nearly 6 times more permeable to oxygen than EVOH32 [27]. It is remarkable to note, however, that the limonene permeance of the M2 multilayer was lower than that of M1, with respective values of 6.94×10^{-11} and 1.51×10^{-10} kg.m^{-2}.Pa^{-1}.s^{-1}, suggesting that the aroma barrier of EVOH increases at higher ethylene contents.

Table 2. Total thickness and permeance to water and limonene vapors and oxygen of the polylactide (PLA), M1, and M2 multilayer films.

Film	Thicknesses (µm)	Water Vapor Permeance $\times 10^{11}$ (kg.m^{-2}.Pa^{-1}.s^{-1})	Limonene Permeance $\times 10^{10}$ (kg.m^{-2}.Pa^{-1}.s^{-1})	Oxygen Permeance $\times 10^{14}$ (m^3.m^{-2}.Pa^{-1}.s^{-1})
PLA	20 ± 1 [a]	92.59 ± 5.82 [a]	2.20 ± 0.17 [a]	11.54 ± 2.02 [a]
M1	137 ± 3 [b]	0.58 ± 0.40 [b]	1.51 ± 0.14 [b]	0.02 ± 0.00 [b]
M2	98 ± 2 [c]	2.32 ± 1.67 [c]	0.69 ± 0.04 [c]	0.97 ± 0.03 [c]

[a–c] Different superscript letters within the same column indicate significant differences among the samples ($p < 0.05$).

In the case of the PLA film, it can be observed that its permeance to water vapor was about 20–200 times higher than that of the two commercial multilayer films, while it was in the same order of magnitude in terms of limonene vapor. This 20-µm PLA multilayer yielded permeance values of 9.26 and 2.20×10^{-10} kg.m^{-2}.Pa^{-1}.s^{-1} for water and limonene vapors, respectively. These vapor permeance values result in equivalent permeability values for a monolayer of PLA of 1.85 × 10^{-14} and 4.42 × 10^{-15} kg.m.m^{-2}.Pa^{-1}.s^{-1}, respectively, which are very similar to those reported in the literature for PLA (1.23 × 10^{-14} and 3.30 × 10^{-15} kg.m.m^{-2}.Pa^{-1}.s^{-1} at 25 °C) [37], confirming that the multilayer was fully based on PLA. The slight reductions observed in the vapor barrier properties in comparison with neat PLA could be ascribed to the plasticization performed on the outer layers to yield a film with enhanced thermosealability. In comparison with its petrochemical counterpart, PET, one can consider that the PLA film yielded lower vapor barrier performance to water but higher to limonene. This is based on the fact that according to the vapor permeabilities values of PET reported in the literature [38], an equivalent 20-μm PET film would provide permeance values of 1.15 × 10^{-10} and 5.85 × 10^{-9} kg.m^{-2}.Pa^{-1}.s^{-1} for water and limonene, respectively. In relation to the oxygen permeance, the PLA film resulted in a value of 1.15 × 10^{-13} m^3.m^{-2}.Pa^{-1}.s^{-1}, which was 2.5 and 1 orders of magnitude higher than in the M1 and M2 multilayer films, respectively. The resultant permeability of the PLA film assuming a monolayer is 2.31 × 10^{-18} m^3.m.m^{-2}.Pa^{-1}.s^{-1}, being very similar to that reported recently for PLA (2.22 × 10^{-18} m^3.m.m^{-2}.Pa^{-1}.s^{-1} at 25 °C and 60% RH) [34] and also 10–15 times higher than that of PET (3.27 and 4.26 × 10^{-19} m^3.m.m^{-2}.Pa^{-1}.s^{-1} at 23 °C and 0% and 75% RH, respectively) [38]. Therefore, the present results confirm that the PLA biopolymer is an intermediate barrier material for vapors but with a low-to-intermediate barrier to oxygen [39]. According to these results, the PLA multilayer film showed significantly higher permeance values to water vapor and, more significantly, oxygen than the commercial multilayers. However, the aroma permeance of the three multilayer films was in the same range, even though the PLA film thickness was considerably lower.

3.4. Optical Properties of Multilayers

The optical properties of the films are shown in Table 3. One can observe that the M1 and M2 multilayer samples, that is, PA6/EVOH32/PA6/LDPE and EVOH48/EVA/EVOH48 /EVA/coPP films, showed very similar values for luminance (L^*), in the 84–86 range, while the PLA multilayer presented a significantly ($p < 0.05$) higher value, that is, 94.9. As for chroma ($C_{ab}{}^*$), the PLA film also showed the highest value (4.3), followed by the M1 (2.6) and M2 (1.9) multilayers. In terms of the hue values ($h_{ab}{}^*$), it can be observed that there was no significant difference ($p > 0.05$) between the PLA and the M1 multilayers, with values ranging from 127 to 132, whereas the M2 multilayer presented a significantly higher value ($p < 0.05$), that is 148.

Table 3. Optical properties in terms of lightness (L^*), chroma ($C_{ab}{}^*$), hue angle ($h_{ab}{}^*$), and internal transmittance (T_i) at 550 nm of the polylactide (PLA), M1, and M2 multilayer films.

Film	L^*	$C_{ab}{}^*$	$h_{ab}{}^*$	*Ti* (550 nm)
PLA	94.9 ± 1.0 ^{a}	4.3 ± 1.0 ^{a}	127 ± 7.0 ^{a}	0.93 ± 0.004 ^{a}
M1	84.6 ± 1.0 ^{b}	2.6 ± 0.4 ^{b}	132 ± 4.0 ^{a}	0.90 ± 0.01 ^{b}
M2	85.6 ± 1.0 ^{b}	1.9 ± 0.2 ^{c}	148 ± 1.4 ^{b}	0.91 ± 0.01 ^{b}

$^{a–c}$ Different superscript letters within the same column indicate significant differences among the samples ($p < 0.05$).

In terms of the internal transmittance (*Ti*) at 550 nm, one can observe values between 0.90 and 0.93 for all the multilayer films. Therefore, all the multilayers showed similar optical properties, being the PLA film slightly but still significantly ($p < 0.05$) more transparent, which can be beneficial for food packaging applications in terms of retail display.

3.5. Physicochemical Properties of Packaged Pork Meat

Figure 5 shows the visual aspect of the pork meat fillets packaged in the different multilayers and control, that is, unpackaged meat wrapped in the PVC cling film during storage in refrigeration conditions of at 5 °C and 48% RH. It can be observed that the control samples suffered the most noticeable changes in color during storage, whereas the samples packaged in the PLA films developed a reddish color at the end of storage.

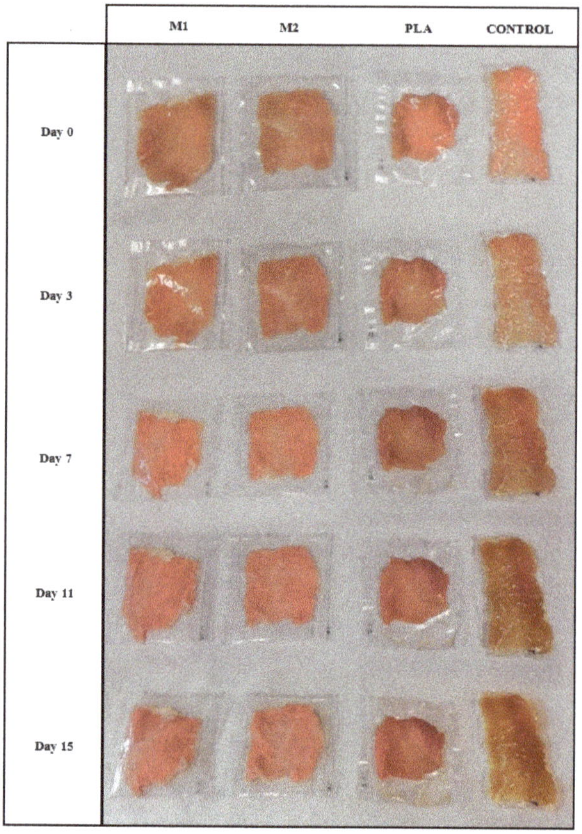

Figure 5. Images of pork meat fillets packaged in the polylactide (PLA), M1, and M2 multilayer films during storage.

Table 4 shows the evolution of pH and weight loss as a function of the storage time. The initial pH of the pork meat samples was 5.49 ± 0.01, which is in the range of the values previously reported for fresh pork meat samples [40,41]. A significant effect ($p < 0.05$) of both storage time and the type of packaging was observed during the analysis of the pH values, being the type of packaging the factor having the most significant influence. During storage, it has been reported that the pH values of meat rise due to the increased content of nitrogenous bases resulting from proteolysis caused by the activity of microorganisms [42]. This behavior was clearly observed in the unpackaged meat, wrapped in PVC film, where pH increased significantly during storage up to 7.66. However, after 7 days of storage, the pH evolution of the meat packaged in the three multilayer materials was significantly different ($p < 0.05$) from that of the control meat, with pH values remaining in the 5–5.5 range. This can be explained by the air-barrier protection offered by the packaging materials, which could slow down the activity of microorganisms causing spoilage and it is also coherent with the lower oxidation level and reduced total microbial counts observed in these samples, as described below. Similar results were obtained by Daniloski et al. [43], who studied the shelf life of pork meat packaged in biaxially oriented polypropylene (BOPP) coated with acrylic/polyvinylidene chloride (BOPPAcPVDC) and biaxially oriented coextruded polypropylene (BOPPcoex) stored under vacuum in refrigeration conditions.

Table 4. Development of pH and weight loss of the pork meat samples packaged in the polylactide (PLA), M1, and M2 multilayer films during storage.

Film	Storage Time (Days)			
	3	7	11	15
	pH			
Control	5.43 ± 0.01 ^{a4}	5.84 ± 0.15 ^{a3}	7.06 ± 0.09 ^{a2}	7.66 ± 0.13 ^{a1}
PLA	5.44 ± 0.02 ^{a3}	5.88 ± 0.05 ^{a1}	5.56 ± 0.02 ^{b2}	5.25 ± 0.04 ^{b4}
M1	5.41 ± 0.02 ^{a1}	5.30 ± 0.01 ^{b2}	5.32 ± 0.02 ^{c2}	5.24 ± 0.01 ^{b3}
M2	5.42 ± 0.03 ^{a1}	5.33 ± 0.01 ^{b2}	5.30 ± 0.01 ^{c3}	5.21 ± 0.01 ^{b4}
	Weight loss (%)			
PLA	1.930 ± 0.300 ^{a3}	3.400 ± 1.160 ^{a3}	8.410 ± 1.500 ^{a2}	14.130 ± 1.300 ^{a1}
M1	0.050 ± 0.004 ^{b3}	0.020 ± 0.010 ^{b4}	0.100 ± 0.010 ^{b2}	1.520 ± 0.800 ^{b1}
M2	0.010 ± 0.001 ^{c3}	0.040 ± 0.010 ^{c2}	0.170 ± 0.110 ^{c1}	0.210 ± 0.100 ^{c1}

Different superscripts within the same column indicate significant differences among samples for the same storage time ($^{a–c}$) or due to storage time for the same sample ($^{1–4}$) ($p < 0.05$).

Furthermore, the weight loss of the pork meat during storage is also reported in Table 4. The so-called M1 and M2 multilayers, that is, PA6/EVOH32/PA6/LDPE and EVOH48/EVA/EVOH48/EVA/coPP films, resulted in a significantly lower ($p < 0.05$) weight loss during storage as compared to meat samples packaged in the PLA film. This weight variation can be related to water loss, which is consistent with the higher barrier to water vapor of the M1 and M2 multilayer films described above. This weight loss can be ascribed to the evaporation of exudate from the packaging, which is a natural event due to the leakage of intramuscular fluids from the cut surface [44]. This weight loss value could not be measured and monitored accurately in the control sample since some loss of exudate occurred during storage. Furthermore, the weight loss values attained herein are in the same range as those reported in previous studies carried out with vacuum-packaged pork chops [45] and beef steaks [46].

Lipid oxidation levels were evaluated by monitoring TBARS formation, which measures the amount of malonaldehyde (MDA) produced by secondary products of polyunsaturated fatty acid peroxidation [47]. The control samples wrapped in the PVC film showed oxidation levels of 1.90 mg MDA/kg at the end of storage. Similar values were found in other studies conducted with refrigerated pork meat at 4 °C [48]. The meat samples packaged in the multilayer films presented significantly lower ($p < 0.05$) TBARS values: 1.19, 0.77, and 0.47 mg MDA/kg for PLA, M2, and M1 films, respectively. The values obtained are consistent with the oxygen barrier properties of the materials described above. Thus, the multilayer film with the lowest oxygen permeance, that is, PA6/EVOH32/PA6/LDPE, resulted in the lowest level of lipid oxidation. In any case, one should consider that off-flavors in pork meat can generally be detected by consumers when the TBARs value is above the threshold of 0.5 mg MDA/kg [49] and that this level was surpassed at the end of storage by all samples except for those packaged in the PA6/EVOH32/PA6/LDPE films.

Figure 6 shows the evolution of the chromatic parameters (L^*, C_{ab}^*, h_{ab}^*) and the total color difference, that is, ΔE, in relation to the initial values (t = 0). The results showed a significant effect ($p < 0.05$) of both storage time and type of packaging on the color properties of the pork meat, being the type of packaging the factor that had the greatest influence. It can be seen in Figure 6a that lightness hardly varied during the 15 days of storage in the pork meat packaged in the M1 and M2 films. In contrast, the meat packaged in the control and PLA films showed a progressive decrease with time. Among the meat samples packaged in multilayers, the highest decrease in L^* detected for PLA can be explained by the higher water permeance of the biopolymer film that, in turn, is correlated with the above-mentioned largest weight loss, which occurs mainly at the food surface. Drying could yield changes in the selective light absorption on meat fillets due to changes in the refractive index of the material and also in the surface concentration of pigments that can affect the C_{ab}* and h_{ab}* values [50]. As described by Faustman et al. [51], oxidation of both lipids and myoglobin (Mb) in meat also leads to discoloration, and these processes are frequently linked since the oxidation of one of these compounds produces chemical species that promote the oxidation of the other. In addition, a lower pH value in pork meat has been associated with greater reflectance, which leads to an increase in lightness and a decrease in the relative amount of the reduced form of Mb. At the same time, a lower pH can be accompanied by a greater susceptibility of muscle pigments to oxygenation and oxidation. Meat yellowness increases due to an increase in the relative amounts of the oxygenated and oxidized forms of Mb, that is, oxymyoglobin and metmyoglobin (MbO_2 and MetMb) at the expense of the reduced form [52].

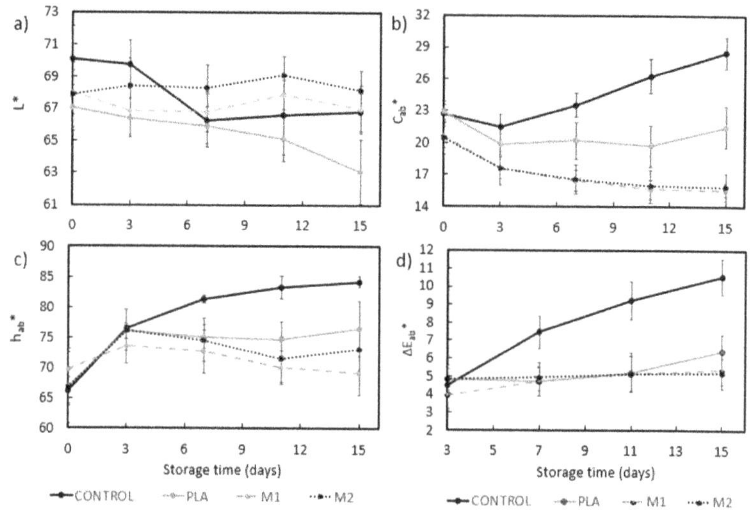

Figure 6. Changes in color parameters of the pork meat samples packaged in the polylactide (PLA), M1, and M2 multilayer films and wrapped in cling film (unpackaged control) during storage in terms of (**a**) Lightness (L*); (**b**) Chroma (C_{ab}*); (**c**) Hue (h_{ab}*); and (**d**) total color difference (ΔE_{ab}*).

In relation to the chroma or saturation (C_{ab}*, see Figure 6b), it can be observed that it slightly decreased in the pork meat packaged in all the multilayer films, showing no significant effect ($p > 0.05$) between the M1 and M2 samples during storage. Samples packaged in all the three multilayer films showed a more saturated color, having no significant changes ($p > 0.05$) during storage. These differences can be mainly ascribed to the light diffraction effect of each film on the meat sample, according to their above-described optical properties. In the case of the unpackaged meat sample (control sample), as shown in Figure 6c, the h_{ab}* values rose significantly ($p < 0.05$) after 11 days of storage, with a more saturated and less red hue. In contrast, the h_{ab}* values of the meat samples packaged in the multilayer films slightly increased during the first 3 days of storage and, then, were kept almost constant during the remaining time of storage. It is also worth noting that the samples packaged in the PLA films developed a redder hue. These observations suggest that the unpackaged meat sample developed a more intense color due to microbial spoilage, while the food samples packaged in the multilayer films better maintained the original color of the fresh meat. In the case of PLA, the red color developed in the meat fillets can be ascribed to an oxidation process, which can be related to the lower oxygen barrier of these films. Finally, as seen in Figure 6d, for up to 11 days, the total color differences during storage of the samples packaged in the multilayer films did not exceed the usual tolerance limit for food products ($\Delta E < 5$) [50], thus indicating better color preservation as compared to unpackaged samples (control).

3.6. Microbial Analysis of Packaged Pork Meat

The main factors determining the shelf life of meat and meat products are the initial composition of the bacterial flora and the method of preservation. Among the microorganisms isolated, LAB represent the dominant group in vacuum-packaged food samples [53]. Figure 7 shows the changes in LAB counts as well as TAC and coliform counts as a function of the storage time.

The microbial analysis showed a significant effect ($p < 0.05$) of both storage time and type of packaging, being the time the factor with the greatest influence in this case. As expected, the control samples were the ones that presented higher microbial counts, though differences were only significant ($p < 0.05$) with respect to the samples packaged with the different multilayer materials after one week of storage. In terms of TAC, which is an important microbiological indicator and the quantitative standard for identifying the conditions and degree of contamination of meat [54], it can be observed that counts increased from approximately 3 log CFU/g to values in the 4.5–3.5 range for the pork meat packaged in the multilayer films during refrigerated storage (Figure 7a). The bacterial growth profile in the meat fillets was also very similar for all the multilayers, remaining nearly constant during the first days of storage and stabilizing after one week with no significant differences ($p > 0.05$) among the films. In contrast, for the unpackaged meat sample, TAC reached a value above 7 log CFU/g. In this regard, the European Commission (EC) Regulation Number 2073/2005 [54] on microbiological criteria for foodstuffs indicates that the maximum acceptable level for TAC in mechanically separated fresh pork meat is 5 log CFU/g [55]. Therefore, this level was not exceeded in the pork meat samples packaged in the multilayers for the whole studied period, accomplishing the requested value of food quality and safety.

(a)

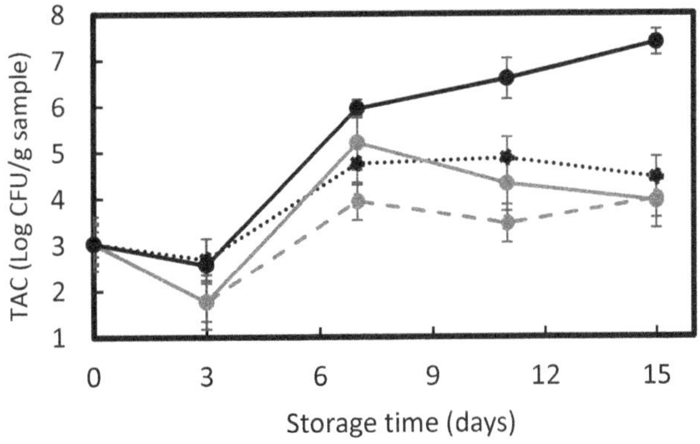

Figure 7. *Cont.*

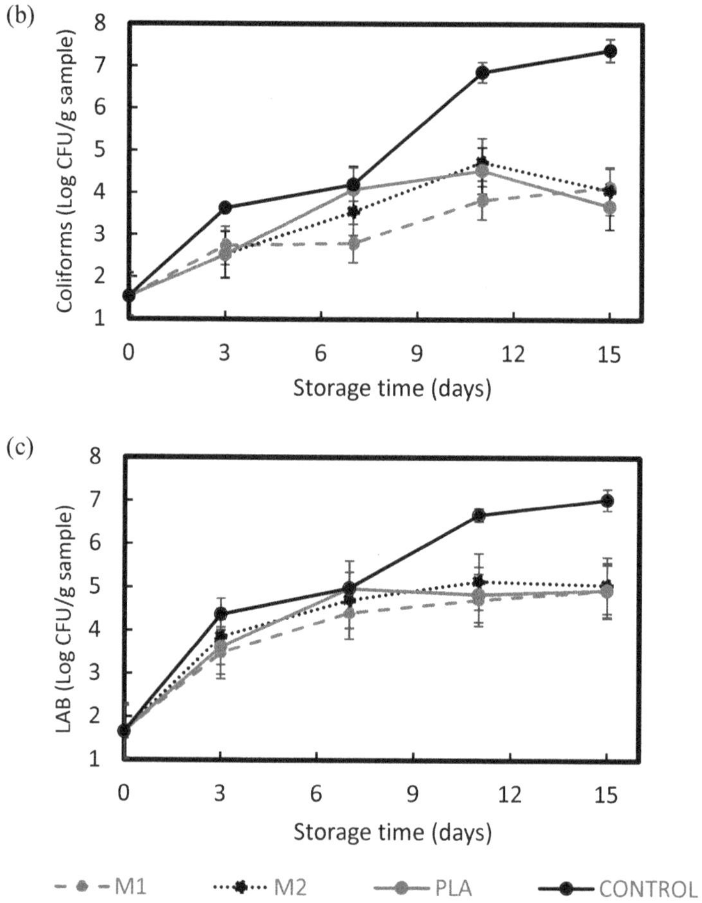

Figure 7. Changes in microbial counts in pork meat samples packaged in the polylactide (PLA), M1, and M2 multilayer films and wrapped in cling film (unpackaged control) during storage: (**a**) total aerobic count (TAC); (**b**) coliforms; and (**c**) lactic acid bacteria (LAB).

Figure 7b shows the total coliform microbial counts, which increased from approximately 1.5 log CFU/g to values nearly 4 log CFU/g in the pork meat samples packages in the multilayer films. It can be observed that microbial counts showed very similar values, showing no significant differences ($p > 0.05$) for the three multilayers during the whole storage period. In the case of the unpackaged meat sample (control), one can observe that the total coliform counts rapidly increased after one week of storage and reached significantly higher values ($p < 0.05$) as compared to samples packaged with the multilayer films. This result correlates to the reported trend in terms of pH shown above. Finally, Figure 7c shows the microbial counts for LAB, which are the dominant group of microorganisms isolated from meat and vacuum-packaged meat products [48]. Results showed that bacterial counts increased in the meat packaged in the multilayer films from 1.3 log CFU/g to values in the 4–5 log CFU/g range after 1 week of storage. Then, bacterial counts remained almost constant during the whole storage period. As reported for the TAC and coliform counts, the growth pattern was very similar in all the pork meat fillets packaged in the multilayers, and no significant differences ($p > 0.05$) were observed in terms of the type of film. In the case of the control sample (unpackaged meat), LAB count progressively increased during storage, being significantly higher than the values reported in packaged samples after one week of storage.

Therefore, the PLA multilayer packaging proved to be effective in terms of preserving the microbiological quality of pork during storage at 5 °C, even though this film showed lower thickness and higher water vapor and oxygen permeances than the commercial multilayers based on high-barrier EVOH materials. This suggests that its medium-barrier to water and aroma can be sufficient to limit and delay microbial growth, whereas its relative low oxygen permeance can also contribute to reducing microbial counts.

4. Conclusions

The negative impact of multilayer food packaging materials on the environment makes it necessary to replace them with biopolymers that are derived from renewable resources and biodegradable. The properties and environmental issues of biopolymers have been widely studied during the last decade but evaluations regarding their actual performance for food preservation applications are scarce. In the present study, PLA multilayer films were thermosealed and applied, for the first time, to package pork meat fillets. The quality and shelf life of the packaged meat were determined and compared to high-barrier commercial multilayer films of PA6/EVOH32/PA6/LDPE and EVOH48/EVA/EVOH48/EVA/coPP. It was concluded that the PLA films exhibited mechanical properties typical of materials used in rigid food packaging, having a moderate barrier to water and aroma vapors and relatively low to oxygen. During the shelf-life evaluation, the PLA packaging led to slight color variations as well as certain dehydration and oxidation phenomena in the cold-stored fresh pork meat samples, particularly after 11 days of storage. However, the pH changes were minimal and similar to those observed in the food samples packaged with the commercial high-barrier multilayer films. In terms of microbial analysis, the PLA film also yielded a comparable performance to the high-barrier multilayers, and the biopolymer films accomplished the requested values of food quality and safety. Therefore, PLA can be regarded as a sustainable alternative packaging material for fresh meat preservation to replace petrochemical polymers that are non-biodegradable and used in multilayers that are extremely difficult to recycle. Nevertheless, the lower water and oxygen barrier of PLA can still restrict its application in products where, for instance, moisture losses or oxidation during storage could be a limiting factor. Thus, high-barrier formulations and structures based on this biopolymer or others will be needed. According to these requirements, future studies will deal with the evaluation of more high-performance bioplastic films for preserving fresh foodstuffs, selecting those that are usually packaged in plastic due to their limited shelf life, whose rapid deterioration represents significant economic losses, and whose market price allows the cost of innovation to be absorbed, such as chicken breast, fish fillets, fresh pasta, or cheese.

Author Contributions: E.H.-G. methodology, investigation, formal analysis, writing—original draft preparation; M.V. formal analysis, validation; supervision, funding; S.T.-G. conceptualization, writing—review and editing, supervision. All authors have read and agreed to the published version of the manuscript.

Funding: This research was funded by the Spanish Ministry of Science and Innovation (MICI), grant number PID2019-105207RB-I00.

Institutional Review Board Statement: Not applicable.

Informed Consent Statement: Not applicable.

Data Availability Statement: Data is contained within the article and also available on request.

Acknowledgments: E.H.-G. and S.T.-G. acknowledge MICI for her predoctoral research grant (BES-2017-082040) and his Ramón y Cajal contract (RYC2019-027784-I). The authors are also grateful to the Packaging Technologies Department of AINIA for the technical support provided during the determination of the multilayer structures. Derprosa™ is also acknowledged for gently providing the multilayer PLA film.

Conflicts of Interest: The authors declare no conflict of interest.

References

1. Robertson, G.L. Packaging and food and beverage shelf life. In *The Stability and Shelf Life of Food*, 2nd ed.; Subramaniam, P., Ed.; Elsevier: Amsterdam, The Netherlands, 2016; pp. 77–106. [CrossRef]
2. Torres-Giner, S.; Gil, L.; Pascual-Ramírez, L.; Garde-Belza, J.A. Packaging: Food waste reduction. In *Encyclopedia of Polymer Applications*; Mishra, M., Ed.; CRC Press: Boca Raton, FL, USA, 2019. [CrossRef]
3. Baroni, L.; Cenci, L.; Tettamanti, M.; Berati, M. Evaluating the environmental impact of various dietary patterns combined with different food production systems. *Eur. J. Clin. Nutr.* **2007**, *61*, 279–286. [CrossRef] [PubMed]
4. Radusin, T.; Torres-Giner, S.; Stupar, A.; Ristic, I.; Miletic, A.; Novakovic, A.; Lagaron, J.M. Preparation, characterization and antimicrobial properties of electrospun polylactide films containing *Allium ursinum* L. extract. *Food Packag. Shelf Life* **2019**, *21*, 100357. [CrossRef]
5. Buntinx, M.; Willems, G.; Knockaert, G.; Adons, D.; Yperman, J.; Carleer, R.; Peeters, R. Evaluation of the Thickness and Oxygen Transmission Rate before and after Thermoforming Mono- and Multilayer Sheets into Trays with Variable Depth. *Polymers* **2014**, *6*, 3019–3043. [CrossRef]
6. Domeño, C.; Aznar, M.; Nerín, C.; Isella, F.; Fedeli, M.; Bosetti, O. Safety by design of printed multilayer materials intended for food packaging. *Food Addit. Contam. Part A* **2017**, *34*, 1239–1250. [CrossRef]
7. Suchek, N.; Fernandes, C.I.; Kraus, S.; Filser, M.; Sjögrén, H. Innovation and the circular economy: A systematic literature review. *Bus. Strategy Environ.* **2021**, *30*, 3686–3702. [CrossRef]
8. Dixon, J. *Packaging Materials: 9. Multilayer Packaging for Food and Beverages*; Report Series; ILSI Europe: Bruxelles, Belgium, 2011; pp. 1–44.
9. Abdin, M.; El-Beltagy, A.E.; El-sayed, M.E.; Naeem, M.A. Production and Characterization of Sodium Alginate/Gum Arabic Based Films Enriched with *Syzygium cumini* Seeds Extracts for Food Application. *J. Polim. Environ.* **2021**, 1–12. [CrossRef]
10. Eltabakh, M.; Kassab, H.; Badawy, W.; Abdin, M.; Abdelhady, S. Active Bio-composite Sodium Alginate/Maltodextrin Packaging Films for Food Containing *Azolla pinnata* Leaves Extract as Natural Antioxidant. *J. Polim. Environ.* **2021**, 1–11. [CrossRef]
11. Groot, W.; van Krieken, J.; Sliekersl, O.; de Vos, S. Production and purification of lactic acid and lactide. In *Poly(Lactic Acid): Synthesis, Properties, Processing and Applications*; Auras, R.A., Lim, L.-T., Selke, S.E.M., Tsuji, H., Eds.; Wiley & Sons Inc.: Hoboken, NJ, USA, 2010; pp. 1–18.
12. Södergård, A.; Stolt, M. Industrial production of high molecular weight poly(lactic acid). In *Poly(Lactic Acid): Synthesis, Properties, Processing and Applications*; Auras, R.A., Lim, L.-T., Selke, S.E.M., Tsuji, H., Eds.; Wiley & Sons Inc.: Hoboken, NJ, USA, 2010; pp. 27–41.
13. Nofar, M.; Sacligil, D.; Carreau, P.J.; Kamal, M.R.; Heuzey, M.-C. Poly(lactic acid) blends: Processing, properties and applications. *Int. J. Biol. Macromol.* **2019**, *125*, 307–360. [CrossRef]
14. Gürler, N.; Paşa, S.; Temel, H. Silane doped biodegradable starch-PLA bilayer films for food packaging applications: Mechanical, thermal, barrier and biodegradability properties. *J. Taiwan Inst. Chem. Eng.* **2021**, *123*, 261–271. [CrossRef]
15. Torres-Giner, S.; Figueroa-Lopez, K.J.; Melendez-Rodriguez, B.; Prieto, C.; Pardo-Figuerez, M.; Lagaron, J.M. Emerging trends in biopolymers for food packaging. In *Sustainable Food Packaging Technology*, 1st ed.; Athanassiou, A., Ed.; Wiley VCH: Weinheim, Germany, 2021; pp. 1–33. [CrossRef]
16. Gerometta, M.; Rocca-Smith, J.R.; Domenek, S.; Karbowiak, T. Physical and chemical stability of PLA in food packaging. In *Reference Module in Food Science*; Smithers, G.W., Ed.; Elsevier: Amsterdam, The Netherlands, 2019. [CrossRef]
17. Dave, D.; Abdel, E.G. Meat Spoilage Mechanisms and Preservation Techniques: A Critical Review. *Am. J. Agri Biol. Sci.* **2011**, *6*, 486–510. [CrossRef]
18. Antoniewski, M.N.; Barringer, S.A. Meat Shelf-life and Extension using Collagen/Gelatin Coatings: A Review. *Crit. Rev. Food Sci. Nutr.* **2010**, *50*, 644–653. [CrossRef] [PubMed]
19. Jeremiah, L.E. Packaging alternatives to deliver fresh meats using short- or long-term distribution. *Food Res. Int.* **2001**, *34*, 749–772. [CrossRef]
20. ASTM. Standard test method for tensile properties of thin plastic sheeting. In *Annual Book of ASTM Standards*; ASTM D882; American Society for Testing and Materials: Philadelphia, PA, USA, 2001; pp. 162–170.
21. ASTM. Standard test methods for water vapor transmission of materials. In *Annual Book of ASTM Standards*; ASTM E96/E96M; American Society for Testing and Materials: Philadelphia, PA, USA, 2005; pp. 406–413.
22. ASTM. Standard test method for oxygen gas transmission rate through plastic film and sheeting using a coulometric sensor. In *Annual Book of ASTM Standards*; ASTM 3985-95; American Society for Testing and Materials: Philadelphia, PA, USA, 2002; pp. 472–477.
23. Andrade, J.; González-Martínez, C.; Chiralt, A. Antimicrobial PLA-PVA multilayer films containing phenolic compounds. *Food Chem.* **2022**, *375*, 131861. [CrossRef] [PubMed]
24. Bonilla, J.; Fortunati, E.; Vargas, M.; Chiralt, A.; Kenny, J.M. Effects of chitosan on the physicochemical and antimicrobial proerties of PLA films. *J. Food Eng.* **2014**, *119*, 236–243. [CrossRef]
25. Siu, G.M.; Draper, H.H. A survey of malonaldehyde content of retail meats and fish. *J. Food Sci.* **1978**, *43*, 1147–1149. [CrossRef]

26. Poisson, C.; Hervais, V.; Lacrampe, M.F.; Kraweczak, P. Optimization of PE/Binder/PA Extrusion Blow-Molded Films. II. Adhesion Properties Improvement Using Binder/EVA Blends. *J. Appl. Polym. Sci.* **2006**, *101*, 118–127. [CrossRef]
27. Maes, C.; Luyten, W.; Herremans, G.; Peeters, R.; Carleer, R.; Buntinx, M. Recent Updates on the Barrier Properties of Ethylene Vinyl Alcohol Copolymer (EVOH): A Review. *Polym. Rev.* **2018**, *58*, 209–246. [CrossRef]
28. Allen, N.S.; Edge, M.; Rodriguez, M.; Liauw, C.M.; Fontan, E. Aspects of the thermal oxidation of ethylene vinyl acetate copolymer. *Polym. Degrad. Stab.* **2000**, *68*, 363–371. [CrossRef]
29. Ortiz-Barajas, D.L.; Arévalo-Prada, J.A.; Fenollar, O.; Rueda-Ordóñez, Y.J.; Torres-Giner, S. Torrefaction of coffee husk flour for the development of injection-molded green composite pieces of polylactide with high sustainability. *Appl. Sci.* **2020**, *10*, 6468. [CrossRef]
30. Quiles-Carrillo, L.; Montava-Jordà, S.; Boronat, T.; Sammon, C.; Balart, R.; Torres-Giner, S. On the use of gallic acid as a potential natural antioxidant and ultraviolet light stabilizer in cast-extruded bio-based high-density polyethylene films. *Polymers* **2020**, *12*, 31. [CrossRef]
31. Alvarez, V.A.; Ruseckaite, R.A.; Vázquez, A. Kinetic analysis of thermal degradation in poly(ethylene-vinyl alcohol) copolymers. *J. Appl. Polym. Sci.* **2003**, *90*, 3157–3163. [CrossRef]
32. Valdés, A.; Martínez, C.; Garrigos, M.C.; Jimenez, A. Multilayer films based on poly(lactic acid)/gelatin supplemented with cellulose nanocrystals and antioxidant extract from almond shell by-product and its application on hass avocado preservation. *Polymers* **2021**, *13*, 3615. [CrossRef] [PubMed]
33. Quiles-Carrillo, L.; Montanes, N.; Pineiro, F.; Jorda-Vilaplana, A.; Torres-Giner, S. Ductility and Toughness Improvement of Injection-Molded Compostable Pieces of Polylactide by Melt Blending with Poly(ε-caprolactone) and Thermoplastic Starch. *Materials* **2018**, *11*, 2138. [CrossRef] [PubMed]
34. Rojas-Lema, S.; Quiles-Carrillo, L.; Garcia-Garcia, D.; Melendez-Rodriguez, B.; Balart, R.; Torres-Giner, S. Tailoring the properties of thermo-compressed polylactide films for food packaging applications by individual and combined additions of lactic acid oligomer and halloysite nanotubes. *Molecules* **2020**, *25*, 1976. [CrossRef]
35. Lagarón, J.M. Multifunctional and nanoreinforced polymers for food packaging. In *Multifunctional and Nanoreinforced Polymers for Food Packaging*; Lagarón, J.-M., Ed.; Woodhead Publishing: Sawston, UK, 2011; pp. 1–28.
36. Massey, L.K. *Permeability Properties of Plastics and Elastomers*; William Andrew Publishing: Norwich, NY, USA, 2003.
37. Quiles-Carrillo, L.; Montanes, N.; Lagaron, J.M.; Balart, R.; Torres-Giner, S. In situ compatibilization of biopolymer ternary blends by reactive extrusion with low-functionality epoxy-based styrene–acrylic oligomer. *J. Polym. Environ.* **2019**, *27*, 84–96. [CrossRef]
38. Gabirondo, E.; Melendez-Rodriguez, B.; Arnal, C.; Lagaron, J.M.; Martínez de Ilarduya, A.; Sardon, H.; Torres-Giner, S. Organocatalyzed closed-loop chemical recycling of thermo-compressed films of poly(ethylene furanoate). *Polym. Chem.* **2021**, *12*, 1571–1580. [CrossRef]
39. Trinh, B.M.; Chang, C.C.; Mekonnen, T.H. Facile fabrication of thermoplastic starch/poly(lactic acid) multilayer films with superior gas and moisture barrier properties. *Polymer* **2021**, *223*, 123679. [CrossRef]
40. Xiong, Y.; Chen, M.; Warner, R.D.; Fang, Z. Incorporating nisin and grape seed extract in chitosan-gelatine edible coating and its effect on cold storage of fresh pork. *Food Control* **2020**, *110*, 107018. [CrossRef]
41. Wang, G.; Liu, Y.; Yong, H.; Zong, S.; Jin, C.; Liu, J. Effect of ferulic acid-grafted-chitosan coating on the quality of pork during refrigerated storage. *Foods* **2021**, *10*, 1374. [CrossRef]
42. Athayde, D.R.; Flores, D.R.M.; da Silva, J.S.; Genro, A.L.G.; Silva, M.S.; Klein, B.; Cichoski, A.J. Application of electrolyzed water for improving pork meat quality. *Food Res. Int.* **2017**, *100*, 757–763. [CrossRef]
43. Daniloski, D.; Petkoska, A.T.; Galić, K.; Ščetar, M.; Kurek, M.; Vaskoska, R.; Nedelkoska, D.N. The effect of barrier properties of polymeric films on the shelf-life of vacuum packaged fresh pork meat. *Meat Sci.* **2019**, *158*, 107880. [CrossRef] [PubMed]
44. Stella, S.; Garavaglia, D.; Francini, S.; Viganò, V.; Bernardi, C.; Tirloni, E.A. Evaluation of the weight loss of raw beef cuts vacuumpackaged with two different techniques. *Ital. J. Food Saf.* **2019**, *8*, 8111. [CrossRef] [PubMed]
45. Cayuela, J.M.; Gil, M.D.; Bañón, S.; Garrido, M.D. Effect of vacuum and modified atmosphere packaging on the quality of pork loin. *Eur. Food Res. Technol.* **2004**, *219*, 316–320. [CrossRef]
46. Bağdatli, A.; Kayaardi, S. Influence of storage period and packaging methods on quality attributes of fresh beef steaks. *CyTA J. Food* **2015**, *13*, 124–133. [CrossRef]
47. Song, N.B.; Lee, J.H.; Al Mijan, M.; Song, K.B. Development of a chicken feather protein film containing clove oil and its application in smoked salmon packaging. *LWT Food Sci. Technol.* **2014**, *57*, 453–460. [CrossRef]
48. Qin, Y.-Y.; Yang, J.-Y.; Lu, H.-B.; Wang, S.-S.; Yang, J.; Yang, X.-C.; Cao, J.-X. Effect of chitosan film incorporated with tea polyphenol on quality and shelf life of pork meat patties. *Int. J. Biol. Macromol.* **2013**, *61*, 312–316. [CrossRef]
49. Sheard, P.R.; Enser, M.; Wood, J.D.; Nute, G.R.; Gill, B.P.; Richardson, R.I. Shelf life and quality of pork and pork products with raised n-3 PUFA. *Meat Sci.* **2000**, *55*, 213–221. [CrossRef]
50. Hutchings, J.B. Instrumental specification. In *Food Colour and Appearance;* Springer: New York, NY, USA, 1999; pp. 199–237. [CrossRef]
51. Faustman, C.; Sun, Q.; Mancini, R.; Suman, S.P. Myoglobin and lipid oxidation interactions: Mechanistic bases and control. *Meat Sci.* **2010**, *86*, 86–94. [CrossRef]
52. Karamucki, T.; Jakubowska, M.; Rybarczyk, A.; Gardzielewska, J. The influence of myoglobin on the colour of minced pork loin. *Meat Sci.* **2013**, *94*, 234–238. [CrossRef]

53. Xu, F.; Wang, C.; Wang, H.; Xiong, Q.; Wei, Y.; Shao, X. Antimicrobial action of flavonoids from *Sedum aizoon* L. against *lactic acid bacteria* in vitro and in refrigerated fresh pork meat. *J. Funct. Foods* **2018**, *40*, 744–750. [CrossRef]
54. Huang, L.; Zhao, J.; Chen, Q.; Zhang, Y. Rapid detection of total viable count (TVC) in pork meat by hyperspectral imaging. *Food Res. Int.* **2013**, *54*, 821–828. [CrossRef]
55. European Commission. Commission Regulation (EC) No 2073/2005 of 15 November 2005 on Microbiological Criteria for Foodstuffs. Available online: https://data.europa.eu/eli/reg/2005/2073/oj (accessed on 12 September 2021).

foods

Article

One-Step Synthesis of Poly(L-Lactic Acid)-Based Soft Films with Gas Permselectivity for White Mushrooms (*Agaricus bisporus*) Preservation

Tao Sun [1], Junxia Bian [1], Yangyang Wang [2], Jian Hu [1], Xueyan Yun [1], Eerdunbayaer Chen [1] and Tungalag Dong [1,*]

1 College of Food Science and Engineering, Inner Mongolia Agricultural University, 306 Zhaowuda Road, Hohhot 010010, China
2 Hohhot Huimin District Center for Disease Control and Prevention, Hohhot 010030, China
* Correspondence: dongtlg@imau.edu.cn

Abstract: Proper packaging can extend the shelf life and maintain the quality of mushrooms during storage. The purpose of this study is to investigate the preservation of *Agaricus bisporus* using copolymer-modified poly (L-lactide-co-butylene fumarate) and poly (L-lactide-co-glycolic acid) (PLBF and PLGA) packaging. Shelf life and quality were evaluated over 15 days of storage of *Agaricus bisporus* at 4 ± 1 °C and 90% relative humidity, including weight loss, browning index (BI), total phenolics (TP), ascorbic acid (AA), malondialdehyde content (MDA), electrolyte leakage rate (EC), and superoxide dismutase (SOD) and catalase (CAT). The results showed that mushrooms packaged in PLBF films exhibited better retention in BI, TP, and AA than those with PLLA, PLGA, or polyethylene (PE) films. They can reduce the rate of weight loss, EC, and MDA, which in turn increases the activity of SOD and CAT. PLBF and PLGA have substantially improved flexibility in comparison with PLLA. They also significantly reduced oxygen (O_2) and carbon dioxide (CO_2) permeability and changed the gas permeability ratio. These positive effects resulted in the effective restriction of O_2 and CO_2 in these packages, extending the post-harvest storage period of white mushrooms.

Keywords: *Agaricus bisporus*; one-step synthesis; poly (L-lactic acid); soft packaging film; O_2/CO_2 atmosphere; food shelf life

check for updates

Citation: Sun, T.; Bian, J.; Wang, Y.; Hu, J.; Yun, X.; Chen, E.; Dong, T. One-Step Synthesis of Poly(L-Lactic Acid)-Based Soft Films with Gas Permselectivity for White Mushrooms (*Agaricus bisporus*) Preservation. *Foods* **2023**, *12*, 586. https://doi.org/10.3390/foods12030586

Academic Editors: Sergio Torres-Giner, Amparo Chiralt and Chelo Gonzalez-Martinez

Received: 21 December 2022
Revised: 17 January 2023
Accepted: 20 January 2023
Published: 30 January 2023

1. Introduction

Agaricus bisporus (button mushroom) is popular in the global food market and accounts for 30% of the world's total mushroom production [1]. Their excellent texture is characterized by their high nutritional value, characteristic flavors, and high protein and fat contents compared to other vegetables [2]. However, the potential market for this type of mushroom is somewhat inhibited by its short shelf life [3], which is 1~3 days at room temperature and 5~8 days under refrigerated conditions [4]. Without the cuticle, high metabolic activity, high respiration rates, and dehydration are responsible for the rapid decay of the mushroom [5]. As it is separated from the water and nutrients (organic matter, minerals) in the culture medium, harvested *Agaricus bisporus* is subject to the influence of the external environment, which leads to a constant depletion of nutrients and deterioration in storage quality [6]. Upon leaving the culture medium, a mushroom begins to respire, a process in which tissues decompose faster to maintain their own physiological metabolism, including breaking down cellular walls and producing reactive oxygen species [7]. Mushroom cell walls are composed of chitin and β-glucan, and as a result of enzyme action, they gradually decompose; firmness is rapidly destroyed, and tissue degeneration occurs more rapidly due to autolysis [8]. Reactive oxygen species and free radicals produced by respiration accelerate the oxidation of phospholipids on cell membranes, which leads to the production of peroxide malondialdehyde (MDA) and electrolyte leakage [9,10]. If the

O_2 content in the environment around the bivalve mushroom is reduced to less than 1% (O_2 < 1%, CO_2 > 5%), the respiration rate is inhibited, which can prevent the accumulation of MDA while reducing oxidative damage to the mushroom [11]. Therefore, to extend the shelf life of *Agaricus bisporus*, it is initially necessary to ensure that it remains in a normal state of activity, and on this basis, to create an environment with extremely low oxygen content inside the package to inhibit its respiration, creating a slower metabolic environment and thus retaining freshness.

Poly (L-lactic acid) (PLLA) is a fully biodegradable polymer derived from sustainable sources. PLLA is becoming increasingly popular as an alternative to synthetic plastics in the packaging sector due to its desirable characteristics of high transparency, high stiffness, high strength, and thermal stability [12–14]. Nevertheless, PLLA's poor gas barrier properties present a challenge and a need for further research in order for PLLA to compete with fossil-based plastics in such essential functions [15]. Many techniques can be used to enhance the gas barrier properties of PLLA. In Ali's study, PLLA was used to prepare PLA-based polyurethane (PLAPU) films containing PCL diol as a softening agent. An acquired blend consisting of PLAPU and PCL at a ratio of 1:3 demonstrated the best mechanical and gas barrier properties [16]. Genovese's group synthesized novel PLLA-based chain-extended triblock copolymers, and the results showed that copolymerization led to better mechanical and barrier properties in comparison with PLLA [12]. Thus, by altering the polymerization structure of PLLA, we can enhance its mass transfer properties, which improves its adsorption properties as well as its O_2 and/or CO_2 barrier properties [17,18]. This will help maintain food quality and safety since the rate of typical food degradation reactions (such as oxidation, spoilage, and deterioration) will be significantly reduced.

Equilibrium modified atmosphere packaging (EMAP) reduces the product's respiration rate, preserving its quality and extending its shelf life [19]. EMAP has become increasingly common in recent years, where products are packaged using selected types of film, and the required atmosphere is formed naturally due to product respiration and diffusion of gases through the film [20,21]. EMAP is richer in CO_2 and lower in O_2 than air, with the potential to reduce respiration rate, spoilage, and physiological changes [22]. EMAP is more adaptable than conventional packaging and is, therefore, a better solution for plants undergoing respiration; as a consequence, researchers have gradually turned their attention to the development of EMAP over the past few years [23]. The study of the effect of EMAP on the quality changes in mushrooms has found that low O_2 levels and/or high CO_2 levels produced by the packaging material have been effective in improving the storage quality of *Agaricus bisporus*, reducing browning production, enhancing antioxidant enzyme activity, and preventing ageing [11,24,25]. For this effect to be achieved, it will be necessary to improve the O_2 barrier properties and O_2 to CO_2 transmission ratio of the materials used for EMAP. Commercial modified atmosphere packaging used in the cryopreservation of *Agaricus bisporus* has been made from nondegradable plastics, which can cause environmental pollution. Researchers are now working on the preparation of EMAPs using degradable plastics [22,26–28].

Accordingly, the development of modified PLLA to prepare EMAP is an opportunity to increase the shelf life of mushrooms. To our knowledge, there has been little research conducted on the application and comparison of modified PLLA films to enhance the shelf life and quality of *Agaricus bisporus*. Due to its low gas barrier and poor flexibility, PLLA is difficult to use directly in food packaging, especially for irregularly shaped foods such as mushrooms. When PLLA is used for mushroom packaging, it usually breaks due to its brittleness, and its barrier properties do not meet the needs of packaging mushrooms. The novel approach we propose is to introduce poly(butylene fumarate)(PBF) or poly(glycolic acid)(PGA) blocks with high barrier properties into PLLA. PBF, which is a biobased unsaturated aliphatic polyester with properties similar to those of the traditional biobased saturated polyester poly (butylene succinate) (PBS), is attracting increasing interest because it has excellent processing characteristics and low production costs. The excellent flexibility of PBF may bring better mechanical properties to PLLA, while its similar structure to PSB also indicates that it may promote the barrier property of PLLA. [29–32]. While the excellent barrier and processing properties of PGA and its composites have been demonstrated to make them suitable for food packaging, their promotion for use in this application has been difficult due to high production costs. [33,34].

In this study, we developed PLLA-based films by a one-step synthesis method that included the addition of small amounts of oligo (butylene fumarate) (OBF) or glycolic acid (GA). Our work investigated the effects of OBF or GA on the properties of PLLA films, including flexibility properties as well as gas-selective permeability of the films. Additionally, their application in the preservation of white mushrooms under refrigeration at 4 °C was studied to determine the differences in commercial value during storage of white mushrooms under different atmospheric conditions created by PLBF and PLGA and the effect of films with different barriers and selective gas permeability on the quality of white mushrooms.

2. Materials and Methods

2.1. Materials

Agaricus bisporus was harvested from the Shengle plantation in Hohhot City Hellinger County. PE film, purchased from the Hohhot Hualian supermarket, was approximately 30 ± 0.5 μm thick. The CDP and OP of the PE film were 5.94×10^{-12} $cm^3/m{\cdot}s{\cdot}Pa$ and 1.97 $\times 10^{-12}$ $cm^3/m{\cdot}s{\cdot}Pa$, respectively.

L-lactic acid (L-LA; 90 wt % in water; Aladdin), glycolic acid (GA; 98%; Aladdin Bio-Chem Technology Co., Ltd. Shanghai, China), butane-1,4-diol (BDO; 99%; Aladdin), fumaric acid (FA; 99%; Sigma-Aldrich (SHANGHAI) Trading Co., Ltd. Shanghai, China), tin(II) chloride dihydrate ($SnCl_2{\cdot}2H_2O$; 99.99%; Aladdin), hydroquinone (99%; Sigma-Aldrich), and P-toluenesulfonic acid (TSA; 99%; Aladdin) were used. Unless otherwise specified, all the other reagents used in this study were obtained from Sinopharm Chemical Reagent Co., Ltd. (Shanghai, China) and were analytical grade.

2.2. Film Preparation

2.2.1. Synthesis of Oligo (L-Lactic Acid) and Poly (L-Lactic Acid)

Lactic acid (700 g) was added into a 2 L reaction flask equipped with a mechanical stirrer and nitrogen inlet. The flask was sequentially degassed by a vacuum pump and purged with high purity Ar 3 times to ensure an Ar atmosphere during the reaction. The system was first heated to 110 °C for approximately 1 h under 400 mbr, then to 150 °C for approximately 2 h under 130 mbr, and then degassed to 40 mbr for 4 h; reduced pressure ensured the removal of water monomers. The products were finally poured out of the flask, resulting in the production of oligo (L-lactic acid) (OLLA).

The synthetic catalyst ($SnCl_2{\cdot}2H_2O$/TSA, 0.5 wt% of gross weight) was added to the OLLA, warmed to 180 °C, and reacted under vacuum for 12 h. After the reaction was stopped, the product was removed from the reactor and cooled to room temperature naturally. The product is the PLLA sample.

2.2.2. Synthesis of Oligo (Butylene Fumarate)

A total of 64.4 g of fumaric acid, 50 g of 1,4-butanediol, and a proper amount of hydroquinone (0.5 wt%) were added into a 500 mL reaction flask equipped with a mechanical stirrer and nitrogen inlet. The flask was sequentially degassed by a vacuum pump and purged with high purity Ar 3 times to ensure an Ar atmosphere during the reaction. Then, the system was heated to 150 °C for approximately 2 h under continuous Ar purge and was subsequently degassed to a pressure of 10 mbr for 4 h. The products were poured out of the flask, and the oligo (butylene fumarate) (OBF) was obtained.

2.2.3. Synthesis of the PLGA Copolymer

PLGA synthesis was conducted as follows. Appropriate masses of OLLA (92 wt%), GA (8 wt%), and catalyst ($SnCl_2{\cdot}2H_2O$/TSA, 0.5 wt% of gross weight) were weighed into a

2-L batch reactor equipped with a mechanical mixer, a nitrogen inlet, temperature control by oil circulation, and a vacuum pump. The reactor's temperature was raised gradually from ambient temperature to 180 °C within 10 min; subsequently, the pressure was reduced stepwise to 30 mbar within 5 min. After a total reaction time of 20 h, the polymer was removed from the reactor and given the designation PLGA.

2.2.4. Synthesis of the PLBF Copolymer

PLBF synthesis was conducted as follows. Appropriate masses of OLLA (92 wt%), OBF (8 wt%), and catalyst ($SnCl_2 \cdot 2H_2O$/TSA, 0.5 wt% of gross weight) were weighed into a 2 L batch reactor equipped with a mechanical mixer, a nitrogen inlet, temperature control by oil circulation, and a vacuum pump. The reactor's temperature was raised gradually from ambient temperature to 180 °C within 10 min; subsequently, the pressure was reduced stepwise to 30 mbar within 5 min. After a total reaction time of 20 h, the polymer was removed from the reactor and given the designation PLBF.

2.2.5. Preparation of Films

PLLA, PLGA, and PLBF samples (0.9 g) were dissolved in chloroform (50 mL) used for film mold, and the films were removed from the mold after the chloroform evaporated and allowed to dry for 30 days.

2.3. *Measurements and Characterization*

2.3.1. Attenuated Total Refection Fourier Transform Infrared Spectroscopy

An IRAffinity-1 spectrophotometer (SHIMADZU (CHINA) Co., Ltd. Shanghai, China) equipped with a diffuse reflectance accessory was used to measure ATR-FTIR spectra with a resolution of 4 cm^{-1} and 64 scans from 700–4000 cm^{-1}.

2.3.2. Nuclear Magnetic Resonance

The ^1H NMR spectra of the polymers were recorded on a 400 MHz Bruker Advance 2B spectrometer in a solvent of deuterated chloroform ($CDCl_3$). The copolymer composition was evaluated with ^1H NMR. ^1H NMR signals of PLGA: 1.57 ($-CH_3$), 4.85 and 4.60 ($-CH_2-$), 5.17 ($-CO\underline{C}\underline{H}(CH_3)O-$). ^1H NMR signals of PLBF: 1.57 ($-CH_3$), 1.75 ($-CH_2-$), 4.22($-CH_2-$), 5.17 ($-COC\underline{H}(CH_3)O-$), 6.88($-CH{=}CH-$).

2.3.3. Gel Permeation Chromatography

Molecular weights were determined using gel permeation chromatography (GPC, Waters Co., Milford, MA, USA). A 1.0 mL/min flow rate of THF was used as the eluent; the column temperature was 30 °C. The polydispersity was also determined using gel permeation chromatography.

2.3.4. Tensile Test

The tensile strength (σ_t, MPa), Young's modulus (*E*, MPa), and elongation at break (ε_b, %) were measured on a tensile testing machine (XLW(EC), Labthink, Jinan, China). According to ASTMD882-09, the film samples were cut into rectangular strips (28 mm × 5 mm) and tested on the machine at a speed of 10 mm/min. Seven parallel samples were taken from each group, and the average test results were taken.

2.3.5. Gas Barrier Properties

Oxygen and Carbon Dioxide Permeability

The O_2 and CO_2 permeability of the films (OP and CDP) was determined by a gas permeability tester (VAC-V2, Labthink, Jinan, China) in accordance with GB/T1038-2000. The test was performed under environmental conditions (temperature: 5 °C, RH: 50%). The transmission rates (TRs) of CO_2 and O_2 were obtained directly from the test results, and the permeability of CO_2 and O_2 ($cm^3/m^2 \cdot d$) was calculated using the following formula.

$$\text{OP/CDP} = \frac{\text{OTR/CDTR} \times D}{\Delta P} \quad (1)$$

D represents the thickness of the films, and ΔP represents the gas pressure difference between the two sides of the film (Pa), which was a standard atmospheric pressure (1.01×10^5 Pa).

2.4. *Application for Agaricus bisporus Packaging*

Agaricus bisporus mushrooms were precooled after being picked under 2 ± 1 °C for two hours before testing and then those mushrooms that were round, of good hardness, white in color, unopened, free of pests, and without mechanical damage were selected for testing. The mushrooms were packaged in PE, PLLA, PLBF, and PLGA bags, with each bag containing 80 ± 5 g of mushrooms. The packaged mushrooms were stored at 4 °C for 15 days, and their quality was tested every three days to evaluate whether they were protected. At the same time, 80 ± 5 g of mushrooms were weighed in each of 6~8 clean trays as a CK group and stored in the same environment as the rest of the packaged group to achieve a similar storage effect. A randomly selected group was tested against the packaged group on each test day.

2.4.1. Headspace Analysis

The gas composition of the bags was measured using a headspace gas analyzer (Type 6600, Systech, Oxford, UK). The instrument was calibrated before the test, and each group was measured 3 times in parallel. The CO_2 and O_2 concentrations in % were displayed directly on the instrument's screen.

2.4.2. Weight Loss

The weight loss rate of the samples was calculated by following formula.

$$Weight\ Loss = \frac{m_1 - m_2}{m_1} \times 100\% \quad (2)$$

where m_1 (g) represents the initial weight of the mushrooms, and m_2 (g) represents the weight on the day of the experimental test. Three bags of mushrooms from each group were measured at a time.

2.4.3. Browning Index

The color of mushrooms during storage was measured using a portable colorimeter (CR-20, Minolta Co., Ltd., OSAKA, Japan). Three mushrooms were randomly selected from each packaged group after calibration using the accompanying standard white plate. As a result of testing the center of the mushroom cap, L (lightness), a (redness), and b (yellowness) values were gathered. The resulting browning index (BI) can be calculated as follows [10].

$$\text{BI} = \frac{100 \times (\text{X} - 0.31)}{0.172} \quad (3)$$

where

$$\text{X} = \frac{a^* + 1.75L^*}{5.645L^* + a^* - 3.012b^*} \quad (4)$$

2.4.4. Total Phenolics and Ascorbic Acid

The total phenolics (TP) of the samples were determined by the method developed by Pirie and Mullins in 1976, where gallic acid was used as a standard solution [35]. The ascorbic acid (AA) content was determined using a 2,6-dichlorophenol indophenol (DCPIP) titration method [36].

2.4.5. Malondialdehyde Content and Electrolyte Leakage Rate

The mushroom's internal tissue (1 g) was ground and homogenized with 10 mL of 100 g L^{-1} trichloroacetic acid (TCA) solution and centrifuged at 10,000× *g* and 4 °C for 20 min to obtain the mushroom extract. The extract was mixed with 4 mL of 0.67 g/100 g thiobarbituric acid (TBA), incubated in boiling water for 20 min, cooled, and centrifuged again to collect the supernatant. The extraction was replaced by 4 mL of 0.67 g/100 g TBA and 4 mL of 100 g L^{-1} TCA solution as the control blank. The absorbance of the supernatant was measured at 532 nm (OD_{532}), 600 nm (OD_{600}), and 450 nm (OD_{450}). The MDA content could be calculated as follows:

$$\mathrm{MDA}\left(\mathrm{nmol\ g^{-1}\ FW}\right) = \frac{[6.45 \times (OD532 - OD600) - 0.56 \times OD450] \times Volume}{Sample\ mass} \quad (5)$$

The electrolyte leakage rate was analyzed following a previously described method [37]. Approximately 2 g of mushrooms was cut into small cubes, washed 3 times with deionized water, and dried. The plates were immersed in 25 mL of deionized water at 25 °C for 30 min. The electrical conductivity (γ0) of the suspension solution was determined immediately using an electrical conductivity meter (DDSJ-308A, Leici Instrument Co., Shanghai, China). The electrical conductivity (γ1) was measured after the electrolyte was boiled for 10 min and cooled to 25 °C. The electrolyte leakage rate was calculated as follows.

$$\gamma(\%) = \frac{\gamma 0}{\gamma 1} \times 100\% \quad (6)$$

2.4.6. Enzyme Activity

Catalase (CAT) activity was measured following Cao and Jiang's method with some modifications [38]. CAT was measured in a reaction mixture of 1.5 mL of 50 mmol/L potassium phosphate buffer (pH 7.0), 0.3 mL of 50 mmol/L H_2O_2, and 0.2 mL of enzyme extract. The absorbance at 240 nm was recorded once every 30 s for 3 min. One unit of enzymatic activity was defined as the amount of the enzyme that caused a change of 0.01 in absorbance per minute and caused changes in the substrate. The specific CAT activity was expressed as U/g on a protein basis.

Superoxide dismutase (SOD) activity was measured according to the method of Cao [38]. The change in absorbance at 560 nm was recorded. SOD activity was expressed as U/g fresh weight (FW), where one unit of SOD activity was defined as the amount of enzyme that caused 50% inhibition of nitroblue tetrazolium (NBT).

2.5. Statistical Analysis

All quantitative data were statistically analyzed and presented as the means ± standard deviations (SD) of three independent treatments. One-way analysis of variance (ANOVA) was performed using IBM SPSS Statistics 22 to compare the means by Duncan's test. Differences among groups were considered significant at $p < 0.05$.

3. Results and Discussion

3.1. Characterizations of Films

3.1.1. Structural and Molecular Characterization of Films

Figure 1 presents the ATR-FTIR spectra of PLLA, PLGA, and PLBF. All three polymers in the spectrum exhibit the carbonyl group (C=O) stretching mode band as part of the PLLA unit at 1750 cm^{-1} [39]. This peak is the amorphous absorption peak of the PLLA ester carbonyl group, indicating that the PLLA unit in all three materials exhibits an amorphous state at this time. It was also observed at 1180 cm^{-1} and 1080 cm^{-1} that there was a robust asymmetric stretch vibration peak of the ether bond (C-O-C) and a symmetric stretch vibration absorption peak. Similarly, the characteristic peak at 1423 cm^{-1} in the PLGA spectra was assigned to the (-CH_2-) bending in GA units [40], while the characteristic peak at 980 cm^{-1} in the PLBF spectra was assigned to the C-H bending on trans CH=CH in PBF units [41]. This result showed that PLGA and PLBF copolymers were successfully synthesized.

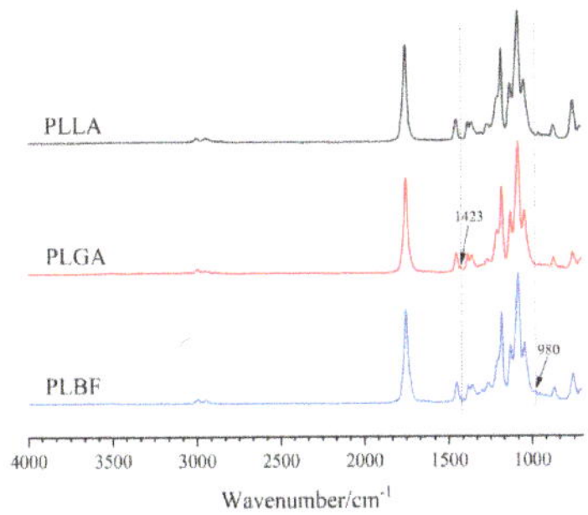

Figure 1. ATR-FTIR spectra of PLLA, PLGA, and PLBF.

The chemical shift values measured in this study are analogous to those reported in the literature for PLGA [42] and PBF [41] copolymers. As seen in Figure 2, the (2H, 4.85 and 4.60 ppm) peaks of GA units are multiple peaks, separated by a wide gap, indicating that in the one-step melt copolymerization of GA and LA, large molecules of random polymers are the main components of polymers, and the different sequences complicate peak splitting [43]. However, in the one-step manufacture of the PLBF copolymer, this phenomenon also exists, indicating that PLBF is also a random polymer. There are several explanations for such complex peak splitting, including the lower weight of GA and BF units and the complex sequences in the molecules in the PLGA and PLBF molecular chains.

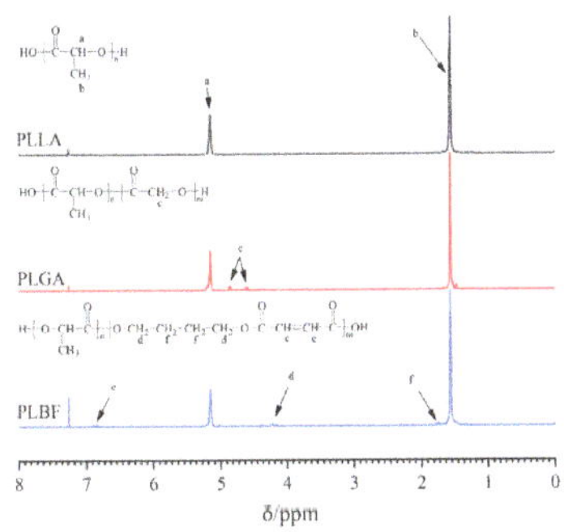

Figure 2. ^{1}H NMR spectra of PLLA, PLGA, and PLBF.

We evaluated the molecular weight, polydispersity (Pd), and relative mass ratio of polymers by GPC and ^{1}H NMR, as shown in Table 1. There is a range of PLLA units between 92.5 and 91 wt%, and all copolymers possess molecular weights greater than 4.0×10^4, indicating acceptable film formation characteristics. There was a slight difference in the percentage of each feedstock unit determined by NMR compared to the feedstock additions during the synthesis, which may be attributed to the PLLA homopolymer present in the systems during the synthesis. However, it is worth mentioning that all copolymers have relatively low Pd values, suggesting a narrow distribution of molecular weights. The above results demonstrated that PLLA, PLGA, and PLBF copolymers with high molecular weights were successfully synthesized in this study.

Table 1. Mechanical properties of PLLA/PLGA/PLBF modified atmosphere breathing membrane.

Sample Name	*x*/LA wt/wt	*x*/LA [a] wt/wt	Mn [b]	Pd [b]
PLLA	—	—	40,375	2.33
PLGA	8/92	7.5/92.5	45,296	2.02
PLBF	8/92	9/91	58,359	1.97

x means GA unit or BF unit. [a] Determined by ^{1}H NMR based on GA unit (2H, 4.85 and 4.60 ppm) and lactic acid (LA) unit (1H, 5.15 ppm) of the PLGA copolymers. Determined by ^{1}H NMR based on BF unit (1H, 6.88 and 4.23ppm) and lactic acid (LA) unit (1H, 5.15 ppm) of the PLBF copolymers. [b] Determined by GPC.—means this is a pure PLLA with no other components present.

3.1.2. Mechanical Properties

The mechanical properties of packaging materials are essential performance parameters and critical performance indicators in *Agaricus bisporus* packaging. Table 2 contains the parameters associated with tensile strength (σ_t), elongation at break (ε_b), and Young's modulus (E) for PLLA, PLGA, and PLBF films. PLLA films are brittle materials with a tensile strength of 40.3 MPa and an ε_b value of 5.7%; however, the ε_b values of the PLGA and PLBF films increased to 121.1% and 139%, respectively, while their σ_t values decreased to 23.3 and 25.6 MPa, respectively. Meanwhile, the E value of PLLA films was 1352 MPa, while the E values of PLGA and PLBF copolymer films decreased to 525 and 559 MPa, respectively, which was 61.2% and 58.7% less than that of PLLA films, showing a tomographic decrease.

Table 2. Mechanical properties of PLLA/PLGA/PLBF modified atmosphere breathing membrane.

Films	E/(MPa)	σ_t/(MPa)	ε_b/(%)
PLLA	1352 ± 16.7	40.3 ± 2.0	5.7 ± 1.1
PLGA	525 ± 8.6	23.3 ± 3.1	121.1 ± 10.4
PLBF	559 ± 6.4	25.6 ± 2.4	139 ± 15.8

Due to the addition of PGA and PBF units, the PLBF and PLGA copolymers have a higher degree of flexibility than PLLA, which is probably caused by a disruption of the regularity of the PLLA molecular chain. This decreases the σ_t values of PLGA and PLBF, consequently increasing flexibility [44]. Furthermore, the PLGA and PLBF films are in an amorphous state after copolymerization, thereby reducing the intermolecular forces, resulting in increased movement of molecular chain segments within the material, resulting in a significant decrease in E values.

3.1.3. Gas Permeability Properties

Gas permeability is a crucial factor influencing the freshness retention of equilibrium modified atmosphere packaging. Table 3 reveals OP, CDP, and $P_{C/O}$ for the copolymer films. It can be seen from Table 3 that CDP, OP, and $P_{C/O}$ for PLLA were 4.51×10^{-12} cm^3/m·s·Pa, 1.50×10^{-12} cm^3/m·s·Pa, and 3.0. In contrast, when GA and PBF were introduced into PLLA, CDP and OP decreased outstandingly, and $P_{C/O}$ increased significantly. It is worth noting that PLBF had the lowest CDP value of 2.08×10^{-12} cm^3/m·s·Pa and the lowest OP value of 0.46×10^{-12} cm^3/m·s·Pa; only 46% and 30% of PLLA, respectively, and PLBF has 1.5 times more $P_{C/O}$ than PLLA.

Table 3. Gas barrier properties of PLLA/PLGA/PLBF modified atmosphere breathing membrane.

Films	CDP (10^{-12}·cm^3/m·s·Pa)	OP (10^{-12}·cm^3/m·s·Pa)	$P_{C/O}$
PLLA	4.51 ± 0.35^{A}	1.50 ± 0.23^{A}	3.0
PLGA	2.55 ± 0.23^{B}	0.69 ± 0.13^{B}	3.7
PLBF	2.08 ± 0.12^{B}	0.46 ± 0.12^{B}	4.5

Note: Different capital letters in the same column with shoulder markers indicate significant difference ($p < 0.05$), same capital letters in the same column indicate non-significant difference ($p > 0.05$).

The decrease in CDP and OP of PLGA films is because the PGA chain segment possesses high gas barrier properties [45]. It should be noted, however, that in PLBF, the PBF chain segment has similar gas barrier properties to PBS [46]. Furthermore, the PBF chain segment contains unsaturated carbon–carbon double bonds and a stable trans double bond structure, which makes PLBF less susceptible to cis-trans isomerization, resulting in PLBF being more rigid and less susceptible to gas passage, thereby promoting a significant improvement in the barrier properties of PLBF. The size of gas molecules can also affect their solubility and diffusion in the polymer; O_2 molecules have a diameter of 0.346 nm [47], while CO_2 molecules have a diameter of 0.280 nm [48]. The large size of O_2 molecules makes it more difficult to pass through PLBF copolymer films, and thus, $P_{C/O}$ is enhanced.

3.2. *Characteristics of Mushrooms*

3.2.1. Headspace Analysis

Figure 3A shows the variation pattern of O_2 inside each package group with storage time. The O_2 concentration of all four package treatments, PE, PLLA, PLGA, and PLBF, showed a sharp decrease in O_2 concentration during 3 days of storage. The oxygen concentration fluctuated between 0.74%–0.38%, 0.59–0.24%, 0.74–0.38%, 0.33–0.15%, and 0.27–0.11%, respectively. Typically, lower oxygen concentrations slow the respiration rate of mushrooms, reducing changes in physiological structure and extending shelf life. In contrast, the respiration of mushrooms was first inhibited in the PLBF package group due to its lowest OP. Due to the relatively close O_2 concentrations in each sample package, the effects on the weight loss of mushrooms may not differ greatly after reaching equilibrium. A similar study noted that the O_2 concentration in the package was less than 1% on the 15th day but still better maintained the quality of edible mushrooms better and prolonged storage to 25 days [49].

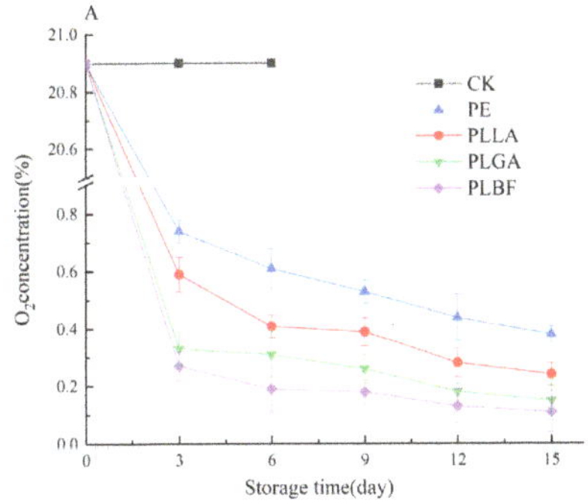

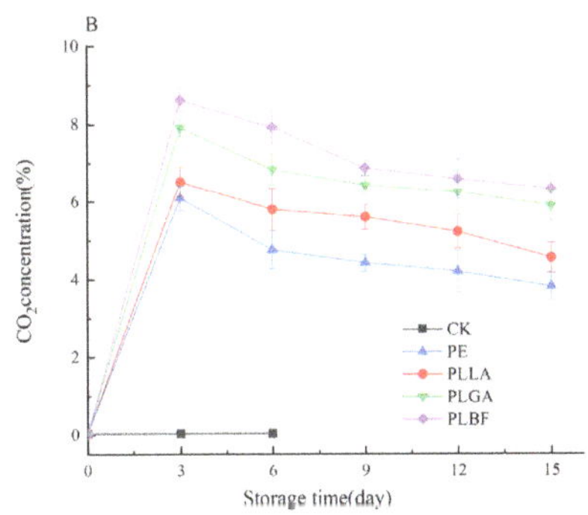

Figure 3. Changes in gas components O_2 (**A**) and CO_2 (**B**) of each packaging group during storage.

As shown in Figure 3B, the CO_2 concentration inside the packaging with PLBF film was higher than that inside the packaging with the other films. This was attributed to the lower CDP of PLLA films incorporated with PBF. In general, a CO_2 concentration lower than 12% could reduce the growth and reproduction of thermophilic bacteria, browning, and respiration rate of mushrooms [7] and help maintain the hardness of mushrooms. Moreover, a high CO_2 concentration of over 2.5% was also beneficial in reducing the opening rate of mushrooms [50]. As the storage time increased (6 days-15 days), the PLBF treatment group still had the highest CO_2 content, fluctuating between 6.33% and 7.92%, and the CO_2 concentration in PLBF was significantly higher than that in the other three treatment groups throughout the storage period. The reasons for this were mainly due to the variability in the CO_2 transmission properties of the different films.

3.2.2. Weight Loss

The weight loss rate during storage can be utilized as an essential indicator to determine white mushroom freshness due to their delicate tissue and high moisture content. Respiration and transpiration processes result in water loss and nutrient depletion, which causes white mushrooms to deteriorate during storage [51,52]. The weight loss rate of white mushrooms packaged in different films is presented in Figure 4. In all samples, weight loss increased as the storage period increased. The CK group could not control the dehydration and transpiration of the mushroom skin due to a lack of film protection, resulting in a mass loss rate of 30.99% on the third day of preservation, which was significantly higher than that of the other packaged groups.

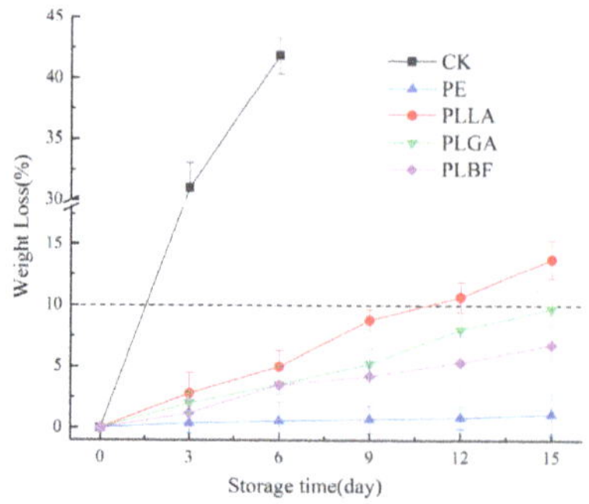

Figure 4. Changes in the weight of *Agaricus bisporus* in each packing group during storage.

The quality loss of mushrooms packaged in PE film was always at a low level, only 1.7% after 15 days of storage, which was significantly lower than that of the PLLA (13.8%), PLGA (9.75%), and PLBF groups (6.78%), which was mainly due to the low water vapor permeability of the PE film, which resulted in a high humidity environment inside the package and reduced transpiration [28,53]. However, after 3 days of storage of *Agaricus bisporus*, water droplets could be found on both the film and *Agaricus bisporus* in the PE group packaging, and with the extension of the storage period, a large number of water droplets condensed to produce condensation, and condensation inside the packaging would promote the growth of microorganisms and accelerate the spoilage of *Agaricus bisporus*.

On the 15th day of storage, no condensation occurred inside the PLGA and PLBF groups, and the weight loss rate of the PLBF group was 6.78%, which was significantly lower than that of the PLGA group (9.75%), indicating that the change in weight loss rate was closely related to the packaging material. In addition to these factors, the change in the weight loss rate of mushrooms during storage was also related to other factors, such as the release of CO_2 due to respiration, transpiration, and the role of microorganisms [50,54]. To a certain extent, packaging with a modified atmosphere could reduce water loss from produce.

3.2.3. Browning Index

Agaricus bisporus is prone to browning after harvest, and the degree of browning increases as the time after harvest increases (Figure 5). It is generally believed that the browning of fresh *Agaricus bisporus* is mainly caused by enzymatic browning, and during storage, white mushrooms turn from white to brown or black due to enzymatic browning of the substrates, which negatively affects their commercial value [55]. Figure 6 illustrates the impact of packaging on BI when mushrooms are packaged in different films. All mushrooms showed an increase in BI during storage, indicating that their whiteness was reduced. Figure 6 shows that the BI of both the CK group and the packaged group increased as the storage period increased. On day 3 of storage, there was a significant difference in BI between the CK group and the packaged group, with the control group having a significantly higher BI than the packaged group. After day 9 of storage, the BI of the PLBF group was significantly lower than that of the PE, PLLA, and PLGA groups, and this advantage was maintained until the end of the storage period. The high BI value in the PE group is probably a result of high relative humidity in the package, which resulted in high bacterial growth, reproduction, and number, and therefore *Agaricus bisporus* substrates were degraded and the oxidation of phenolic substances by tyrosinase produced brown substances [56].

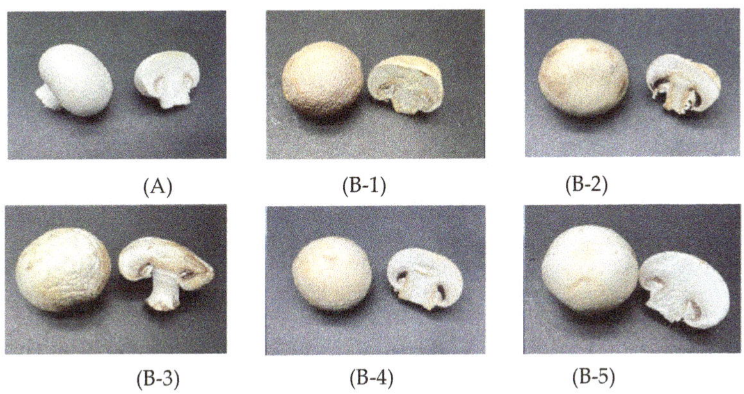

Figure 5. Surface and section images of *Agaricus bisporus* of CK (**A**) in 0 days and CK (**B-1**), PE (**B-2**), PLLA (**B-3**), PLGA (**B-4**), and PLBF (**B-5**) at 15th days.

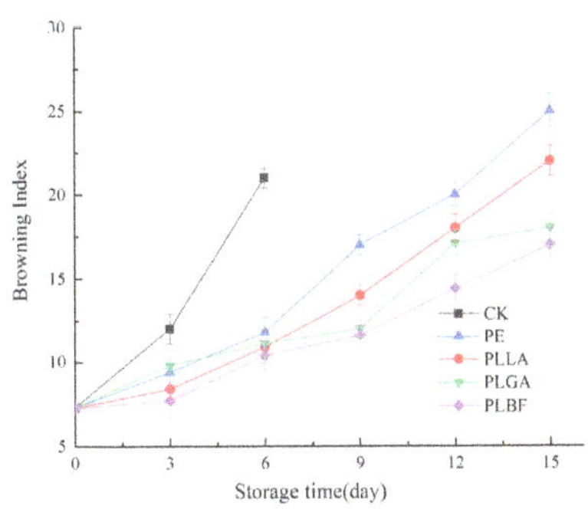

Figure 6. Changes of BI value of *Agaricus bisporus* in each packing group during storage.

3.2.4. Total Phenolics and Ascorbic Acid

Phenolic compounds are major antioxidant components in mushrooms and other plants. They are important secondary metabolites that scavenge free radicals and chelate metals and reduce lipoxygenase activity, which is essential in resisting adverse external environmental stresses, internal oxidative stresses, and pathogenic bacterial growth infestation. Oxidation of phenolics is the main factor in the postharvest browning of *Agaricus bisporus*, which adversely affects its commercial value [57,58]. As shown in Figure 7, mushrooms' total phenolic (TP) content was reduced differently in different packages. The TP content in the control group was the lowest at the end because polyphenols might convert to quinones by polyphenol oxidases and react with oxygen to form brown-colored compounds [59].

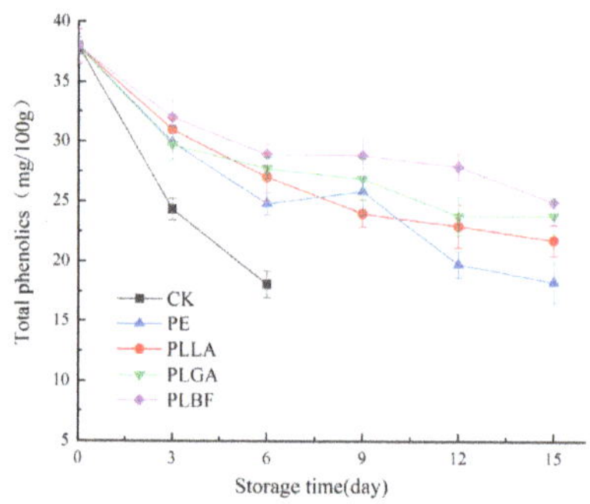

Figure 7. Changes of total phenolic content of *Agaricus bisporus* in each packaging group during storage.

After 6 days, the TP content was reduced by 52.34%, 34.64%, and 28.80% in the CK, PE, and PLLA groups, respectively, compared to the initial TP content, while the PLGA and PLBF groups had significantly higher TP contents than the other groups. Throughout the storage period, the decreasing trend of phenolics in the samples of the PLBF group was smoother and consistently higher than the other packaging groups. This indicates that the low O_2 and high CO_2 gas environment formed inside the PLBF package can effectively inhibit the activity of polyphenol oxidase, thus reducing the oxidation of polyphenols and inhibiting the decrease in TP content during storage of *Agaricus bisporus*.

Ascorbic acid (AA) is a powerful endogenous antioxidant, that plays a vital role in removing reactive oxygen species from fruits and vegetables, and is related to the senescence of mushrooms. As shown in Figure 8, the AA content of all treatment groups showed a decreasing trend during storage, among which the AA content of the CK group decreased rapidly and was significantly lower than that of the other four treatment groups throughout storage. The AA content in the PE, PLLA, and PLGA treated groups decreased rapidly to 13 mg/100 g, 15 mg/100 g, and 18 mg/100 g, exhibiting a decrease of 69.04%, 62.2%, and 57.1% at the end of storage, respectively. The AA content of the PLBF group decreased relatively slowly with the extension of storage time, and the AA content of this group was higher than that of the other three packaged groups throughout storage and remained at 22.01 mg/100 g at the end of storage. The loss of AA was better inhibited in the PLBF treatment group, which may be related to the higher CO_2 content in the package, and the higher concentration of CO_2 in the bag could form hydrogen bonds between the cell wall and polar groups, preventing the expression of certain enzymes and AA reduction [19]. Furthermore, AA can reduce reactive oxygen species (ROS) through the ascorbate-glutathione cycle or react with reactive forms of oxygen such as ozone to inhibit tissue senescence and browning [60], and the rapid loss of AA in the three packaging groups PE, PLLA, and PLGA may be related to the rate of reactive oxygen species production in these three groups.

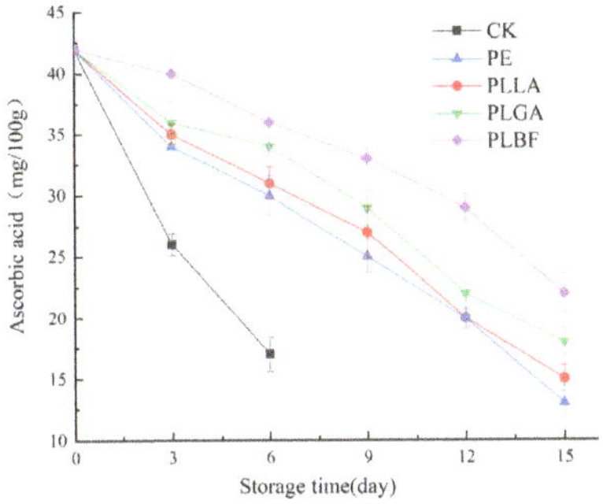

Figure 8. Changes of total ascorbic acid of *Agaricus bisporus* in each packaging group during storage.

3.2.5. Malondialdehyde Content and Electrolyte Leakage Rate

Electrolyte leakage(EC) and malondialdehyde (MDA) content indicate the extent of damage and membrane lipid peroxidation in *Agaricus bisporus* cell membranes [61,62], which is indicative of the quality of *Agaricus bisporus* as is evident from Figures 9 and 10, which show that the electrolyte leakage and MDA of all samples raised gradually during storage. The explanation for this is that the quality of *Agaricus bisporus* continues to decline with storage time, with soft wilting and browning, leading to oxidation of the lipids and efflux of cytoplasm from *Agaricus bisporus*.

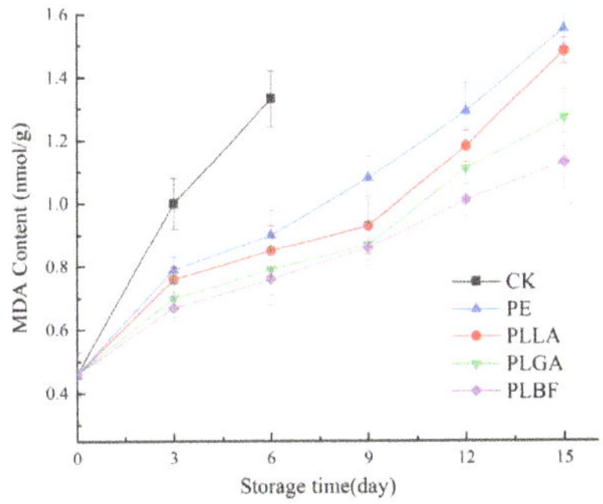

Figure 9. Changes of MDA of *Agaricus bisporus* in each packing group during storage.

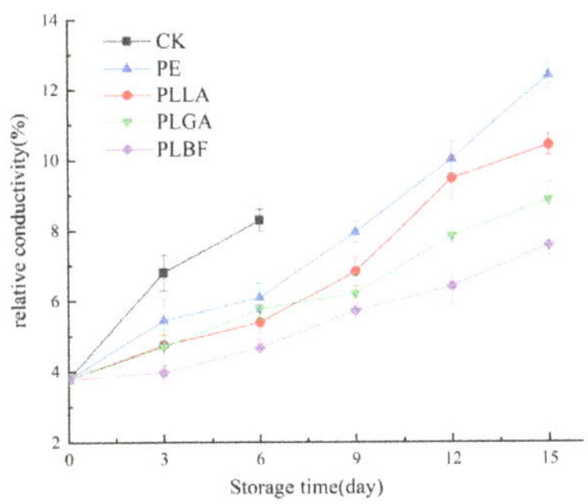

Figure 10. Changes of relative conductivity of cell membrane of *Agaricus bisporus* in each group during storage.

On the 6th day of storage, the relative conductivity and MDA content of the CK group increased to 8.3% and 1.33 nmol/g, respectively. The relative conductivity of PE and

PLLA increased to 12.4% and 10.4% on the 15th day of storage, while the MDA content increased to 1.55 nmol/g and 1.48 nmol/g, respectively. Because an outer packaging does not safeguard CK, while the PE and PLLA groups have a high O_2 content inside the bags, the metabolically active cells of *Agaricus bisporus* are under stress from the gaseous environment, which leads to lipid oxidation within the cell, further disrupting the integrity of the cell and leading to an increase in relative conductivity. In addition, the hyperoxic environment also accelerates the oxidation of TP and AA, leading to a sharp rise in MDA.

On the 15th day of storage, the relative conductivity of PLGA and PLBF was 8.8% and 7.6%, and the MDA was 1.27 nmol/g and 1.13 nmol/g, respectively, markedly inferior to the other groups. This can be attributed to the fact that the low O_2 and higher CO_2 environment created by the PLGA and PLBF groups effectively inhibited the respiratory metabolism of *Agaricus bisporus*. It is worth mentioning that PLBF maintained the lowest relative conductivity and MDA values compared to the other groups, which indicates that the storage atmosphere created by PLBF delays the oxidation of *Agaricus bisporus* membrane lipids while effectively maintaining the integrity of *Agaricus bisporus* cells.

3.2.6. Enzyme Activity

The activities of catalase (CAT) and superoxide dismutase (SOD) enzymes are intimately related to the degree of damage of the organism, which can directly affect the quality of *Agaricus bisporus* [11]. Antioxidant enzymes such as CAT and SOD play a crucial role in antioxidant defense during the storage of fruit and vegetables [9].

The changes in CAT and SOD enzyme activities are summarized in Figures 11 and 12. CAT and SOD enzyme activities showed an overall increasing trend at the beginning of storage, and the enzyme activity decreased as the storage time increased. This can be explained by the inability of the mycelium of *Agaricus bisporus* to be supplied with water and inorganic and organic nutrients after harvest. They resulted in the badlands' stress of the SOD and CAT enzymes, which may have contributed to the increase in CAT and SOD enzyme activities [63]. In the later stages of storage, the quality of *Agaricus bisporus* dropped, and the activity of CAT and SOD enzymes continued to decline.

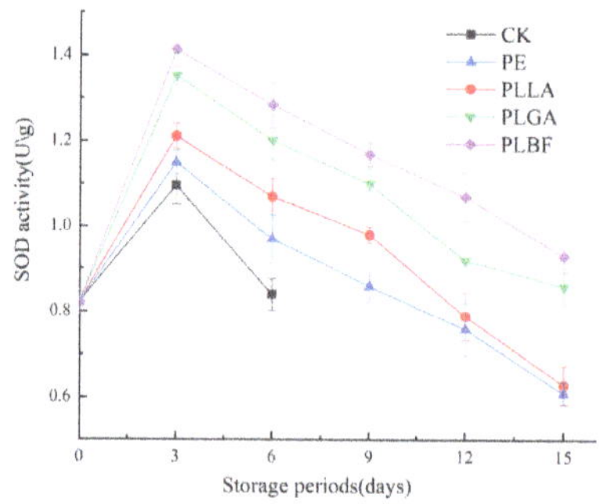

Figure 11. Changes of SOD activity of *Agaricus bisporus* in each packing group during storage.

The CAT and SOD activities of PLBF were higher during storage. On the 15th day of storage, the CAT activities of the PE, PLLA, and PLGA groups were only 13 U/g, 15 U/g, and 20 U/g, and the SOD activities were only 0.61 U/g, 0.64 U/g, and 0.86 U/g, respectively, which were substantially weaker than those of the PLBF group. The CAT and SOD activities of the PLBF group were as high as 28 U/g and 0.93 U/g, respectively. These manifestations indicate that the gas fraction formed by PLBF (0.11% O_2 + 6.3% CO_2) can effectively maintain the activity of CAT and SOD, slow the oxidation and ageing process of *Agaricus bisporus*, and maintain the excellent quality of *Agaricus bisporus*.

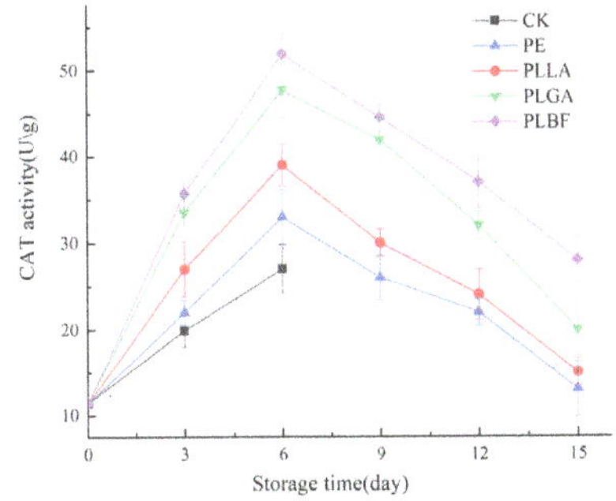

Figure 12. Changes of CAT activity of *Agaricus bisporus* in each packing group during storage.

3.2.7. Relevance Analysis

Correlation analysis between the gas composition in the package and the quality of *Agaricus bisporus* was performed. As shown in Table 4, there was a close relationship between the gas composition in the package and the storage quality of *Agaricus bisporus*. In the results obtained, the O_2 concentration in the bag was highly significantly correlated with the TP of *Agaricus bisporus* ($p < 0.01$) and with the BI and EC of *Agaricus bisporus* in the bag ($p < 0.05$). The CO_2 concentration in the bag was highly significantly correlated ($p < 0.01$) with BI and significantly correlated with TP, AA, MDA, EC, and SOD of *Agaricus bisporus*. In addition, O_2 inside the package had a strong correlation of −0.924, 0.930, −0.875, and −0.901 with AA, MDA, CAT, and SOD of *Agaricus bisporus*, respectively, and CO_2 inside the package had a strong correlation of 0.931 with CAT of *Agaricus bisporus*. These results further suggest that regulating gas composition inside the pouch by changing the OTR and CDTR of PLLA-based films can further regulate the changes in postharvest physiological and biochemical indices, sensory traits, and antioxidant enzymes of *Agaricus bisporus* and improve its postharvest quality and shelf life of. This is because the TP and AA of *Agaricus bisporus* are easily oxidized [64], and higher O_2 concentrations in the storage environment of *Agaricus bisporus* lead to higher BI, EC, and MDA and lower CAT and SOD [65,66]. Higher CO_2 concentrations and lower O_2 concentrations are more favorable for maintaining the quality of *Agaricus bisporus* [11].

Table 4. Correlation analysis between gas composition in packing and postharvest quality of *Agaricus bisporus*.

	O_2	CO_2	WI	BI	TP	AA	MDA	EC	CAT	SOD
O_2	1	−0.972 *	−0.507	0.987 *	−10.00 **	−0.924	0.930	0.989 *	−0.875	−0.901
CO_2		1	0.296	−0.996 **	0.969 *	0.962 *	−0.984 *	−0.984 *	0.931	0.978 *
WI			1	−0.377	0.511	0.184	−0.158	−0.398	0.082	0.094
BI				1	−0.984 *	−0.944	0.966 *	0.987 *	−0.904	−0.957 *
TP					1	0.926	−0.929	−0.990 **	0.878	0.897
AA						1	−0.989 *	−0.970 *	0.993 **	0.954 *
MDA							1	0.968 *	−0.978 *	−0.987 *
EC								1	−0.936	−0.935
CAT									1	0.942
SOD										1

Note: ** represents extremely significant correlation ($p < 0.01$); * represents significant correlation ($p < 0.05$).

4. Conclusions

PLLA film cooperative polymerization with OBF or GA showed a positive preservation effect on the cold storage of *Agaricus bisporus*. Compared with the control, PE, and PLLA film packaged groups, copolymer-modified PLLA films could not only decrease the rate of weight loss, browning index, and electrolyte leakage rate of *Agaricus bisporus* but also decrease the contents of MDA. Meanwhile, TP and AA increased in the PLGA and PLBF groups. Additionally, PLLA films incorporated with OBF increased the activity of SOD and CAT in mushrooms during storage. The results showed that, in two copolymer-modified PLLA films, PLBF packaging had lower OP and CDP. It effectively reduced the weight loss rate to less than 7% at the end of storage. In addition, it created the optimal gas composition for *Agaricus bisporus* preservation, 0.27–0.11% O_2 and 6.33~7.92% CO_2, which significantly reduced the respiration metabolism, membrane lipid peroxidation, and postharvest senescence of *Agaricus bisporus* ($p < 0.05$). It maintained the highest level of market acceptability ($p < 0.05$). The shelf life of *Agaricus bisporus* in PLBF reached 15 days. This study provides a feasible solution to reduce postharvest senescence and maintain the quality of *Agaricus bisporus*.

Author Contributions: Conceptualization, T.D.; methodology, J.H. and E.C.; validation, J.H. and E.C.; formal analysis, T.S., J.B. and Y.W.; investigation, J.B.; resources, J.H.; data curation, J.B. and Y.W.; writing—original draft preparation, T.S.; writing—review and editing, T.D. and X.Y.; visualization, T.S.; funding acquisition, T.D. and X.Y. All authors have read and agreed to the published version of the manuscript.

Funding: This work was supported by National Natural Science Fund Project (21965026); Science and Technology Program of Inner Mongolia Autonomous Region (2021GG337); Youth Science and Technology Talent Development Project of Inner Mongolia Autonomous Region (Innovation Team) (NMGIRT2310).

Data Availability Statement: Data is contained within the article.

Conflicts of Interest: The authors declare that they have no known competing financial interest or personal relationships that could have appeared to influence the work reported in this paper.

References

1. Sun, B.; Chen, X.; Xin, G.; Qin, S.; Chen, M.; Jiang, F. Effect of 1-methylcyclopropene (1-MCP) on quality of button mushrooms (*Agaricus bisporus*) packaged in different packaging materials. *Postharvest Biol. Technol.* **2020**, *159*, 111023. [CrossRef]
2. Yan, M.; Yuan, B.; Xie, Y.; Cheng, S.; Huang, H.; Zhang, W.; Chen, J.; Cao, C. Improvement of postharvest quality, enzymes activity and polyphenoloxidase structure of postharvest *Agaricus bisporus* in response to high voltage electric field. *Postharvest Biol. Technol.* **2020**, *166*, 111230. [CrossRef]
3. Zhang, K.; Pu, Y.-Y.; Sun, D.-W. Recent advances in quality preservation of postharvest mushrooms (*Agaricus bisporus*): A review. *Trends Food Sci. Technol.* **2018**, *78*, 72–82. [CrossRef]
4. Wu, W.; Ni, X.; Shao, P.; Gao, H. Novel packaging film for humidity-controlled manipulating of ethylene for shelf-life extension of *Agaricus bisporus*. *LWT* **2021**, *145*, 111331. [CrossRef]
5. Chang, C.-K.; Cheng, K.-C.; Hou, C.-Y.; Wu, Y.-S.; Hsieh, C.-W. Development of Active Packaging to Extend the Shelf Life of *Agaricus bisporus* by Using Plasma Technology. *Polymers* **2021**, *13*, 2120. [CrossRef]
6. Gholami, R.; Ahmadi, E.; Farris, S. Shelf life extension of white mushrooms (*Agaricus bisporus*) by low temperatures conditioning, modified atmosphere, and nanocomposite packaging material. *Food Packag. Shelf Life* **2017**, *14*, 88–95. [CrossRef]
7. Zhang, L.; Liu, Z.; Sun, Y.; Wang, X.; Li, L. Combined antioxidant and sensory effects of active chitosan/zein film containing α-tocopherol on *Agaricus bisporus*. *Food Packag. Shelf Life* **2020**, *24*, 100470. [CrossRef]
8. Hu, Y.-H.; Chen, C.-M.; Xu, L.; Cui, Y.; Yu, X.-Y.; Gao, H.-J.; Wang, Q.; Liu, K.; Shi, Y.; Chen, Q.-X. Postharvest application of 4-methoxy cinnamic acid for extending the shelf life of mushroom (*Agaricus bisporus*). *Postharvest Biol. Technol.* **2015**, *104*, 33–41. [CrossRef]
9. Jiang, T.; Zheng, X.; Li, J.; Jing, G.; Cai, L.; Ying, T. Integrated application of nitric oxide and modified atmosphere packaging to improve quality retention of button mushroom (*Agaricus bisporus*). *Food Chem.* **2011**, *126*, 1693–1699. [CrossRef]
10. Gao, M.; Feng, L.; Jiang, T. Browning inhibition and quality preservation of button mushroom (*Agaricus bisporus*) by essential oils fumigation treatment. *Food Chem.* **2014**, *149*, 107–113. [CrossRef]
11. Chen, C.; Chen, W.; Dai, F.; Yang, F.; Xie, J. Development of Packaging Films With Gas Selective Permeability Based On Poly (butylene Adipate-co-terephthalate)/Poly (butylene Succinate) and Its Application in the Storage of White Mushroom (*Agaricus Bisporus*). *Food Bioprocess Technol.* **2022**, *15*, 1–16. [CrossRef]
12. Genovese, L.; Soccio, M.; Lotti, N.; Gazzano, M.; Siracusa, V.; Salatelli, E.; Balestra, F.; Munari, A. Design of biobased PLLA triblock copolymers for sustainable food packaging: Thermo-mechanical properties, gas barrier ability and compostability. *Eur. Polym. J.* **2017**, *95*, 289–303. [CrossRef]
13. Jiang, Y.; Yan, C.; Wang, K.; Shi, D.; Liu, Z.; Yang, M. Super-toughed PLA blown film with enhanced gas barrier property available for packaging and agricultural applications. *Materials* **2019**, *12*, 1663. [CrossRef]
14. Qiu, S.; Zhou, Y.; Waterhouse, G.I.; Gong, R.; Xie, J.; Zhang, K.; Xu, J. Optimizing interfacial adhesion in PBAT/PLA nanocomposite for biodegradable packaging films. *Food Chem.* **2021**, *334*, 127487. [CrossRef]

15. Hosseini, S.F.; Kaveh, F.; Schmid, M. Facile fabrication of transparent high-barrier poly (lactic acid)-based bilayer films with antioxidant/antimicrobial performances. *Food Chem.* **2022**, *384*, 132540. [CrossRef]
16. Ali, F.B.; Kang, D.J.; Kim, M.P.; Cho, C.H.; Kim, B.J. Synthesis of biodegradable and flexible, polylactic acid based, thermoplastic polyurethane with high gas barrier properties. *Polym. Int.* **2014**, *63*, 1620–1626. [CrossRef]
17. Yuk, J.S.; Mo, E.; Kim, S.; Jeong, H.; Gwon, H.; Kim, N.-K.; Kim, Y.-W.; Shin, J. Thermoplastic Superelastomers Based on Poly (isobutylene)-graft-Poly (l-lactide) Copolymers: Enhanced Thermal Stability, Tunable Tensile Strength, and Gas Barrier Property. *Macromolecules* **2020**, *53*, 2503–2515. [CrossRef]
18. Marano, S.; Laudadio, E.; Minnelli, C.; Stipa, P. Tailoring the Barrier Properties of PLA: A State-of-the-Art Review for Food Packaging Applications. *Polymers* **2022**, *14*, 1626. [CrossRef]
19. Cheng, P.; Yun, X.; Xu, C.; Yang, Y.; Han, Y.; Dong, T. Use of poly (ε-caprolactone)-based films for equilibrium-modified atmosphere packaging to extend the postharvest shelf life of garland chrysanthemum. *Food Sci. Nutr.* **2019**, *7*, 1946–1956. [CrossRef]
20. Mistriotis, A.; Briassoulis, D.; Giannoulis, A.; D'Aquino, S. Design of biodegradable bio-based equilibrium modified atmosphere packaging (EMAP) for fresh fruits and vegetables by using micro-perforated poly-lactic acid (PLA) films. *Postharvest Biol. Technol.* **2016**, *111*, 380–389. [CrossRef]
21. Matar, C.; Salou, T.; Hélias, A.; Pénicaud, C.; Gaucel, S.; Gontard, N.; Guilbert, S.; Guillard, V. Benefit of modified atmosphere packaging on the overall environmental impact of packed strawberries. *Postharvest Biol. Technol.* **2021**, *177*, 111521. [CrossRef]
22. Wang, L.; Zhou, Y.; Wang, Y.; Bu, H.; Dong, T. Changes in cell wall metabolism and flavor qualities of mushrooms (*Agaricus bernardii*) under EMAP treatments during storage. *Food Packag. Shelf Life* **2021**, *29*, 100732. [CrossRef]
23. Tumwesigye, K.S.; Sousa, A.; Oliveira, J.; Sousa-Gallagher, M. Evaluation of novel bitter cassava film for equilibrium modified atmosphere packaging of cherry tomatoes. *Food Packag. Shelf Life* **2017**, *13*, 1–14. [CrossRef]
24. Zalewska, M.; Marcinkowska-Lesiak, M.; Onopiuk, A.; Stelmasiak, A.; Półtorak, A. Modified atmosphere packaging for extending the shelf life of fresh Agaricus bisporus. *J. Food Process. Preserv.* **2018**, *42*, e13839. [CrossRef]
25. Sami, R.; Elhakem, A.; Alharbi, M.; Benajiba, N.; Almatrafi, M.; Abdelazez, A.; Helal, M. Evaluation of antioxidant activities, oxidation enzymes, and quality of nano-coated button mushrooms (*Agaricus bisporus*) during storage. *Coatings* **2021**, *11*, 149. [CrossRef]
26. Cheng, P.-f.; Liang, M.; Yun, X.-y.; Dong, T. Biodegradable blend films of poly (ε-caprolactone)/poly (propylene carbonate) for shelf life extension of whole white button mushrooms. *J. Food Sci. Technol.* **2022**, *59*, 144–156. [CrossRef] [PubMed]
27. Yun, X.; Wang, Y.; Li, M.; Jin, Y.; Han, Y.; Dong, T. Application of permselective poly (ε-caprolactone) film for equilibrium-modified atmosphere packaging of strawberry in cold storage. *J. Food Process. Preserv.* **2017**, *41*, e13247. [CrossRef]
28. Qin, Y.; Liu, D.; Wu, Y.; Yuan, M.; Li, L.; Yang, J. Effect of PLA/PCL/cinnamaldehyde antimicrobial packaging on physicochemical and microbial quality of button mushroom (*Agaricus bisporus*). *Postharvest Biol. Technol.* **2015**, *99*, 73–79. [CrossRef]
29. Zheng, L.; Wang, Z.; Li, C.; Zhang, D.; Xiao, Y. Novel unsaturated aliphatic polyesters: Synthesis, characterization, and properties of multiblock copolymers composing of poly (butylene fumarate) and poly (1, 2-propylene succinate). *Ind. Eng. Chem. Res.* **2012**, *51*, 14107–14114. [CrossRef]
30. Naghavi Sheikholeslami, S.; Rafizadeh, M.; Afshar Taromi, F.; Shirali, H. Crystallization and photo-curing kinetics of biodegradable poly (butylene succinate-co-butylene fumarate) short-segmented block copolyester. *Polym. Int.* **2017**, *66*, 289–299. [CrossRef]
31. Zheng, Y.; Liu, X.; Liu, F.; Xue, J.; Zhou, J.; Huo, H.; Li, L. Crystallization of isomorphic poly (butylene succinate-co butylene fumarate) biopolyester: Single crystals and ring-banded spherulites. *Polym. Test.* **2018**, *68*, 379–387. [CrossRef]
32. Zheng, Y.; Pan, P. Crystallization of biodegradable and biobased polyesters: Polymorphism, cocrystallization, and structure-property relationship. *Prog. Polym. Sci.* **2020**, *109*, 101291. [CrossRef]
33. Ellingford, C.; Samantaray, P.K.; Farris, S.; McNally, T.; Tan, B.; Sun, Z.; Huang, W.; Ji, Y.; Wan, C. Reactive extrusion of biodegradable PGA/PBAT blends to enhance flexibility and gas barrier properties. *J. Appl. Polym. Sci.* **2022**, *139*, 51617. [CrossRef]
34. Regubalan, B.; Manibalan, S.; Pandit, P. *Polyglycolic Acid-Based Bionanocomposites for Food Packaging Applications*; Elsevier: Amsterdam, The Netherlands, 2022; pp. 153–164.
35. Pirie, A.; Mullins, M.G. Changes in anthocyanin and phenolics content of grapevine leaf and fruit tissues treated with sucrose, nitrate, and abscisic acid. *Plant Physiol.* **1976**, *58*, 468–472. [CrossRef]
36. Song, H.; Yuan, W.; Jin, P.; Wang, W.; Wang, X.; Yang, L.; Zhang, Y. Effects of chitosan/nano-silica on postharvest quality and antioxidant capacity of loquat fruit during cold storage. *Postharvest Biol. Technol.* **2016**, *119*, 41–48. [CrossRef]
37. Kaya, C.; Kirnak, H.; Higgs, D.; Saltali, K. Supplementary calcium enhances plant growth and fruit yield in strawberry cultivars grown at high (NaCl) salinity. *Sci. Hortic.* **2002**, *93*, 65–74. [CrossRef]
38. Cao, J.; Jiang, W.; Zhao, Y. Experiment Guidance of Postharvest Physiology and Biochemistry of Fruits and Vegetables; China Light Industry Press, Beijing, China, 2007; pp. 84–87.
39. Yun, X.; Li, X.; Cheng, P.; Pan, P.; Dong, T. Controllable Poly (L-lactic acid) Soft Film with Respirability and Its Effect on Strawberry Preservation. *Polym. Sci. Ser. A* **2021**, *63*, 77–90. [CrossRef]
40. Chen, S.; Zhang, X.; He, M.; Li, J. Degradation of PGA, prepared by reactive extrusion polymerization, in water, humid, and dry air, and in a vacuum. *J. Mater. Res.* **2020**, *35*, 1846–1856. [CrossRef]
41. Zheng, L.; Wang, Z.; Li, C.; Xiao, Y.; Zhang, D.; Guan, G.; Zhu, W. Synthesis, characterization and properties of novel linear poly (butylene fumarate) bearing reactive double bonds. *Polymer* **2013**, *54*, 631–638. [CrossRef]

42. Wang, L.; Zhang, Z.; Chen, H.; Zhang, S.; Xiong, C. Preparation and characterization of biodegradable thermoplastic Elastomers (PLCA/PLGA blends). *J. Polym. Res.* **2010**, *17*, 77–82. [CrossRef]
43. Lu, D.D.; Yuan, J.C.; Lei, Z.Q. High molecular weight biodegraded poly (lactic acid-glycolic acid-ε-caprolactam) copolymer: Direct polycondensation of lactic acid, glycolic acid and ε-caprolactam using Sn (II)-organic anhydride as catalysts. *Polym. Adv. Technol.* **2009**, *20*, 536–540. [CrossRef]
44. Coudane, J.; Van Den Berghe, H.; Mouton, J.; Garric, X.; Nottelet, B. Poly (Lactic Acid)-Based Graft Copolymers: Syntheses Strategies and Improvement of Properties for Biomedical and Environmentally Friendly Applications: A Review. *Molecules* **2022**, *27*, 4135. [CrossRef]
45. Magazzini, L.; Grilli, S.; Fenni, S.E.; Donetti, A.; Cavallo, D.; Monticelli, O. The Blending of Poly (glycolic acid) with Polycaprolactone and Poly (l-lactide): Promising Combinations. *Polymers* **2021**, *13*, 2780. [CrossRef] [PubMed]
46. Peñas, M.I.; Pérez-Camargo, R.A.; Hernández, R.; Müller, A.J. A Review on Current Strategies for the Modulation of Thermomechanical, Barrier, and Biodegradation Properties of Poly (*Butylene succinate*) (PBS) and Its Random Copolymers. *Polymers* **2022**, *14*, 1025. [CrossRef] [PubMed]
47. McKeown, N.B.; Budd, P.M.; Msayib, K.J.; Ghanem, B.S.; Kingston, H.J.; Tattershall, C.E.; Makhseed, S.; Reynolds, K.J.; Fritsch, D. Polymers of intrinsic microporosity (PIMs): Bridging the void between microporous and polymeric materials. *Chem. A Eur. J.* **2005**, *11*, 2610–2620. [CrossRef]
48. Echeverría, J.C.; Estella, J.; Barbería, V.; Musgo, J.; Garrido, J.J. Synthesis and characterization of ultramicroporous silica xerogels. *J. Non-Cryst. Solids* **2010**, *356*, 378–382. [CrossRef]
49. Jafri, M.; Jha, A.; Bunkar, D.S.; Ram, R.C. Quality retention of oyster mushrooms (*Pleurotus florida*) by a combination of chemical treatments and modified atmosphere packaging. *Postharvest Biol. Technol.* **2013**, *76*, 112–118. [CrossRef]
50. Guillaume, C.; Schwab, I.; Gastaldi, E.; Gontard, N. Biobased packaging for improving preservation of fresh common mushrooms (*Agaricus bisporus* L.). *Innov. Food Sci. Emerg. Technol.* **2010**, *11*, 690–696. [CrossRef]
51. Pourbagher, R.; Abbaspour-Fard, M.H.; Sohbatzadeh, F.; Rohani, A. Inhibition of enzymes and Pseudomonas tolaasii growth on *Agaricus bisporus* following treatment with surface dielectric barrier discharge plasma. *Innov. Food Sci. Emerg. Technol.* **2021**, *74*, 102833. [CrossRef]
52. Wang, T.; Yun, J.; Zhang, Y.; Bi, Y.; Zhao, F.; Niu, Y. Effects of ozone fumigation combined with nano-film packaging on the postharvest storage quality and antioxidant capacity of button mushrooms (*Agaricus bisporus*). *Postharvest Biol. Technol.* **2021**, *176*, 111501. [CrossRef]
53. YE, J.-J.; LI, J.-R.; HAN, X.-X.; Zhang, L.; JIANG, T.-J.; Miao, X. Effects of active modified atmosphere packaging on postharvest quality of shiitake mushrooms (*Lentinula edodes*) stored at cold storage. *J. Integr. Agric.* **2012**, *11*, 474–482. [CrossRef]
54. Qu, T.; Li, B.; Huang, X.; Li, X.; Ding, Y.; Chen, J.; Tang, X. Effect of peppermint oil on the storage quality of white button mushrooms (*Agaricus bisporus*). *Food Bioprocess Technol.* **2020**, *13*, 404–418. [CrossRef]
55. Lin, X.; Sun, D.-W. Research advances in browning of button mushroom (Agaricus bisporus): Affecting factors and controlling methods. *Trends Food Sci. Technol.* **2019**, *90*, 63–75. [CrossRef]
56. Xu, D.; Gu, S.; Zhou, F.; Hu, W.; Feng, K.; Chen, C.; Jiang, A. Mechanism underlying sodium isoascorbate inhibition of browning of fresh-cut mushroom (*Agaricus bisporus*). *Postharvest Biol. Technol.* **2021**, *173*, 111357. [CrossRef]
57. Haminiuk, C.W.; Maciel, G.M.; Plata-Oviedo, M.S.; Peralta, R.M. Phenolic compounds in fruits-an overview. *Int. J. Food Sci. Technol.* **2012**, *47*, 2023–2044. [CrossRef]
58. Sánchez, C. Reactive oxygen species and antioxidant properties from mushrooms. *Synth. Syst. Biotechnol.* **2017**, *2*, 13–22. [CrossRef] [PubMed]
59. Cheung, L.; Cheung, P.C.; Ooi, V.E. Antioxidant activity and total phenolics of edible mushroom extracts. *Food Chem.* **2003**, *81*, 249–255. [CrossRef]
60. Lin, S.; Chen, C.; Luo, H.; Xu, W.; Zhang, H.; Tian, J.; Ju, R.; Wang, L. The combined effect of ozone treatment and polyethylene packaging on postharvest quality and biodiversity of *Toona sinensis* (A. Juss.) M. Roem. *Postharvest Biol. Technol.* **2019**, *154*, 1–10. [CrossRef]
61. Liu, H.; Chen, F.; Lai, S.; Tao, J.; Yang, H.; Jiao, Z. Effects of calcium treatment and low temperature storage on cell wall polysaccharide nanostructures and quality of postharvest apricot (*Prunus armeniaca*). *Food Chem.* **2017**, *225*, 87–97. [CrossRef]
62. Chen, L.; Zhou, Y.; He, Z.; Liu, Q.; Lai, S.; Yang, H. Effect of exogenous ATP on the postharvest properties and pectin degradation of mung bean sprouts (*Vigna radiata*). *Food Chem.* **2018**, *251*, 9–17. [CrossRef]
63. Sharma, P.; Jha, A.B.; Dubey, R.S.; Pessarakli, M. Reactive oxygen species, oxidative damage, and antioxidative defense mechanism in plants under stressful conditions. *J. Bot.* **2012**, *2012*, 217037. [CrossRef]
64. Vunduk, J.; Kozarski, M.; Djekic, I.; Tomašević, I.; Klaus, A. Effect of modified atmosphere packaging on selected functional characteristics of *Agaricus bisporus*. *Eur. Food Res. Technol.* **2021**, *247*, 829–838. [CrossRef]

65. Sun, J.; Jiang, X.; Chen, Y.; Lin, M.; Tang, J.; Lin, Q.; Fang, L.; Li, M.; Hung, Y.-C.; Lin, H. Recent trends and applications of electrolyzed oxidizing water in fresh foodstuff preservation and safety control. *Food Chem.* **2022**, *369*, 130873. [CrossRef]
66. Ali, S.; Khan, A.S.; Malik, A.U.; Anjum, M.A.; Nawaz, A.; Shah, H.M.S. Modified atmosphere packaging delays enzymatic browning and maintains quality of harvested litchi fruit during low temperature storage. *Sci. Hortic.* **2019**, *254*, 14–20. [CrossRef]

MDPI
St. Alban-Anlage 66
4052 Basel
Switzerland
www.mdpi.com

Foods Editorial Office
E-mail: foods@mdpi.com
www.mdpi.com/journal/foods

www.ingramcontent.com/pod-product-compliance
Lightning Source LLC
Chambersburg PA
CBHW040745250926
51850CB00007BB/1417
9783725801121